D1661404

My Path To Peaceful Energy

By Roderich W. Graeff

My Path To Peaceful Energy

By Roderich W. Graeff
rwgraeff@yahoo.com

First Edition November 2010

ISBN 1451591861
9781451591866
LNCC

Cover
Water color painting 1948 by Roderich W. Graeff
Designed by Gisela Hoffmann, Judith T. Hermann

Contents

This book is dedicated to

Rolf Schleu
Gert Wigand
Andreas Zacharias-Langhans

my high-school classmates, who were killed during the last weeks of the Second World War,
like millions of others at the time.
Also to those, who continue to be killed in wars all over the world.

Their lives being taken away totally senseless.

And to the students and their professor at the University of Munich,

who spoke out against Hitler's tyranny
non-violently
under the sign of the White Rose:

Sophie Scholl, age 22, decapitated on February 22, 1943
Christoph Probst, age 24, decapitated on February 22, 1943
Hans Scholl, age 25, decapitated on February 22, 1943

Professor Kurt Huber, age 50, decapitated on July, 12, 1943
Alexander Schmorell, age 26, decapitated on Juiy 13, 1943
Willi Graf, age 25, decapitated on October 12, 1943

And to all who work for peace

Acknowledgements

Typically, a book has one author. But in its path to creation a multitude of people are involved, most of them extremely helpful. Without their participation in my interest and work in Non-Violence and The Second Law of Thermodynamics, my Path to Peaceful Energy would not have seen the light of day.

Thank you for all your help, every one of you!

Preface

Why does one write a book? Why did I write this book?

Originally I wanted to describe my experiments in physics, experiments with results which are a world sensation. At least I thought so and frankly, I still think so. These results change our thinking about the most basic law of life, the Second Law of Thermodynamics. It opens the possibility to create usable energy out of a heat bath with the help of gravity. Maybe one day it will help us to reduce the burning of fossils fuels with its consequence of global warming.

In 2003 a Munich professor wrote to me: If you are right, this would be a sensation. A Nobel prize-winner judging if a paper I had written could be published in his physics magazine said, "....for a publication you have to rewrite it totally. But I don't want to discourage you. Maybe it contains a diamond in the rough...". But these two are the only ones in now more than ten years who offered a favourable commentary. The many other professors of physics or engineering whom I approached in Germany, the USA and other countries listened for 15 minutes and then offered this advice: We can't say that there is anything wrong with your experimental results nor with your theoretical interpretation, but your statements about the Second Law are wrong. Again, we can't say why, but the Second Law has never been proved wrong. Better concern yourself with something more worthwhile.

So I had to write this book because the specialists, the scientists who I thought would be interested and predestined to repeat my experiments and discuss the consequences declined to look at or even to publish them. But when I started I realized very quickly that this Second Law had fascinated me throughout my life, beginning with my first experiments at age 11. And very quickly I realized the connection to the whole areas of energy, how we use it, how we generate it and how we even start wars to keep it for ourselves or to get it from outside our own borders. Only 20 years ago an American president gave as a reason for invading another country that we had to defend "our" oil. This pre-emptive invasion of another country was called a "war of defence".

My first "war of defence" I had experienced as a 15 year old boy as member of the auxiliary anti- aircraft force in my home town of Hamburg, trying to shoot down the British and American bombers. This experience led me to start a foundation "Gewaltfreies Leben" – Violence-free Living. Circability was the aim of my work there, trying to get society to perform only those actions which the following generation could undo if they realized their parents had followed a wrong path.

The Second Law, violence-free living, killing is out, green energy, alternative, violence free methods to alleviate violence, Circability: It sounds as if I am trying to cover many different topics. But for me they are one, all interconnected. Reader, don't give up. Concentrate on the parts which you feel close to, which touch a cord in you, those which might encourage you to work for a more peaceful future of our children, for the children around the whole world.

Roderich W. Gräff
Koenigsfeld, Germany

October 13, 2010

Chapter 1
Children's Evacuation, my First Perpetuum Mobile

Autumn, 1939. My mother surprised us four children: I am a slim boy of 11 years, my sister Hilde, two years older, my 9 year old brother Gernot, and not to forget the darling of the family, Ingeburg. Tomorrow we will go downtown and buy a pair of shoes for all of you. Great astonishment. New shoes we would get normally only at Easter or Christmas.

On the next day Adolf Hitler made an announcement on the radio: Beginning at 4 o'clock that morning we are shooting back at invading soldiers at our border to Poland! We were at war. Three days later Great Britain and France declared war on Germany. To me, it was clear: Germany had to defend herself against the attacking enemy! My parents, the people on the street, everybody showed a very subdued mood. There was no sign of any great excitement which supposedly was expressed in Germany and France at the outbreak of the First World War, which I read about in my history books. It was only years later when my father admitted that he knew what was coming. As an officer of the reserve he had been informed that mobilization would be announced on the following day. So that was the reason for the unusual purchase of four pairs of shoes.

Of course we children were excited. War, what did that mean? We lived in Hamburg, a city with more than one million people in Northern Germany, lying on the river Elbe with a large harbor, not far from the North See. Living in a suburb of Hamburg, Hoheneichen, our family enjoyed a single family house with a large, beautiful garden. Our spacious living room opened up to a terrace where we enjoyed breakfast on Sundays. A few steps further down was a lawn big enough for playing crocket and soccer.

The author as a three year old, 1930

What did I feel at this moment holding this big ball in my hands? Was this a ball to me, or the world? Did this feeling influence me 15 years later to start to think about the radiative heat exchange between the surfaces of spheres? Is our future already established at that young age as some people believe our future life is in God's hands? Or is it fixed by the laws of physics, totally fixed but not predictable? Does it leave us with something what we call FREE WILL?

1997

Or did I have then this worried look realizing that at the time of my 70th birthday I would have to share this world with two additional little boys, another 6 billion people all together, instead of only 2 billion in 1930?

Our big garden was also used for growing vegetables and flowers. Each of us children had our own little plot, about 1 square meter or 10 square feet large, where we could grow water cress, radishes and salad. 4 apple trees, 3 trees with Morella Cherries, one with sweet cherries and 3 pear trees supplied a bountiful harvest of fruit which filled many jars of compote and jams. We children would pull our little wagon filled with apples and pears over 2 miles to a little juicery where a week later we would receive many dozens of glass bottles filled with wonderful tasting juice which were served on Sundays at our family dinners and which lasted all through the winter. Soft drinks were practically unknown. Only for birthday parties we got as a special treat a glass with some syrup and filled up with sparkling water.

I had experienced a peaceful and happy childhood. I had not encountered any form of real violence up to this point in my life. During elementary school I often missed class being sick with all the typical childhood diseases one would get at that time. But school was easy for me. I had been the best pupil in my elementary school class and just for one year in 1936, the best in each class were allowed to skip 4th grade and enter the Gymnasium. So suddenly I was the youngest one in my new class. I was not one of the stronger boys but I was fast and quick. With cleverness and dexterity I managed and survived rare fights with some of my older class mates without any major problem.

Playing soccer during recess I bumped into an older student supervising us. He caught me, grabbed me at the neck and pushed me head first into the schoolyard which was covered with slag. A bigger stone cut my skin at the base of my nose. I bled somewhat and ran to a teacher. He just put a tape across the bottom of my nose covering up some dirt. My mother sympathized with my agitation but I don't recall that any discussion with my teacher about this incident or any other follow up took place. For many years thereafter, there was a dark spot under the skin of my nose reminding me of this incident.

The citizens of Hamburg were asked to prepare themselves for possible air raids. We received sturdy paper bags which we had to fill with sand and distribute around the house, 2 or 3 for each room. If an incendiary bomb should fall into one of our rooms we were supposed to put one of these bags on top of the bomb which then would burn a hole into the bottom of the paper bag and the sand would extinguish the fire. But we should be careful if the bomb had a red line around it. This would mean that they would explode after the fire started. What to do then was not explained. Luckily we never had to use them in our own home. But when bumping the bags by accident with our shoes they tore, which provided a constant battle trying to keep the sand from distributing all over the room.

To avoid becoming buried in our cellar by our own house collapsing from a nearby bomb hit, we were advised to dig a trench in our lawn. It was located 10 meters away from our house, 4 meter long, 2 meter deep and 1 meter wide with an earthen stairway at each end which angled into our bunker. (1 meter equals about one yard). These two stairways made our bunker accessible from both sides. We received requisition papers allowing us to get wooden planks from the local building supply yard to cover our refuge. On top of these we put a layer of soil and grass sod. My sister Hilde sewed some make-believe curtains which we stuck with wire to the vertical earth walls, pretending to cover windows.

Location of the trench which we dug in the lawn before our home
From left to right: My mother, the author, my sister Hilde, my brother Gernot, my father

Three days later sirens sounded the first air raid warning. It happened during the afternoon, my father was at work, my mother shopping. Playing in the street we raced home and straight into our house cellar, totally forgetting our garden bunker. There we tried to put on the new gas masks we had just received. After half an hour the "All Safe" signal was given. Our garden bunker was totally forgotten. Who would want to go to such a muddy and moist hole and stay there for hours, especially in the middle of the night? We never used it, it slowly filled with rain water and not too long thereafter we filled it up again with the soil we had worked so hard to dig out. Workers came with a strong wooden beam which was installed in one of our cellar rooms, selected to become our air raid cellar. The beam was placed against the roof of this room, supported by three strong posts. Now we had to paint big white arrows on the side of our house pointing to the cellar window through which our air raid cellar could be reached most easily. This should show a rescue party the way to find us and dig us out in case the house had collapsed on top of us.

My mother and my father enjoying the sunshine on our terrace, 1940, with a white arrow painted on the house wall

Soon thereafter the sirens sounded every few days, both day and night time. In class, looking out of the windows quite often, I would suddenly see how a number of tethered balloons fastened to a thin steel cable were slowly rising into the sky. They were supposed to reach their operation height higher than the approaching bombers in order to protect with their steel cables the inner city around the Alster, a lake located in the middle of the inner city. Across its narrowest part a railroad bridge was used by all trains connecting the areas north of Hamburg with the south. This strategic bridge was further protected by artificial fog which was released every time bombers approached and the balloons rose. After a year or two this policy was stopped, being deemed ineffective. In spite of this and in spite of the fact that the inner city of Hamburg in later years was bombed to rubble, this bridge, the Lombard bridge, amazingly

survived throughout the war completely intact.
But this was early in the war. During the first years of the war bombs came down quite rarely. Actually we children sometimes hoped for an air raid warning to happen during the night, especially if on the following day in school a test in math or Latin was scheduled. If the air raid finished with the All Clear signal before midnight, the normal school opening would be delayed by just one hour.

If the All Clear signal sounded before one o'clock in the morning, we had to show up by ten o'clock. But, if the English bombers fulfilled our hopes to leave Hamburg only after one o'clock in the morning, school for the next day was completely canceled.

During these early months of the war the anti-aircraft guns were rarely fired. During the summer of 1940, the first bomb was dropped close to us (probably by mistake) within our suburb and exploded harmlessly in a yard between three houses. Nobody was hurt but the people living in the surrounding homes were put in a hospital for a few days for observation. How times would change!

As the war continued, the sirens would sound more and more often. To reduce these many interruptions of everybody's routines a new signal was introduced, a "Pre-Raid" warning. Instead of varying the sound of the sirens continuously between high and low they would do this just 3 times with staying on the high tone just three times for a somewhat longer period. Now everyone would know that there was no immediate danger right then but enemy planes were approaching. Once they got close the real alarm was sounded and we knew, it was time to enter our cellar.

If an alarm happened during the day or in the evening hours every one of the family who happened to be home would gather in the study of my father around the only radio we owned. The antenna inlet of the radio could be connected by a single wire to a blank metal part of the telephone. Wonder of wonders: Through this wire-telephone we would get air raid announcements. A serious mans voice would interrupt a continuous... beep....beep....beep tone : " A few enemy airplanes are 50 kilometers north of the city flying in a south westerly direction. No major raid is expected at this time." Or scarier:" Strong enemy forces are approaching the mouth of the river Elbe in a southerly direction. A major raid has to be anticipated. Take cover now."

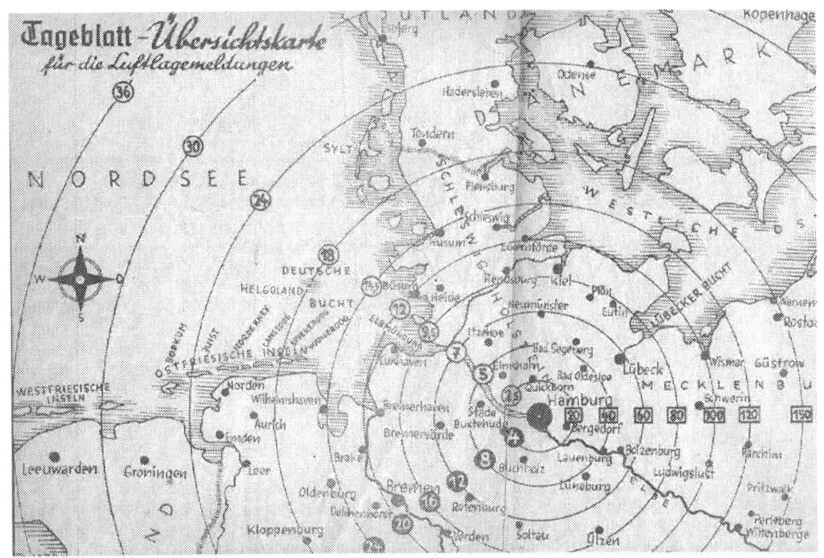

The local papers printed these maps showing circles around Hamburg indicating the number of minutes it would take the bombers to reach the city.

The worst especially for us children was to be wakened up so frequently during the night. Two wooden bedsteads with two beds on top of each other stood in our air raid cellar in Hamburg, ready to be used by my two sisters, my brother and by me. Still half asleep we would leave our nice warm beds and slump down on the wooden slabs of these beds. Before joining us in the cellar my father would try to fill our bath tub in our upstairs bathroom with water to be used in case the house caught on fire. But as thousands of other Hamburgers tried to do the same the water pressure would fall down to zero. Only a few drops appeared. Once the noise of the airplane motors or the anti-aircraft fire came closer my father would also come down into our cellar. If the outside noise became too threatening the open water faucet would be forgotten. Twice it happened: half an hour later the water pressure came back, the bath tub filled, it overflowed, and the water would run down from the second floor over the stairway into our living room and through the floor of the living room into our cellar. Drip, drip, drip...Years after the end of the war huge darkened circles of paint on the upper wall in our living room would remind us of those long nights.

But even without bombs falling the nights could be quite scary. The explosions of the anti-aircraft guns made a terrible noise especially when we could hear the noises of the British bombers getting closer. It sounded like the bursting anti-aircraft shells were exploding right over our heads. Still, these nights created one advantage. There would be a chance the next morning to find nice parts of the bursted anti-aircraft shells, splinters out of glistening metal pieces, maybe 5 cm, 2 inches long and a cross section of 1 square centimeter. Even in the middle of the night whenever the All Clear siren would sound my brother Gernot and I would race up to my fathers study. This was located in an extension built by the owner of the house, an artist. The whole room was covered by a double glass roof to provide optimal lighting for his paintings. Whoever managed to get there first had the best chance to find these shrapnel parts. He had to look only if there was a hole in the glass roof, as sure enough below there would be one of our treasured metal pieces. Worse, of course, this created many small holes which remained in the wire glass. My mother had to put up more and more containers on the floor below them to collect the water dripping in whenever it rained.

It happened in 1941. In order to escape these continuous sleep interruptions together with thousands of other school children from Hamburg I was sent to southern Germany, which was still a very peaceful place at that time. I would live with my grandparents in Freiburg. Here I would hardly notice that we were at war. When the fighting with France had ended, I stood waiving on the sidewalk as the garrison from Freiburg moved back into town in a small parade. We would jump onto the military wagons which quite often were still pulled by horses and begged from the homecoming soldiers for French coins or helmets.

The author in 1941, 13 years old, standing in Freiburg at the entrance to his grandparents home, Jakobistr. 29. In its cellar I performed my first experiments trying to create a Perpetuum Mobile.

I went to the local high school. Here they had started in fifth grade with Greek and with Latin as a second foreign language in the seventh grade, just the opposite as it was taught in my school in Hamburg. As a result I could shine in Latin, having had it already for four years, and had a wonderful excuse for my bad grades in Greek, as I first had to make up for the missed two years.

In early summer the class made an excursion to the town of Kehl, located on the edge of the Rhine River. We were allowed to walk to its banks. Over on the other side lies France, our teacher explained. This is what a foreign country looks like? As a thirteen year old boy, I never had visited a foreign country, and I was surprised. Over there on the other side of the river everything looked just the same as here in Germany! Apparently I had assumed that there would be greater differences between the appearances of different countries.

We were told it happened just about here where our brave soldiers first crossed the Rhine in rubber dinghies in a hail of enemy fire. Quite a few of them sank with their young soldiers. Only very few of these were able to swim to the shore. Most of them drowned as they were pulled down by the weight of the ammunition fastened to their uniform belts. I wondered: Was this a meaningful way to live and to die?

The Black Forest area in Southwestern Germany is a very beautiful area with rolling mountain tops covered by pine trees. On your hikes you will pass through small picturesque villages and valleys, typically with a little stream gurgling on its downwards path. Was it on this excursion that I saw one of the many waterwheels still running in this part of the world? Was it this waterwheel which gave me the idea to design a Perpetuum Mobile?

And then it happened in Physics class. The teacher demonstrated to us how to lift water out of an aquarium. He used a rubber hose, filled it with water, and closed both ends of the hose with his fingers. He put one end into the water of the aquarium, lowered the other end over the edge of the aquarium wall, and let go with his fingers. To my great astonishment the aquarium water flowed upwards through the hose across the wall of the aquarium, and down on the other side into a second basin.

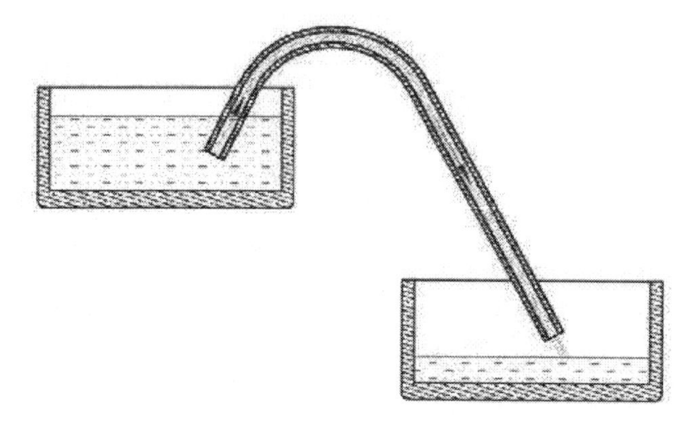

Thoughts raced through my head. How was it possible that water could flow upwards? Would it be possible to install a water wheel below the end of the hose where the water flowed out of the tubing to drive a wheel, a grain or a saw mill? Why would one not arrange the water wheel above the water level of the aquarium so that the water would return to the basin where it came from? Then this process could continue endlessly. This way it would be possible to produce energy free of charge. I had invented a Perpetuum Mobile!

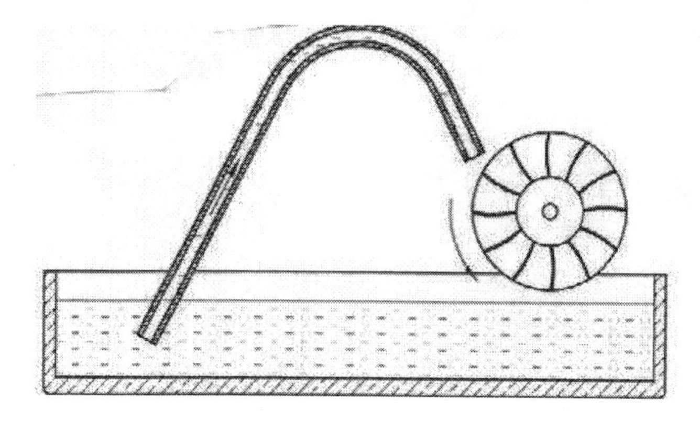

I could hardly wait for the end of the day's last lecture. I raced home and into the cellar room which was used as a laundry room. With a half-filled water bucket and a rubber hose I started my experiments. After a while I could repeat what the teacher had shown us in the experiment. Water was flowing out of the bucket across its upper edge onto the cellar floor. A great feeling of success flowed through me. Now I only had to lift the end of the hose where the water flowed out above the water level of my pale, and my invention would be perfect.

To my great surprise I noticed very quickly that the water stopped running as soon as the end of the hose was lifted above the water level. Lunch interrupted my experiments. With the last bite still in my mouth, I ran back into the cellar. Jumping down the stairway the solution struck me: Yes, the end of the hose where the water came out had to be below the water level. But to keep running may be the lower end had to be in a space filled with air. And in this air space I would have to mount my water wheel. And this air-filled space I had to arrange within the bucket! Nothing was simpler than that. I raced back up into the kitchen and got a metal funnel. Turning it upside down I pushed it into the open end of the hose. I filled the hose with water squeezing both ends tight with my fingers, and pushed the funnel upside down into the water bucket. Then I let go. After a few tries I seemed to be successful. It looked to me that water was really running through the hose into the funnel. As the hose and the funnel were not transparent I could not actually see water running through the hose but I got this impression because air bubbles appeared across the edge of the funnel.

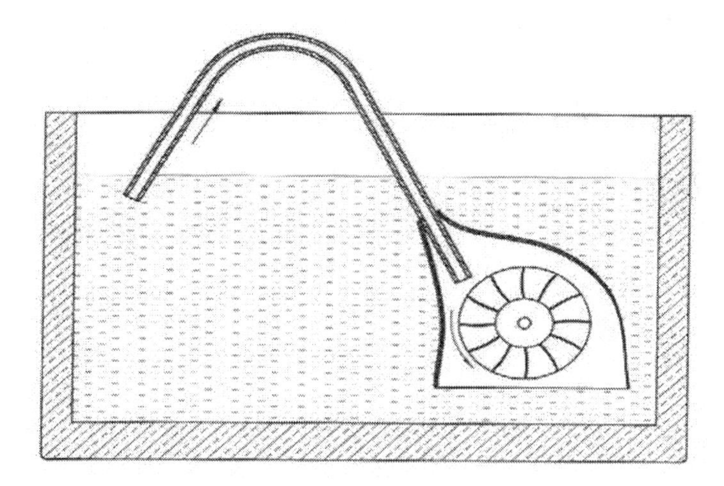

But after a short while the air bubbles stopped coming. Apparently the funnel had filled with water.

What to do now? I went to my grandfather, a pathologist, and reported to him my problem. In his experience, he told me, some lard put onto the edge of the funnel would prevent air from passing. Defeated, I left him. How could I, in the middle of a war, put my hands on some lard? I crawled into the kitchen and confided my problem to our cook, Anna. She did not see this as a problem at all. She went into the kitchen store room and came back with a spoonful of lard. Our strong-willed Anna had grown up on a farm, and visiting there on weekends she typically returned with some edible treasures.

I put the lard on the edge of my funnel. But in spite of a number of additional tests I have to report that I was not successful. My invention just did not work as planned. Not long thereafter I had to abandon this project. Summer vacation started and suddenly our whole class was sent for six weeks to various villages in the Black Forest in order to help the farmers with their harvest.

Chapter 2
Farmer's Helper; We Search for the Potato Bug; In an Emergency I Will Hit Hard

A small train pulled by a steam locomotive took my classmate, Eckhard Hölscher, and me to the village of Gündelwangen, located 8 km from Titisee, deep in the Black Forest. The mayor welcomed and delivered me to the farmer family Bühler, and Eckhard to another farm. During these times of war when most young farmers had to fight as soldiers we were supposed to help during our summer vacations with the harvest for the next 6 weeks. In this manner my whole 7th grade class was sent to farmers in other small nearby villages.

Family Graeff visits farmer Bühler while I was working there. Front row from left to right: My brother Gernot, my father, my sister Ingeburg, farmer Bühler and his wife, Paul, the author.

CHAPTER 2

The farmer and his wife looked skeptically at this 13-year-old city boy, thin and somewhat pale, who was supposed to assist them with their daily chores. Both sons of the farmer had been drafted into the armed forces. In their place a young, 20 year- old Pole Paul worked with them. Today one would consider his work as forced labor. I don't know if he came voluntarily to Germany hoping to improve his chance to survive the war, or if he had been forced against his will. He certainly made a content and happy impression. He was treated like a son of the family, was an industrious worker, and we got along with each other very well. Each of us occupied one of the rooms belonging to the absent sons.

A wonderful time began. Ever since I was a young boy, I had wanted to become a farmer. As soon as Germany regained her former colonies which she had lost after the First World War, I thought, I would start a farm in Africa. Now I was being taught how to clean out the cow stable and feed the animals, and I was allowed to help with the milking. We were in the middle of the hay season. In the morning, I spread out the grass freshly cut by hand by the farmer and Paul to dry in the sun. In the evening, I would rake it together and pile it into little heaps. The following day, providing the weather was favorable, I again spread these little piles as equally as possible across the field for further drying. Our village was located at a height of about 2000 feet or 700 meters and the days were short.

Farmer Bühler sharpening his scythe, an every day task.

During the night dew would collect on the ground so this process had to be repeated for several days until finally the hay was dry enough to be transported into the barn. Every day there was something new. More importantly, it was wonderful that I could eat as much as I wanted of the hearty farmers' food.

My farmer, like all other farmers in our village, owned neither a tractor nor horses. To pull the heavily loaded hay wagons, two of his cows, and sometimes even four, were hitched to the hay wagon. In the field I was allowed to stand in front of the cows and order them to pull the hay wagon ahead another ten meters to the next pile of hay for further loading. The huge cows obeyed me. I was surprised how they followed the commands of this small 13-year-old boy.

Hay wagon pulled by two cows

Once a week, each Wednesday afternoon, we would meet with the other farmers and their families next to the various potato fields of the village. In groups of 10 we would walk between the potato rows, carefully inspecting each potato plant. Our task was to find potato bugs. There were rumors that the enemy had dropped them from airplanes in order to damage our potato crop. We were shown pictures of this yellow and brown striped insect. Would this be called "biological warfare" today? The Wednesday afternoons provided a nice interruption to our daily work, but we never found even one of these little creatures.

One day my farmer and I rode on an empty hay wagon to one of his outlying fields. From the road we pulled into the courtyard of the town hall as farmer Bühler had to talk to the mayor. I was positioned just ahead of the two cows, holding in my left hand the bridle of the left cow, and in my right hand a whip. "Watch the cows. I will be back in ten minutes." And my farmer disappeared into the building.

It was just before noon. The sun burned mercilessly on me and the cows. The biting insects became worse and worse. I was stung all over, especially on my uncovered legs. Similarly the cows had a problem, swinging their tails trying to get rid of their tormentors. Suddenly, I was pulled out of my dreams. As though following an order both cows started to move ahead. I pulled on the reins, and cried out, "Halt!" But the rein had gotten caught under the hitch. As hard as I pulled, the cows felt nothing. After two or three steps they began to run, and I was pulled along. They started to gallop in the direction of the road with the wagon behind them and after 2-3 steps I could not keep up with them. How could I stop them? What could I do in this emergency?

I still had the whip in my right hand. Everything happened so quickly, there was no time for me to think. Somehow I lifted my right hand with the whip high in the air and let it come down as hard as I could on the behind of the galloping cow. Now the cows really had a reason to run! I was forced to let go of the reigns and the cows with the wagon behind them passed me. At this moment the door of the town hall opened and my farmer reappeared. I froze! Do I look at the farmer or at the disappearing cows? A big crash! Just before reaching the road, one wheel got caught by a big stone marker. The wagon tipped over. The cows stopped in their tracks. The hay wagon, which was lying on its side, must have broken into 1000 pieces. The farmer ran to us and took the whip out of my hand. What might he have thought when he opened the door of the town hall and saw how I hit the rear of his galloping cow as hard as I could? I wished I could sink into the earth and hide. It was all my fault and now his wonderful hay wagon lay totally destroyed on the ground.

But now a wonder occurred! Farmer Bühler showed me how I could help him straighten out the hitch along with one axle and the two wheels still fastened to it. We moved back about four meters and arranged there the back axle. Next, two thick wooden planks about four meters long were put on top of the axles. Then we lifted the two ladders and arranged them next to the wooden planks connecting the upper beams of the ladders across the wagon with the cross beams having two iron rings which fit over the upper beams of the ladders. With some chains which we collected from the grass the whole contraption was stabilized. Wonder of wonders, in just a few minutes the hay wagon stood reassembled before me in its old beauty. I could hardly believe trust my eyes. Instead of having to bring home a collection of broken wooden beams and parts, I realized that all the various pieces, in fact the whole wagon, had survived this catastrophe without any damage.

Picture of a typical hay wagon, reassembled.

I was greatly impressed by this design. I was further impressed in the following days when the two ladders constituting the sides of the wagon were replaced by two wooden planks. It was converted in just a few minutes into a wagon for carrying dung. A week later we mounted a dung barrel onto the two axles, and the dung wagon was ready. Maybe this experience was one of the reasons why, after finishing my time at school, I decided not to become a farmer but an engineer?

Looking back I remember vividly how I had hit the cows with the whip with all my might. Already then I had realized that my hitting had made the whole situation only worse. Why did I hit the cows? Had I done this instinctively out of a feeling of powerlessness, with the realization that I had lost the control of the situation? How often does this happen in our lives as married partners, as parents, as members of society, as policemen, soldiers, or statesmen? Does it happen when we do not know how to proceed in a difficult situation and when we become really fearful? Do we then start to hit with our fists, with words, with economic power, or with bombs, depending which of these means of power happens to be available to us at the time? Do we forget so easily our basic intention to act non-violently?

As long as the weather did not demand otherwise, on Sunday afternoons we didn't work. My Perpetuum Mobile discussed in Chapter 1 was still on my mind. I sat down and tried to describe it in a letter addressed to my Uncle Volker. He had been drafted as an electrical engineer and was working in Gotenhafen, a harbor city in East Prussia, trying to design improved torpedoes. Once ejected from a submarine, the torpedo was supposed to follow the sound of the screw of the warship, trying to hit and sink it. Of course, this letter to my uncle is long gone. But I saved his answer to me, dated October 6, 1942:

Dear Roderich!

I think it is great that you concern yourself with technical problems during your work as a farmer's helper. But I find it less pleasing that you still dream about a perpetuum mobile. You can be totally convinced that there is no solution for this. Relatively often I have been in contact with inventors of a perpetuum mobile, and I have always had the same experience. These inventors know the basic facts of mechanics and realize that using these facts they cannot reach their goal. Therefore they think up very complicated designs which finally they themselves cannot understand any more, but they still keep hoping to somehow reach their goal.
It is the task of an engineer to reduce these complicated designs to basic simple forms so that the impossibility of the process becomes obvious. In your case the basic problem is this:
From an upper container water flows into a lower one. The flowing water drives a water wheel, and this water wheel is supposed to drive a pump to bring the water back into the upper container. It is totally unimportant that you provide additional design features. The expected effect of the air chambers is just an illusion. It just complicates the system and leads to the assumption that the design could run by itself.
You might also forget the pump and water wheel as unnecessary. What you are trying to achieve is the following: In a closed tube on one side, water shall stream downwards, and on the other side it shall flow back to the top. This would constitute the simplest solution to your task, as losses through friction would be the lowest. But our daily experience teaches us that such a water circulation, even if it could get started somehow, would come to a halt very shortly. If you add pumps, water wheels, air chambers, and valves, the circulation will stop much earlier as much more friction has to be overcome.

He then proposed in his letter, that I should concern myself with simpler and more practical problems: How would my farmer try to lift a heavy stone onto his wagon? Could he pull it up on a slanted board? What would be the force of the stone pulling on the rope?

How did this letter affect me? I was pleased that my uncle had written me two pages, but I was really disappointed because I did not want to hear that basic principles would always show that a perpetuum mobile could not work. I wanted to know the specific reasons why my wonderful invention did not perform and was perhaps not as wonderful as I thought. For instance, could it be that the pressure in my air chamber was too high for the water to flow into it from above?

The final sentence in my uncle's letter stated:

You can study these practical questions like lifting a stone by very simple experiments on small models. I believe that such tests will give you more joy than working on a perpetuum mobile which never can function.

Did he think I would lose my interest and joy in thinking about a Perpetuum Mobile? How wrong he was!

Chapter 3
Child Soldiers Protect their City; Firestorm in Hamburg July, 1943. Temperature Distribution in the Atmosphere above a Burning City

February, 1943. The battle of Stalingrad had been raging on for months. Suddenly an announcement was made on the radio: Stalingrad has fallen to the enemy! A terrible defeat? No, the German defenders had fought to the last man, a great heroic victory! In reality, 100,000 soldiers running out of food and ammunition had surrendered and were marched off to work in Siberian prisoner of war camps. 90,000 of them died there, only 10,000 survived and were allowed to return to their homes in Germany in 1955, ten years after the end of the war.

I was 15 years old and listened a few days later to the radio when the propaganda minister Goebbels asked a crowd sitting before him in a huge hall in Berlin: "Do you want Total War?" And the crowd exploded, screaming "Yes!" Of course I don't know what the participants were feeling at that moment. I would have also screamed an enthusiastic "Yes!" It was not clear to me what "Total War"" meant. I just felt you don't give up defending your country just because you lost a battle. You renew your efforts to reach final victory and work hard at it in all aspects of your life.

Frankly, today, following the tragedies which occurred on September 11, 2001 strange feelings come over me when I hear the President of the mightiest country in the world talking about revenge, "a crusade" and the possibility of declaring war on the "evil ones." It saddens me when I realize that the vast majority of his citizens enthusiastically follow his lead.

In 1943, what did this total war mean to me? A few days later on February 15, along with my classmates in the 9th grade, I received an induction order.

Sixth grade class in 1938 of the Gymnasium Johanneum in Hamburg, Germany

We were to report to duty one week later at the site of an anti-aircraft battery in a suburb of Hamburg, our home town. Very suddenly we 15 and 16 years old boys were pulled out of our life as school children into the world of soldiers. Even so this picture shows our class five years earlier it gives an impression how we looked at that time. I am the boy sitting on the chair 2 rows below the twin boys in the sailor suit. The German teacher on the left and the music teacher on the right both demonstrate with their pin on their lapels that they are members of the National Socialist Party. Practically every teacher had to be a member but we students knew quite well who of our teachers felt close to the ideology of the times or who dared to be somewhat critical.

Out of these 30 class mates in the picture three of them died as soldiers during the last month of the war. One of them was my best friend, Gert Wigand, sitting on the chair next to my right, wearing a jacket, and Rolf Schleu, leaning his knee against my elbow. The third one, sitting in his white shirt in the row between the teachers, was Andreas Zacharias-Langhans.

I dedicate this book to you, Gert, Rolf und Andreas. And I dedicate it to Hans and Sophie Scholl and their four friends, who risked their lives in working non-violently underground under the sign of the White Rose against Hitler's Tyranny. On the day we were drafted to the anti-aircraft, February 15, 1943, on this day they were arrested. During the first week of our training learning how to shoot, they were tried, convicted and decapitated.

The two classmates sitting in the picture next to me got killed in this terrible war, and I survived! Was it determined already then that this would be our fate?

We were not part of the regular Air Force but organized as an auxiliary corp. We received specially designed blue uniforms. Through our work some of the Air Force soldiers manning the anti-aircraft guns would be freed to fight the war in Russia.

Due to space limitations our barrack was sitting with one side on a bank of a side arm of the river Elbe, making necessary these five posts holding up the other side of our home. We kept wondering if one day we would find ourselves in the river Elbe due to a bomb falling close by.

We were housed in wooden barracks within the battery of six 88mm guns, later upgraded to 105mm. Our school teachers tried to come to us each morning to teach the normal school subjects. As we soon found out, however, our school sessions were often interrupted by alarm sirens, which made us run to our battle stations waiting for the English or American bombers to come close enough that we could shoot at them before their bombs fell on us.

Look at your 15-year-old son or friend and you will know how we looked when we 40 schoolboys arrived at our battery in Barmbek, a northern suburb of Hamburg. We learned that we would fight alongside 40 German aircraft soldiers and surprisingly, 40 Russians. These were prisoners of war who supposedly had volunteered to fight with us against the West. Had they joined because they felt it increased their chances to survive the war? I never found out the real motives of these young Russian men.

The author, 15 years old, can be found in the top row, third from the right, summer 1943.

While the German soldiers did the heavy work, like hand loading the heavy shells into the guns, we were trained to aim the guns and to mount the radar and fire control units. From cavities in the protective wall surrounding our guns, the Russians were responsible for bringing enough ammunition to our firing guns.

I was proud to be able to serve my country. The tasks we had to perform we grasped quickly and easily. Most alarms and air raids took place not during the day, but throughout the night. This was the hardest for us not to be able to sleep through the night but to be awakened two or three times almost every night. We would run to our battle stations and after several hours of waiting or shooting, we could return to our barracks, fall into our bed bunks and into a deep sleep, just to be awakened again a little later.

While we were inspected by some higher up officers we Air Force Helpers (Luftwaffenhelfer) stand to attention with our helmets around our gun. In the background before the bunker filled with ammunition one can see 2 of the Russian prisoners belonging to the crew of this gun.

Today we discuss if forced sleep deprivation performed on political prisoners constitutes torture or not. At the time we felt to fulfill a worthwhile task, a patriotic duty. Looking back today, how stupid, how senseless it all was, that war, any war, past or future. Are we not able to learn and to create a warless future for our children?

For many years I have struggled with how to share my experiences during those years between 1943 and 1945. How can you describe the horror when more than 40,000 people of Hamburg, mostly women and children, burned to death within just a few hours during one night? For anyone who survived this inferno it is nearly impossible to describe it meaningfully. Somebody who has not experienced it cannot grasp such a happening, so what would be the point? For anyone with a real interest in this disaster you can read about it in books like "The Night Hamburg Died" [26] or Dresden or Tokyo.

Personally, the experience of the bombing raids in July of 1943 led me to become a strong believer in non-violence. I don't wish anyone to have to live through a similar experience. But how can the next generation and those following learn from such a terrible happening if those who experienced it, don't tell anyone about it? It is very hard to talk about it, and I realize it might be similarly difficult for you to read about it.

The worst air raid for Hamburg happened the night of July 27-28, 1943. It was the first night the English bombers used millions of small metallic film strips. The first planes arriving threw them out all over the city and suddenly our radar installations were blinded. On our radar screens, where we normally saw clear signals indicating the location of a plane, we only saw a picture of what looked like snow on our electronic screen. Our battery, as well as all others around Hamburg, was totally helpless. Instead of firing at a target which we had located with the help of our radar, we could only shoot blindly and senselessly into the air.

We did not know what had happened. Only the next day did we found lying on the ground thousands of these thin foil strips, and slowly we understood. These thin foils reflected our radar signals, making them totally ineffective. Here are some of them which I picked up 67 years ago. The scale is in centimeters:

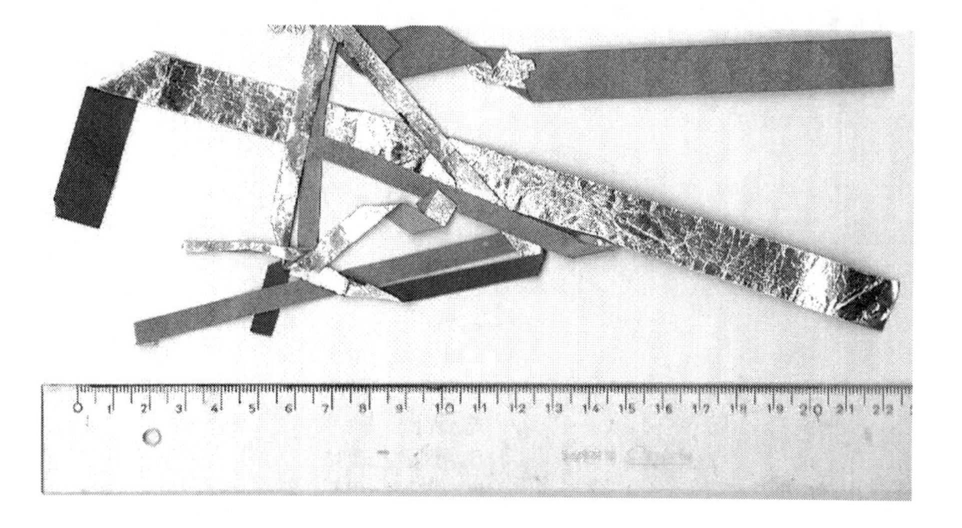

But morning didn't seem to arrive. Around ten a.m. we still looked at a dark sky filled with ash from the big fire storm which had devoured great parts of my city.

The next morning the new day did not seem to arrive as usual. Instead of getting light a strange misty darkness remained throughout the daylight hours. All we could see of the sun was a dark red ball which slowly moved across the sky. This condition lasted for two days until the dust and ash slowly settled. A strange, sweetish smell surrounded us for weeks on end reminding us of the thousands of people who had burned to death.

Here are some of the narratives of survivors as reported by my father in his book "Der Tod im Luftangriff". (Death in an Air Raid) [27]:

Page 18; By A man, 40 years old:

"Since the beginning of the alarm I was in our air raid shelter. Every 5 minutes I checked the surrounding areas. Around 2 o'clock a bomb hit very close. Our house and those around us were burning. We seemed to be totally surrounded by flames. Even the pavement seemed to be burning. Our cellar had slowly filled with a policeman and some civilians, men, women and children, altogether about 20 people. On the street roared a fire storm. The flames blew in different directions, sometimes forward, sometimes backward. The cellar got hotter and hotter; outside the heat was nearly unbearable. It seemed like you were breathing flames. Together with some other people we run back and forth trying to find a place free of fire. The hair of an older lady next to me caught fire from the sparks in the air. Soon the houses started to collapse. A woman next to me fell down. I tried to pull her up, but after two or three steps she fell again. We tried to reach a waterway. A few meters next to me many people were running, falling, never to stand up again. Finally we reached the water....."

Page 26; A woman, 35 years old:

"On July 27 my sister and I stayed with friends as our house had been destroyed in the raid 3 days earlier. At midnight the alarm sounded. After one hour of quiet, bombs began to explode. The whole house seemed to shake. We heard calls for help so we broke the wall to the next cellar filled with neighbors and their frightened children. One man went outside. Coming back he reported: the house across the road is on fire. And shortly thereafter: Our bedroom upstairs is on fire. We tried to extinguish the fire but soon ran out of water. .... It must have been around 3 o'clock when all the houses around us were burning. It became unbearably hot. A wind storm raced through the streets, with streaks of sparks

and burning amber all around us. We returned to our cellar, totally helpless. My sister, in desperation said: "Now we will burn to death!" If we are going to die it doesn't matter where, let's make a run for it. We found some water with which we moistened our clothing and a cloth which we wound around our head. We climbed through a window onto the street; holding onto each other's hands we fought the fire storm, trying to find a safer place. Other people, covered with blankets ran through the fire back and forth like they had lost their minds and were insane. I lost one of my shoes. No time to look for it. Behind the wall of a small building we collapsed. I nearly fainted. More people came running, bleeding from their wounds. Mothers were calling for their children; children for their parents. I felt terrible as I wanted to help, but couldn't. My eyes hurt terribly. I worried that I would become blind. Finally we made it to a cellar of a hospital where I received a salve for my eyes. By now it was 4 o'clock in the morning when we collapsed on a bed. At seven a call: The roof of the hospital is on fire! We run out. The fire storm had subsided. With houses still burning and dead people lying around, we started on our slow and long walk out of the city."

More than 40,000 people died a horrible death that night. Many were found in the cellars of their houses. It took months to open them; some of them were too hot to enter even after weeks had passed. One section of Hamburg, the borough Hammerbrook was totally destroyed as it was the centre of the firestorm. As it would take months to take out all the dead people a wall was erected around this section of the city.

A wall built around the borough of Hammerbrook, Hamburg, Germany after it was totally destroyed in a firestorm in July 1943 [31]

The year 1943 was the same year in which the wall in Warschau was build around the Jewish part of the city. Today we just had the 20th anniversary of the fall of the wall in Berlin, built to keep the East Germans from trying to leave their own country. We still have a wall between South and North Korea and we are building new ones between the United States and Mexico, between Saudi Arabia, Yemen and Iraq, on the southern edge of Europe to keep out people trying to escape the misery of Africa and between Israel and the occupied west bank and Gaza strip, and between there and Egypt, just to mention some of them. Whenever we feel we are confronted with unsolvable problems we build walls trying to keep people in or out. What a disgrace it is!

Had the world come to an end? Even nature was affected. The local temperatures had dramatically increased, as the fires continued to burn. This created an atmosphere which confused the chestnut trees standing along the remaining major roadways. They wondered if spring had come again and started to sprout new leaves, in late July!

And what has this part of history to do with "Gravity Machine" and my "Search for Peaceful Energy"? Well, there is one very direct connection. My father, a pathologist, had the task of investigating why people had died during that horrible night. After the war he wrote a book about his findings "Tod im Luftangriff" (Death In Air Raids) [27]. When I leafed through this book a few weeks ago, to my great surprise I found on page 40 a description of the temperature gradient in the air above the burning city. From a plane from the weather service which had been sent up, at a height of 4500 meters it measured a decrease in temperature of 0.8 degrees centigrade or 1.4 degrees Fahrenheit per 100 meters of altitude. Now, about 67 years later, I, the son of this pathologist, am trying to measure the temperature distribution in vertical columns of gases. I realize that the values measured on that fateful day above the burning city of Hamburg are the same which meteorologist's measure today when they determine the air temperatures of the atmosphere all over the globe. And these temperature gradients in air, gases and other media are a major topic throughout this book.

Why do we have wars? Isn't the basic cause very often the real or the imagined need for more and cheaper resources of food, water, and especially energy? If we could find a new source of energy which would not damage our environment and which would be available equally to all humans on this earth, wouldn't that be a great step towards a more peaceful human community?

Chapter 4
The Divided Living Room; A Renoir; The Stolen Tree; Solar Energy on the Window Sill.

Spring, 1945. The enemy came closer and closer, the British and the Americans from the west, the Russians from the east. Was the war lost? There were rumours that negotiations were going on. German soldiers would join the British and the Americans, and together we would stem the Bolshevists flood coming from the east and occupying all of Europe. For me, a 17-year-old boy, it was difficult to imagine that after all this hardship Germany could lose the war.

Most of my classmates had left the auxiliary corps of the anti-aircraft and had become real soldiers in the regular army fighting somewhere along the borders of Germany. I had volunteered to become a reserve officer in the air corps because I wanted to be a fighter pilot trying to shoot down those incoming bombers. I was sent to a training camp to learn to fly a glider. After three weeks I ended up in a military hospital with a very large abscess on the back of my neck. The doctors declared that my heart showed serious problems, and I really should be sent home to recover. So one day in January 1945, in order to be checked if I was fit to become a soldier I stood side by side with another man who looked to me like he was 70. We were standing to attention naked next to each other in a huge, empty and unheated ballroom. It was a scene for a movie. I wondered: Were these two naked people the last reserves Germany could muster to win the war?

Soon a military doctor appeared and declared that the old man surely could serve in the *Volkssturm*, an organization of old men and boys which was supposed to defend Hamburg against the approaching enemy. Looking at me the doctor decided that with my heart problem I was not fit to become a soldier but was also well-suited to join the Volkssturm, the Peoples Army, and to learn how to shoot at tanks with a *Panzerfaust*, a shoulder-fired rocket grenade, in order to save my city from the enemy.

Soldier training with a shoulder fired rocket grenade for defending against tanks.

In the coming weeks I was trained to master this weapon, but providence was kind to me. I was lying at home in bed with diphtheria and a high fever when it was announced over the radio: Germany had surrendered, the war was over.

Slowly the news sank in. The war was lost. Would peace come, what would it bring? My father came to my bed and told me that during the First World War, German soldiers were supposed to have cut off the hands of civilians in Belgium, as the Allies had declared. Of course after the war it was established that this was just one of many propaganda lies. Now, in May of 1945, he said the occupation forces claimed to have found not only in the Eastern occupied territories of Poland, but also in Germany, concentration camps where humans had lived and died under horrible conditions. My father doubted this could be true, but if it

turned out to be, all he had to say was "poor Germany." The future of Germany, my future, too, would be very, very bleak.
It was not long before the first one-page newspapers were printed, confirming these reports with terrible stories and pictures. It was too horrible to imagine. Could this have happened under Hitler, who so heroically fought for the rebirth of Germany? Could it be that the SS and the Gestapo performed these deeds without his knowledge? What a shock for a 17-year-old. Was the whole past of my growing up in Germany based on deceit and lies? Was there nothing any more one could believe in?

But life went on. During the war for more than 5 years, practically during all my teenage years, you went to bed at night, and then to the cellar, and you shuddered under the noise of the anti aircraft and sometimes of the exploding bombs. And you wondered: Will I be alive the next day? Now at least you could go to sleep without worry and stay asleep throughout the whole night. A new experience! How wonderful!

Food and fuel for heating your home or apartment were scarce. So each day would be spent trying to find something to eat, and something with which to create some warmth in your home. We separated one-third of our living room from the rest of the house by bringing the support beams out of the cellar. They had originally been installed to hold up the cellar roof in case the house got a direct hit and would collapse. Now we used them in our living room to support our huge rug hanging in a vertical position, thus creating a wall. This reduced living space also housed the furnace, which we tried to keep heated during part of each day. We also moved my parent’s two beds into this space which now would function both as our living room and their bedroom.

Most German cities lay in ruins. Millions had lost their homes through the air raids. Other millions were refugees or expellees from the eastern parts of Germany which now had become part of Poland. All these people had to find a space to live. So two families were assigned to move into our home. One family had come from the East and the other had been bombed out in Hamburg. Our family of 6 had to squeeze into 3 rooms instead of the 7 rooms we had enjoyed formerly.

Still, our family was lucky. My parents, two sisters, my brother and I had all survived the war, and we were at home together, not prisoners of war like many of my classmates. Quite a few of my school comrades did not come home at all.

They were killed during the last days of the war, including my best friend Gert Wigand.

So called Nissen-Huts supplied by the British to provide living space for some of the bombed out Hamburger's [31]

In the evenings we were visited quite often by new friends, Mr. and Mrs. Krüger. They had been bombed out in Berlin, where Wolfgang Krüger had previously owned and managed a publishing house. Whenever my parents wanted to contact them, I was sent over to the neighbouring house where the authorities had relocated them. The Kruegers lived in just one room with a closet, a small table, two beds and a stove in the corner for heating and cooking. But three of their four walls were covered by huge paintings. All were originals, especially French paintings from the 19th century. I vividly recall a huge Renoir hanging above their beds which left me speechless. I simply marvelled at these paintings which were just about the only possessions the Krueger's had been able to save.

To do something useful and to learn how to repair cars, I worked in a downtown garage. As a worker, I received an additional food ration, about 100 grams of sausage per week. My mother let me buy and eat it all by myself. The 100 grams of sausage could be purchased for two coupons of 50 grams each. So instead of asking for 100 grams of sausage on my way home I would first stop at a certain butcher shop, wait to be served by the woman rather than the man who worked there also, and ask for 50 grams of sausage. It was difficult for her to accurately cut such a small amount from a thick sausage so, looking at this pale, thin young man, she typically sliced a piece weighing 70 or even 80 grams. This procedure I would repeat at the next shop, and in this way, was able to convert my two times 50 gram coupons into 150 grams of sausage.

Everything was rationed, including electricity. The garage where I was working was getting electricity for only 24 hours every other day. So we would work throughout one day and the following night, after which we could rest for 24 hours. It was still dark when at six o'clock in the morning I would walk to the suburban train station to take a train downtown to work. Still half asleep, I noticed an older man passing me on the other side of the street. With an axe in his right hand, he was pulling a small tree behind him. It was rather obvious that he had cut down this tree in the public park next to the station. I had also looked longingly at these trees, as I'm sure many of our neighbours had. But somehow we all felt that these trees belonging to all of us should survive. Suddenly I was wide awake. Wasn't this older man Wolfgang Krüger, owner of a publishing house in Berlin, living with a Renoir above his bed, going out in the wee hours of the morning to steel a small tree in the public park? Surely I could not be mistaken. I nearly went across the street to say hello to him, but thought better of it.

At home we had only cold running water. On Saturdays the stove in the bathroom was lit. There was a basin on top of it for heating water and to wash our clothing. The water, heated with wood from our garden, was transferred to our bath tub and we took turns for our weekend bath. I wondered: was there some other way to find energy for heating, cooking, or making warm water for washing? Couldn't the sun be used somehow?

With this in mind, I went to the kitchen and found a baby bottle. I filled it with water, put a thermometer into the opening, insulated the back, and set it close to the window in our living room. It was cold in the room, maybe 16 C or 60 F, but outside it was a nice sunny day with the sun's rays falling through the windows oriented to the South. And wonder of wonders, the thermometer slowly showed an increase in temperature. It was not long before it exceeded 20 C. It took nearly an hour to reach 23 degrees C. It was late afternoon when I admired a maximum temperature of 32 C or 92 F. I was excited. I had produced a quarter litre of warm water just by using the sun!

Producing warm water with the help of the sun in a baby bottle

Was this a realistic way to produce peaceful energy? I certainly did not foresee at that moment, to what degree my professional life would be intertwined with the study and use of solar energy.

Chapter 5
The Joys of Student Life; Coal and Folding Chairs: Is this Stealing? Who are these Quakers?

Summer, 1945. The war had ended. Germany was defeated, but there was peace. The country was divided into four occupation zones. Hamburg was located in the British zone in the north, the Americans were in central and southeastern Germany; the French in the southwest; and the Russians in the east. Twenty-five percent of the people in the British, American and French zones were German refugees from the East. They had to be housed in a country where to a large degree most towns and cities had been destroyed by bombs. Food was very scarce. In order to get your monthly ration cards every grown up had to collect building blocks out of the rubble, clean them of mortar with a hammer, and pile them up by the sidewalks in neat piles.

I, 17 years old, was happy that the war had ended. What a pleasure to go to bed at night and to be able to sleep undisturbed until the next morning. Some of my class mates had been killed during the last month of the war as soldiers and quite a few more of them were held in the west and the east as prisoners of war. It took another 2 to 4 years until finally they were let go from France, Belgium, Great Britain and even all the way from Siberia. But I was alive and wanted to make plans for my life. What profession should I choose? Could I go to a university? Would universities open up again? Or was this made impossible by the

Morgenthau Plan, which proposed to convert Germany from an industrial into an agricultural country? Hamburg University was in shambles and closed. Nobody knew if ever it would function again. But there were two cities in Germany which had been spared from the bombing: Heidelberg in the American and Göttingen in the British zone. Would they open their universities? No trains were running yet. I was allowed to travel only in the British zone, but Göttingen was only 150 miles away. Out came my bike; off I went.

My uncle Jürgen, a physiologist and professor at the university, lived in Göttingen. He was helpful. We went to see Professor Pohl, who headed the Physics Department. Would the University open again? He did not know. Should I study physics? He was not very hopeful for the field of physics for Germany. Agriculture might be a more promising subject. But then he suggested that I talk to Professor Walther, a professor of applied mathematics. He had just arrived from the Institute of Technology in Darmstadt, a town in the American Zone, a short distance south of Frankfurt. He had come to Göttingen to find out if this university had any plans to re-open.

Professor Walther told me about the total destruction of the Institute of Technology in Darmstadt. Maybe two lecture rooms were still usable. Though he didn't know if they would ever be able to fully re-open, he didn't want to give up. In the autumn there was a plan to offer a two-month introductory course for returning soldiers. If I was interested in joining and could find a way to get from the British into the American Zone, he offered the couch in his living room for me to sleep on the first days after my arrival. Yes, I told him, I would try to come.

October, 1945, my parents wished me good luck and let me head into an uncertain future. By now a few trains had started to run connecting Hamburg to the South. They consisted of open freight cars which had carried coal from the Ruhr area to the north and were now returning empty going south. So one day I began the long wait at the main station of Hamburg with hundreds of other people. Finally, an endless freight train pulled into the station and stopped on track five. We helped one another to climb up and over the side of a box car, open at the top. Our back-packs and my bike were lifted in. Finally the steam engine whistled and the hissing locomotive started out into the night.

It was a ride with many stops on the open track, and it was quite cold. How could we avoid freezing to death, shivering in this open freight car with the cold winter wind blowing? Could I find some way to warm us? At one of the stops I climbed out and next to the tracks found some wooden sticks, some tree branches and even some pieces of coal which apparently had fallen from a passing coal train. In the middle of the wooden floor of our box car laid a metal plate, and on this plate I lit a fire. Soon the train started up again. Rambling through a number of railway stations and passing by the two-story high towers from which railway workers controlled signals and switches, we wondered if the railroad officials would stop the train thinking it was on fire. But we didn't care. Slowly we got warm and felt alive again. Suddenly one of the riders thanked me profusely and offered me a present, a cigarette. This was a great surprise, because cigarettes were currency which you could exchange for food. To part with a cigarette really must have meant quite a bit to this man. I was a non-smoker, always having exchanged the few cigarettes I had gotten on ration cards for bread. But I could not refuse this heart-felt present. And so, that night in this open box car rolling south on the tracks between Uelzen and Celle, I smoked the second cigarette in my life.

The train took us to Hanover. From there I rode my bike south, sometimes hitching a ride on a truck. With great luck I managed to cross the border from the British Zone to the American without getting caught. Finally I arrived in Darmstadt, Professor Walther and his wife were surprised but happy to see me. They offered me a place to sleep on the living room floor, as the couch was already taken up by Rudi Przybilla, a young man from the eastern part of Germany whom they also tried to help in his new life as a student.

The next day he and I walked floor to floor and door to door in those houses which had survived the bombing on the outskirts of Darmstadt, asking the owners if they had a room to rent. I finally found a small room under the slanted roof of a five-story building. It became my home for the next five years. It contained a bed, a closet, a chair, and a small table. A cast iron stove could be used for heating and cooking. At the end of the corridor there was a small room with a sink and a cold water faucet and a toilet which was shared with another family living next door. Rudi finally found a room in another building nearby.

But what could we use to heat our rooms? Near the main railroad station at the switching yard with its many railroad tracks, Rudi and I had seen a long line of locomotives which had been blown up and destroyed by low flying fighter planes. Perhaps their tenders contained some leftover coal? We planned to check it out the next evening. Curfew started at 6 p.m., but around 5 p.m. it got dark enough for us to investigate without being seen by railroad workers. The first two tenders were empty, but the third one still had a layer of coal. The next night would be it. Rudi and I, each with a burlap sack, crossed the railroad tracks closing in on our locomotive. In the darkness we kept falling over the wires which ran parallel to the tracks and were used to move the switches and to control the signals. Some electric lights on high posts were periodically turned on and off illuminating the switching yard. We would freeze and then throw ourselves to the ground in a dark spot hoping that we would not be discovered. Finally, we reached our tender and filled our sacks with large pieces of coal. Heavily loaded we made it back to our rooms.

Now I could heat and cook my meals. Was life not wonderful? But was this a peaceful way to get energy? It dawned on me only years later that I had been a thief, stealing coal which did not belong to me. I remembered Wolfgang Krüger and the stolen tree. I had joined his ranks.

Together Rudi and I began organizing our new life as potential students. We were unbelievably happy to get this opportunity to learn. The results of the war were only too obvious all around us. Darmstadt had been destroyed by a number of nightly air raids during the last half year of the war. The inner city, including most of the Institute of Technology, lay in ruins. Only the chemistry department had survived, but it had been sequestered by the American Army.

TH Darmstadt: Part of the Department of Mathematics [25]

I still vividly remember our first lecture. Dressed in old army clothes we went to room 202, one of two lecture halls which still had a roof. On one wall a blackboard was fastened; otherwise the room was totally empty. In the cold temperatures we kept our gloves on. The professor entered and stood in his winter coat in front of us to begin his first lecture. His breathing showed up as white steam. There were no desks, no chairs, no chalk, no pencils, no paper, no books, but we all were very happy during this, our first lecture.

How could we improve our situation? One of our fellow students had an idea. Not far from us in the center of the city there used to be an open air restaurant. Like all the other buildings, it now lay in ruins. But he remembered that garden chairs were usually stored in the cellar. Maybe they were still there, spared from the fire storm which had destroyed that area. Let's look! So a group of us started out, and he was right. We could find our way through the rubble into the cellar and triumphantly returned one hour later, each of us carrying a metal folding chair. Lecture room 202 started to have some furniture, 22 salvaged garden chairs. What a surprise it was for our professor the next morning!

"Der weisse Turm" – the White Tower – near the center of Darmstadt [25]

The building where we had confiscated our folding chairs was located next to the castle in the center of Darmstadt. It has long since been rebuilt, but whenever I pass it these days I am flooded with memories. Yes, this was the place where we found the chairs under the rubble of a building, when it still was surrounded by ruins, the ruins which comprised the center of Darmstadt. And I can't help but remember the people who huddled in their cellars while this destruction took place. This terrible bombing happened on September 11th, not in 2001 but in 1944. Until that date Darmstadt had been spared from any serious destruction. Some citizens might even have dared on that day to listen to the BBC and heard that Darmstadt was on the list to be bombed very soon. During that terrible night in September 1944, 350 bombers approached the undefended city and delivered their deadly load. Here is a report of an eye witness [25]:

We went into the cellar, hoping that after a certain time we could return upstairs. But when we heard over the wire radio (Drahtfunk) that bombers would fly from Frankfurt in a southerly direction ,more bombers from Oppenheim towards the east and another group from Mannheim and Heidelberg towards the north, then we knew that now we would be the target.

A short while it remained calm. But then it happened, explosion after explosion. The light went out; we sensed a thick cloud of dust in the cellar, even with flashlights practically nothing could be seen any more. We sat or lay very close together on the floor. Bottles of Seltzers water which we always had in the cellar were very helpful. The detonations of bombs would not come to any end.
My father who suddenly ran out once more came back with the news that our house must have been hit but it did not yet burn. Then our small apprentice Annie called out: I believe it is burning! I can smell smoke! Soon we all noticed the very strong smell of a fire. We had to try to get out. The stairway was still there but covered with stones. We tried the cellar towards the back but a dense cloud of sparks entered through the cellar window.

From now on I can report only what happened to me. I ran up the cellar steps into the entrance hall but flames drove me back. I returned to the cellar, always calling for my relatives and the small Annie. In spite of my confusion it was clear to me that only through the flames we could find safety. I ran up the stairway again and noticed that suddenly a door had blown out as I assumed through air pressure. As I learned later in reality my father had opened it with an axe. From the store one could see the burning Elisabeth Street. But to reach this street one had to be brave enough to race through the store room which was already burning. An unknown man helped me to move a shelf standing in the way. I helped two or three women out of the cellar always calling for my relatives and the small Annie. The heat became unbearable and the fear of the women so overpowering that I could only scream: I will run now through the flames and if

I succeed, just follow me! I moistened a blanket in a pale of water standing in the cellar, pulled it around me and started to run. In the street everything was burning. I ran across wooden boards, all burning. Then suddenly a racing firestorm caught me and nearly pushed me down to the street. With the last energy I reached the Palais Garden. I ran to the middle finding there a wooden shack. There I stood in the glowing heat, around the Palais Garden everything was burning. The shack was filled with people with small children. The heat became more and more unbearable. A young man offered to look for some water. After a long time he returned, at least we were able to moisten again our clothes.

Suddenly some wood before the shack broke out into flames and soon somebody called, "the shack is burning!" We had to leave - but where to? All the houses around the garden and the Palais itself were burning, nowhere a path to be seen. From my loved ones I did not see anybody. I assumed that maybe I alone had escaped this hell. Within me everything was dead and had died. There was no pain, just a simple dullness. I left with the other people, and then a miracle happened: Suddenly I stood in the middle of the hell in the Palais Garden before my mother! We were deeply shaken but our questions about the others' fate nobody could answer.

But now our martyrdom had just started. Climbing over dead bodies and moaning people we reached the wall of the Palais Garden. We pressed ourselves close together to get some protection from the raging storm with all the sparks. My woolen blanket had been torn away from me by the gale like a piece of newspaper. My mother had lost hers too. I lay down halfway on top of her in order to protect her from the sparks. Then a bush next to us started to burn. We crawled to a big tree around which with the help of a man we put up great metal sheets which we found there and which gave us some protection. But this lasted only for a short while. Pieces of wood and other debris lying around our feet started to burn. We had to go on.

We crawled back to the wall with our heads always close to the soil. I got the shivers and had to throw up. Once I had lost the total volume of my gall bladder I felt better. Then the wall got so hot again that we had to crawl back to the middle of the garden. There we were in the middle of hell. Once ice cold air hit us then again it was glowing hot each time hitting us with burning ashes whenever another house around us started to collapse. We were totally occupied just to put out the sparks hitting us. Astonishing and really deafening was the

CHAPTER 5

Totenstille – dead silence – which surrounded us in the Palais Garden In spite of so many people present. We did not hear any sound, no moaning - nothing. Only from the cellars in the Peter-Gemeinder-Street terrible screams and calls for help were heard. But it was impossible to help anybody.

By now It was 3 o'clock in the morning. The storm calmed down slowly and the heat became bearable. About the gruesome pictures which we had to see when we searched for something to cover us I will not talk. I wish that we will be able to forget them soon. Our eyes were burning to such a degree that we hardly could see any more.

Slowly the fires became smaller. Our house had totally collapsed. Around 5 o'clock in the morning we decided to leave this hell. We held tight to each other and climbed across debris and charcoaled bodies through the still burning streets toward the Marienplatz. There we found the same picture as in the Palais Garden. We tried to reach the house of a friend, but their house was in flames. Now we stumbled along the Heidelberger Straße: Not one house which was still whole or did not burn. Totally exhausted a man in a heavily damaged house invited us in. He gave us some wine and we sat for an hour on his sofa. Later we went to the Marien Hospital to get some help for our half blind eyes. There we saw only pictures of the horror ....

Why do we still have wars in this world after experiencing all this misery? Why do they escalate, engulfing not only the military but more and more the civilian population? And why do vast majorities of the population, especially those on the stronger side, the "winning" side, accept this? Would I, who as a 15-year-old boy had been drafted into shooting at bombers which were dropping bombs on me, would I, now as an adult, be drawn again into a similar situation, perhaps contributing to the war effort as the engineer I hoped to become?

In 1939 after the war started President Roosevelt sent a message to the warring nations of Europe including Great Britain and Germany: Think of the innocent civilians! Don't bomb open cities! Initially it seemed that both nations heeded this advice, only military objectives were selected as targets. But this slowly changed. Each side, of course, blamed the other one for having started the bombing of civilian homes. "We" as it was reported, "only retaliated."

As the war became worse, it expanded to more countries. Could an increase in bombing bring a quicker end to this slaughter? Eventually, the destruction of the houses of the civilian population along with other civilian targets became accepted as an official aim of the war effort. This destruction would prevent the civilians from working in the factories producing arms, and they might become demoralized and lose the will to support the war effort. Surely it was clear that this plan would destroy not only the buildings, but also many of the people living and working there. As it turned out, many thousands would die in a city like London, 10,000 in a small town like Darmstadt, more than 40,000 in my home town in Hamburg, and at least 100,000 in Tokyo. Of course, we say we don't like to do this, but for the greater good and in order to finish a terrible war, we have come to accept what is now called "collateral damage." Questions of legality and morality have been left behind.

I was studying to become an engineer. Had engineers been asked to develop the best methods for burning down an entire city - engineers like I was hoping to become? Yes, carpenters and electricians, architects and engineers in the United States had been called to construct small sections of a street with typical Japanese and German homes in order to develop the most efficient methods for burning them down. They did their task well. Did they realize what they were doing? Probably.

Other engineers designed two types of incendiary bombs. One was filled with a rubber-like material containing phosphorus. Upon impact this material dispersed in all directions, and the phosphorus burst into thousands of little flames capable of igniting bedding, paper, and curtains. A second type exploded and sent small particles of magnesium flying in all directions, burning with such a high temperature that wooden beams would be ignited. And then somebody proposed that when dropping incendiary bombs one should mix in 30% exploding bombs, to make sure that the Germans or the Japanese or the British people would not come out of their cellars to fight the fire. This combination of bombs then could hopefully create thousands of small fires which would combine into a huge fire storm engulfing an entire city.

The first time this policy succeeded on a large scale took place on July 26th and 27th, 1943, in Hamburg. It was planned and performed under the name of Gomorrah! On an even larger scale we humans succeeded in Dresden and Tokyo. If I had been given the task of planning or executing these deeds, would I have gone along or would I have refused to participate? Dear reader, how would you have reacted either then or now if asked to help with such a plan? Most of us never get touched so closely. But are we not all paying our taxes today making possible the bombings in villages and cities all over the globe accepting "collateral damage"? And if we ever think about it we claim to do it as a necessity to reach a greater goal of security, justice, freedom and democracy in accordance with our truth?

September 11, the night Darmstadt died in 1944, has gotten a new meaning for us. On September 11, 2001, nineteen young men, some of them engineers, had worked quietly for years pursuing one goal. They would sacrifice their own lives for what they believed in. On that day they hi-jacked four airplanes. They knew they were going to die by attacking the World Trade Center Towers in New York and the Pentagon in Washington. They must also have realized that, in this process, they would kill not only themselves but also many innocent people. But this they apparently had accepted in order to reach their greater goal. They joined the ranks of those who are willing to die for their "truth" taking along with them many innocent people.

Let us return once again to the Darmstadt of January 1946 when I arrived there to study to become an engineer. As students all of us had to spend one day each week to help to clear the university area of stones and rubble. As citizens of Darmstadt we had to spend another day helping to clean the streets and sidewalks. Only if we presented a paper showing that we had worked for a full day would we receive our food ration cards. That didn't leave much time to study.

One day at the entrance to room 202, I noticed a strange message posted between offers to exchange books for light bulbs. It said that a group of American Quakers were trying to put up some wooden shacks for German refugees from Hungary. Would any student be willing to help during the next weekend?

I had no idea what Quakers were, but I was fascinated. A group of American civilians had come to help us Germans, the defeated enemy? American soldiers still were not allowed to fraternize or even talk to us. And yet, at the same time a group of Americans had apparently come not to judge or condemn us, but to help put up some housing for refugees. So I went there and worked for an afternoon with them. I had learned some Latin and Greek at school, or should I say during my years in the anti-aircraft battery, but hardly any English. Conversing with these men and women was therefore very limited, as they didn't speak any German. But I was extremely moved by this experience. Was this part of an answer to the unanswered questions I had asked myself about dictatorship and democracy, communism and Christianity, war crimes and peaceful action? Did these Quakers represent a kind of deeply human, peaceful energy, free of violence?

Chapter 6
Crossing Borders, I Am Getting Shot, really?

It is Sunday morning August 23, 2009. I am in my office in Germany, surrounded by my physics experiments and listening to classical music from a little radio in the background. From behind my computer, where I am sitting, I can look out of four different windows, all of them revealing the green meadow and bushes growing in the garden around the house. It is beautiful here in the outskirts of our little village of Burgberg in the middle of the Black Forest. "Burg" means castle and "Berg" means mountain. Our house lies on the top of a small mountain and yes, we do have a ruin of a Burg, a castle destroyed about 500 years ago. It feels to me like paradise, surrounded by beauty, peace and quiet.

Today I plan to work on this chapter, CROSSING BORDERS; I AM GETTING SHOT; REALLY? So I think back in time, to the day of June 25, 1946, when a young French soldier shouldered his rifle and ordered me to go ahead into a deep wood, far away from any other human being. Was he going to shoot me to cover up his deed of stealing my boots, a knife, my only watch and a picture of my sister? This happened 63 years ago -- a long time. I am 81 years old now, so in order to stimulate my memory I put some books on the table next to me like "THE NIGHT HAMBURG DIED" [26], and "Die Brandnacht - THE NIGHT OF THE FIRE" [25], documents about the destruction of Darmstadt on

CHAPTER 6

September 11, 1944, and "DIE FLAKHELFER - THE ANTI-AIRCRAFT HELPERS", a book about an auxiliary corps of 15 to 16 year old school boys during the Second World War. These books remind me of my encounters with death, violent death. I have just re-read my Chapter 3,

"Child Soldiers Protect their City; Firestorm in Hamburg July, 1943"

which I wrote seven years ago. I am astonished with what detachment I recalled my time in the anti-aircraft, more like a historian, not like a 15 year old boy who had participated in this horror.

Two months ago I took a train ride from Burgberg, in southern Germany, over 500 miles north to Hamburg, the city I grew up in and tried to defend against the British and American bombers during the last two years of the war. And when the train reached the outskirts of Hamburg and entered the suburb Hamburg, where our battery was located in 1944, my heart started to pound noticeably. I realized I had not come to terms with my experiences during those war years, not with the closeness to death, not of crossing this border. The reason for my present journey was a reunion with my old school comrades which we try to organize every two years. These are the guys I went to school with and lived so closely together with for nearly two years in the anti-aircraft.

During the six hour train ride north I read through class letters we wrote to each other between 1945 and 1947. In these letters we reported what had happened to us in the last years of the war as 17 and 18 year old soldiers, and later as ex-soldiers or prisoners of war. We wrote from Great Britain, Belgium, France and even Siberia, where a few had to work for another two to four years after peace was restored. Three of our class of 22 had stopped writing before the war was over; they had died in the fighting. While we met in the home of one of my classmates I read to them some excerpts from these letters. I was very much astonished: When I read very specific and dramatic experiences they had written about shortly before or after the war had ended, they did not remember these incidents now. How was that possible? Had they really forgotten the incidents or had they just suppressed them?

Why does my heart beat so hard whenever I get close to Hamburg? Is it the terror or closeness to death which I experienced there as a teenage boy, experiencing death all around me or, much worse, by wondering if I would die tonight or even in the next minute? Like my comrades, I had not come to terms with those experiences.

So why write about it now?

In normal times 1945 my comrades and I would have graduated from our high school by taking a difficult examine called “Abitur”. You have to take tests, written and oral, in all of the subjects which were covered during the 8 years of secondary school. In normal circumstances this would have taken place in the spring of 1945. Because of the war this didn’t happen. Most of us had become real soldiers, some were prisoners of war, and three of our classmates had fallen.

Whenever a major graduation anniversary comes around, 40, 50, or 60 years later, we are invited to participate in the graduation ceremonies of the present class of graduates. We sit through the speeches and music played by the school orchestra and that is it. When the last two anniversaries approached, I asked my classmates if we shouldn’t write down some of the experiences about the times when we were the age of these young graduates living together in the anti-aircraft. It had been a bad experience for all of us, for our home city Hamburg, for our families, for the whole of Germany and for many countries involved in this Second World War. Had we not learned something? What could we share with this young generation in order that they might avoid the mistakes we, our parents and our grandparents had made? And each time after some discussion I was told “No, this idea is pointless”. Nobody would listen to us. Most of us had not even been able to talk to our children about it. To try now would be useless; each generation would have to make its own experience.

I disagreed, but ultimately nothing happened. We sat through another round of speeches and music. Speechless! I felt we were speechless because we didn’t want to think about it, we didn’t want to talk about it with any outsider, because admitting it or not, we had not come to terms with it. And that included me.

This book: “My Path to Peaceful Energy!” originally was supposed to present my experiments in physics, which I began in 1997. My results seem to confirm that the ideas Lochschmidt announced in 1867 were correct when he wrote:

…………"Thereby the terroristic nimbus of the Second Law is destroyed, a nimbus which makes that Second Law appear as the annihilating principle of all life in the universe, and at the same time we are confronted with the comforting perspective that, as far as the conversion of heat into work is concerned,

mankind will not solely be dependent on the intervention of coal or of the sun, but will have available an inexhaustible resource of convertible heat at all times" [3].

If the results of my experiments really prove eventually to be correct the world might have a new source of energy independent of burning coal or oil and independent of using atomic energy. Slowly our world is becoming aware of this problem in this age of global warming. It really would be a peaceful energy. How far away are we from this vision in the reality of our present life! I still remember when the first president George Bush, regarding the first Iraq war announced on television: "...this war is necessary as we have to defend **our** oil..."

Our oil? The oil lying underneath the land mass known as Iraq? After two wars in Iraq, we count more than one million Iraqis dead and three million more Iraqis as refugees. Internal strife sets Iraqis against Iraqis, and the country and its infrastructure lies in ruins. Trying to defend **our** oil led again to terrible violence. And I believe experiencing violence in my teenage years was a major factor in my interest in a source of peaceful energy. It led me to develop a new ethnical model I call CIRCABILITY which I will describe in a later chapter. Its basic meaning consists of the idea that we should perform only those tasks which the next generation could undo in case they realize that our actions were a mistake. Killing of any person for any reasons is out, as this would constitute an action which could not be undone. The same would apply to the burning of oil and gas, as we cannot recreate these resources. And the use of atomic energy would not be CIRCABLE as the waste material generated by any atomic power plant will radiate for thousands of years.

Therefore this book covers not only my experiments in physics but my encounter with violence, often generated by our ever increasing need for more energy. This book is not written by a professor who studied the connection of violence and energy for years and then finally wrote a scholarly book about it. It is written by somebody who experienced various forms of violence as a young boy and a young man and was very much affected by it. And writing about it I found that these various incidents affected me to a different degree. Its seems to me that whenever one involved the direct meeting with another human in some kind of violence I remember such an incident more vividly, even if it was not connected with a death of a person. So in the following section I will write about some of those which my memory brings back to me.

It was before the firestorm nights of July 1943, about two months earlier. Every two or three weeks we would get a one night leave from the anti aircraft and were allowed to visit our families with an overnight stay. Would I experience another air raid at home without having to run to our alarm stations? Following one of those nights where a number of bombs fell, my father asked me to join him before I had to return to my battery. He was always very uneasy driving alone in his car. As he was not mechanically inclined, he felt that he would be helpless if something should go wrong with his car. He thought I might be useful in such a situation. On this morning he wanted to drive downtown to investigate a subway station were the previous night a bomb had killed a few people.

It was a beautiful summer's day with a cloudless blue sky when we drove downtown. We parked close to the subway station, which was located in the middle of a wide street. An air raid warden stood guard at its entrance. My father was a professor of pathology and had been drafted as a reserve medical military officer. One of his tasks was to study the causes of death of people in air raids. Both of us being in uniform, we were allowed to enter the subway station.

Climbing down the steps to the platform it was getting darker and darker, there were no lights were burning. Coming out of the bright sunshine it took my eyes quite some time to adjust to the darkness. Finally we reached the platform. It looked like we had entered into hell. There were two holes in the ceiling, each about a meter in diameter, where one could see blue sky and where the sunlight entered, creating two bright spots on the platform. An electrical floodlight running on a noisy generator lit a small area in the station. Once our eyes got accustomed to these stark contrasts between light and dark we saw it: The whole station, on the platform and on the tracks, was filled with mountains of dead bodies. After a while one could visualize individual bodies amongst the heaps, mainly of women and children, many with their cloths half torn off. Under the bright electrical light a doctor in a white gown was examining the body of one of the dead women, checking her ear drums. It was very apparent that this subway station had been used as an air raid shelter and had received a direct hit from two bombs. They had crashed through the roof and exploded on the platform crowded with all the people now lying dead in these huge mountains of human flesh.

Once we had fathomed the situation my father sent me out of this inferno, he would follow soon as a colleague of his was already working here. I left and climbed up the stairway, squinting into the bright sunlight. The air raid warden at the top of the stairway smiled at me and said: "Well, did you get yourself a pound of meat?" He must have seen my shocked expression when he added: "Oh, I am sorry. I did not know that you had a relative down there!" I still wonder today what made this warden utter those words, was it his way of dealing with such a monstrous happening?

It must have been around that same time when my younger brother Gernot lived temporarily with our Uncle and Aunt in Göttingen, 100 miles south of Hamburg, in order to escape the air raids happening in our home town. One day he was looking out of a second story window with our Aunt Hilde watching American bombers passing by in the distance to a target farther away. But they were under attack by German fighter planes. What excitement for a 13 years old boy! Suddenly one of the bombers let out a smoke screen behind it and was gaining speed, falling towards the ground. My brother jumped up and down and screamed: "Der Feind ist besiegt, der Feind ist besiegt!"- the enemy has been defeated! At that moment Aunt Hilde raised her right arm and gave him a big slap on the cheek "Don't you know that there are human beings in there?" she called out. She was one of the few who kept her humanity alive and acted on it even in such a situation. What she did was a very brave deed to have done at that time. If this incident had been reported to authorities, Aunt Hilde could have gotten into deep trouble.

The worst air raids on Hamburg happened under the code name Gomorrah from July 25 through August 3, 1943. In four night raids by the British and two day raids by the Americans, 2500 bombers delivered 8500 tons of explosives and incendiary bombs. That equates to about 10 pounds of explosives for each man, women and child of the 1.5 million inhabitants of the city. One of the raids had created a firestorm in the southwest of the city, the areas of Hamm, Hammerbrook and Rothenburgsort. Our battery had been spared. Two nights later the alarm had sounded again. Very soon it became clear: This time our section of the city was going to be the main target. The "Christmas Trees", as we called the colorful bright lights, slowly sinking hanging on their parachutes were placed into the sky by the first British planes called Pathfinders. They marked the four corers of the area where the following bombers would unload their deadly freight. It indicated the area all around us for the bombers to deliver their bombs.

As in the first night, our radar failed during the first minutes of the attack. Our six guns could only fire at a target whenever the search lights caught one in their beams. This did not happen often, so most of the time our guns just fired wildly in the general direction of the aircraft sound. This night I was part of the group which was supposed to translate directional data coming from other radar stations situated in batteries around the city, translating these data into angles and distances useable by our own fire coordination station. The older soldier in charge and we four boys were standing around the Malsi-table, which was named after the inventor of the instruments we were supposed to be using. But no data came from these stations as again all the British bombers had thrown out millions of staniol strips making our radar stations totally useless. Since no data arrived we were without a task. We just could listen to the firing of our own guns and the explosion of bombs, sometimes closer and sometimes further away.

Then it happened. It sounded like an approaching siren, getting louder and louder. With the increasing intensity the sound and pitch got higher and higher. We all felt that in the next moment we would get hit by a huge bomb. Of course everything happened very fast. Even so, time seemed to have stopped. My three comrades jumped under the Malsi-table trying to get some protection. It was a wooden platform a few centimeters thick and with a diameter of 1.5 meters, standing on three wooden legs. My mind told me right away that this flimsy table would provide no additional security. So I just bent over the top of the table and waited for the explosion. The explosion came. It was not only a deafening noise but a pressure wave hit our little barracks and our eardrums. I still can recreate this kind of noise in my ears by pulling the muscles under tension somewhere within my ears. The earth shook violently like in a big earthquake. I had the desire to crawl into the earth for protection. Our barrack shook but didn't collapse; only some of our wooden planks mounted outside of our windows to block out our light were blown off. Paint and dust fell on us from the ceiling and the walls.

Then the normal sound from our guns returned. We were alive. Of course all of this had happened in very few seconds, but it seemed like an eternity. My next thought went to the soldiers working the search light, which was located about 20 meters west of our barracks in the direction of the big explosion. All these poor people must be dead now, I thought. I also knew this had not been a normal explosive bomb but an air mine, one of those monsters weighing more than a ton and because of their weight being used very rarely in the bombing raids. My comrades crawled out from under the Malsi-table. I don't think we said a word, we just dusted ourselves off. We could hear a screaming voice apparently coming from the lieutenant commanding the fire control center situated 10 meters east of us, opposite the direction of the search lights. "Turn off your search lights!" he screamed. He must have thought that these lights had been the target for this air mine. As it turned out the soldiers were still alive and the search light was still intact. The mine had not exploded that close to us, but a full 200 meters away. It had destroyed a restaurant and a number of little garden houses. When we later past by the garden houses an old couple lay dead on the ground next to one of them.

We just had started to breathe normally again when the officer in charge of communication walked in. "All of our communication to the guns is interrupted! Do you still have a working line to them?" No, we didn't. "They shall stop firing as there are no enemy bombers over the city anymore!" "I could run down and tell them" I proposed and I was told to proceed. So I left the protection of our barracks, which of course constituted no real protection and ran towards the still firing guns. The barrels glowed dark red in the night sky, having shot that many rounds of ammunition. I had no idea where I could find the person in charge so when I got close I screamed between two salvos towards the gun embankments "Stop the fire! No enemy bomber anymore above the city!" And wonder of wonders somebody at the closest gun had heard me and he repeated my scream. At each of the other guns somebody repeated to the next "Stop firing!" and really, the guns fell quiet. I thought I had successfully completed my mission and turned around. But after just a few steps I could clearly hear aircraft motors in the sky and very quickly the guns started to fire again. Then I noticed shadowy figures running from the gun encampments, 20 meters across a meadow to little bunkers containing reserve shells. They took a shell on their shoulders and carried them to their guns. I wasn't needed anymore at my usual station so I joined them. Each shell weighing roughly 70 pounds or 32 kg was a pretty good load for this slim 15 year old boy. But I managed in that excitement to load one on my shoulder and to walk towards the nearest gun encampment. Getting closer to the firing guns the noise from their firing became deafening, especially since I had no cottonwool in my ears or any other ear protection. They shot a salvo about every 15 seconds. But in between I could hear the sizzling noise of falling incendiary bombs nearby ending with the noise like "plumbs" when they hit the meadow and were buried underground.

This was pretty scary. And then a somewhat louder hissing noise announced the arrival of a more normal sized bomb, not an air mine as no siren was sounding. So I threw myself on the ground with my body over the shell trying to protect it so it would not be hit by a splinter from the incoming bomb causing it to explode. The bomb exploded not too far away but it wasn't half as bad as the mine 20 minutes earlier. It was still very, very scary. Could I get up and proceed with my shell?

This bomb missed us by about 200 meters

And then something very strange happened. Suddenly I knew that there existed a personal God, there existed a personal God who was protecting me. He would take care of me, there was nothing to fear. And suddenly I had no problem standing up amongst all of the noise, putting the shell back on my shoulder and fearlessly delivering it to the firing guns. I would now walk upright, not hunched down. I would walk back and forth within all of this chaos, falling bombs and firing guns. I was totally fearless as I was protected by God.

Half an hour later, practically out of ammunition and with bombers now really gone, a strange stillness existed. Once in a while a bomb with the delayed fuse exploded and you could hear the crackling of the fires burning in the buildings around us. As we had no ammunition left we were allowed to go to those buildings and help the people to save their furniture or assist in the fire fighting.

This intense feeling of the existence of a personal God which was protecting me didn't last very long. Already during the next raid, the normal apprehension and fears came back. I even forgot this incident for many years. I remembered it suddenly many years later when I became active in the peace movement in Germany and United States. I started to question the whole meaning of this incident. How strange would it be if God would tell a 15 year old boy to feel protected and safe so he could fearlessly carry shells to guns firing at bombers and their crews? This seemed to be a complete contradiction.

I had to get back to my station now as it was my duty to draw on a piece of paper the flight path of the bombers we had fired upon.

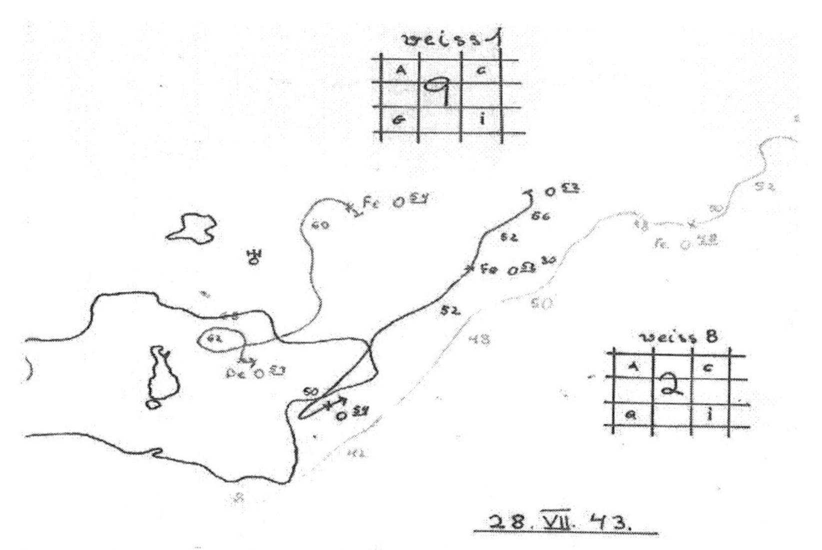

Sample of a flight pass drawing for the night of July 28, 1943. The inner circle indicates Alster Lake in the center of the city, the surrounding line the outskirts of Hamburg. Our battery was located in the circle, the upper flight path indicated for a bomber we shot at.

So while my comrades had to remain at our station, I went next door to our sleeping barrack to fetch my drawing utensils. I stepped down the three steps leading to my locker and surprise -- half of it was gone! In the small circle of my flash light 8 feet away I could see a bomb sticking out of the ground! Of course I could only see the upper part occupying the lower section of my locker, maybe two feet of it. It had a diameter of about one foot. I stopped in my tracks. I tried to breathe very shallow, trying to avoid any vibrations to the floor. Should I slowly and very carefully withdraw? I realized that if the bomb exploded right then I would be dead anyway. So I went closer to the bomb trying to listen for the ticking of a watch, indicating a time bomb. No, everything was quiet. So very slowly I withdrew and reported my find. The area was cleared and we were free to leave the battery to help fight the fires in the neighborhood. We had no ammunition left and more would not arrive until the next morning.

These were the buildings we tried to save, three weeks after the raid.

The arrows indicate were I was balancing on the still standing vertical walls with a water hose trying to oust the fires burning below. At that time only the upper floor was burning. Meanwhile some of my comrades ran through the apartments on the lower floors, throwing bedding, carpets and any clothing we could find out of the windows. Radios, stoves and chairs we carried out. We piled these, together with the stuff which had been thrown out of the windows, in the middle of the street so it would hopefully not catch on fire. The desperate residents of the apartments, mostly old people, women and children, tried to recover what was left of their personal belongings. It was a surreal situation in the middle of the night, illuminated by the burning buildings. This was not one or two buildings, but innumerable ones, and this was the same desperate and hopeless situation, street after street!

When morning broke we had to return to our battery. What had happened to the bomb in my locker? Specialists of a bomb removal team had arrived and defused and removed my bomb. As it turned out it had not been a normal explosive bomb but a special bomb which was supposed to explode high in the sky releasing colored markers to show the ensuing bombers were to release their deadly load. It being a dud had probably saved our battery from a worse fate. Was I saved twice? If he bomb had exploded high in the sky telling the bombers to unload their bombs directly on top of us? Or a second time when the bomb did not explode in my locker 6 meters away from me?

We thought we had put out the fire in the house we tried to save when we left early the next morning. But the fires kept smoldering and when the water ran out they eventually destroyed all the buildings. Only empty shells were left in this section of the city as I found out three weeks later when I took the pictures on the previous page.

All this changed radically when the war came to an end in April of 1945. How wonderful: I could go to bed in the evening and sleep through the night undisturbed by any alarm bell or the sound of an air raid siren! But these sounds did not leave me completely. They are still buried deep inside me. Even now, 65 years later, I freeze when I suddenly hear a siren in the distance!

Peace! Life became more normal, slowly, step by step. In the latter part of 1946 travel became somewhat easier. Leaving Hamburg we didn't have to climb into or onto the top of freight trains anymore. Now the steam locomotive slowly entering the station would pull a strange collection of old passenger cars behind it. The platform was overcrowded with people, most of them having a large amount of luggage. It was clear that there was not enough space in the compartments for everybody. So each of us waiting there was wondering: Who will be left behind? Once the train came to a stop the struggle to get close to the doors began. If you didn't join the pushing and shoving, you might as well give up right away and return home. Most people were carrying along all of what was left of their belongings. The best strategy for getting on the train would be that the father, being the strongest of the family, would try to fight his way into a compartment and then open a window towards the platform. If the family acted quickly enough they would succeed in handing him all of their belongings and then the children. Finally the mother of the family would be pushed up with the help of bystanders, through the window and into the compartment.

Having gone through this 'storming the train' situation a few times, I had adopted my own method. Standing on the edge on the platform above the tracks when the train slowly moved in, I would jump onto the tracks, quickly cross them and climb over to the other side, which was a smaller platform not accessible to passengers but used by railroad workers moving luggage or the mail onto the train. Once the train came to a stop, I was the only person on that side of the compartments and therefore was the first to enter.

Inside each compartment typically ten people would squeeze onto the two benches designed for eight. In the space between the benches another four to six people would stand. The passageway leading to the compartments was also filled with people and their luggage. Another two or three people filled the restrooms. After about twenty minutes it would become clear who could leave with the train and who would be left behind. The outside doors were pushed shut and slowly the train left the station.

In 1946 the northern British and the American zones were combined into one, while the southern French and the Russian zones remained separate. The combination of the northern two zones made travel easier because we were allowed to move between these two zones without a special permit. Traveling from Hamburg in the British zone to Darmstadt in the American Zone, the train would cross into the American zone in Bad Hersfeld. Typically we would arrive in the middle of the night. The train would stop and everybody had to get off the train, taking their entire luggage with them. On the platform American soldiers would check everyone's identity papers and go through some of our suitcases, backpacks and boxes.

Even today, more then sixty years later, I cannot forget one incident that happened at one of those stops in the middle of winter. It was around two o'clock in the morning. The wind was blowing snow across the unprotected platform and everybody quickly became very cold. After about an hour of checking our papers and searching through our luggage, the signal was finally given. We were allowed back onto the train. The struggle to get back into the compartments was not too bad as all of us had arrived with this train. So we knew everybody would be able to squeeze in. The doors started to close, we were ready to leave. But looking out onto the platform I saw a family consisting of 3 children, their parents and grandparents. They must have been one of those refugee families trying to find a place somewhere to start again with their daily lives. "All on board" sounded from the loudspeaker.

Obviously they were late when the family started to move towards the train to climb in. But there was a problem. A soldier still wanted to search through a big cardboard box belonging to the grandfather. The box was held together by multitude of strings. Grandfather had taken off his gloves, trying to untie the valuable strings but with his half frozen fingers progress was minimal. The "All on board" sounded again. Half of the family was now inside and grandma was hesitating between grandfather and the rest of the family already on the train. Finally she also climbed in. She was the last one to enter when all doors were closed. I started to wonder what would happen next. It was obvious that the grandfather would not be able to open his cardboard box before the train would leave. Would he stay behind with the last goods the family owned and let his family go? How would they ever meet again with no mail service or telephone connections, probably not having any relative or friend in this region?

One could feel the tension of the situation as all the people behind the windows in the train watched. What would happen next? The grandfather had to choose between saving the last family belongings or joining his family. At that moment the soldier pulled out a knife bend down to the cardboard box and with two or three quick strokes cut the strings. This was a very flimsy cardboard box and was held together only by this multitude of strings. With the strings gone it just opened up and exposed the contents in a little heap on the floor of the platform. We all realized right away the old man had no chance to reassemble his belongings in any kind of enclosure before the train would start moving. The choice was very clear. Jump on the train and stay with your family or stay behind to save your belongings.

It was a ghostlike picture, the dark night filled with snowflakes and a station illuminated by only a few electric lights. Was the grandfather crying? His head turned back and forth looking at his family on the train slowly starting to pull out of the station and his heap of belongings. And he stayed on the platform next to the soldier with his gun and with his knife.

I keep wondering what the soldier thought at that time. Or what he thought in future years about his actions at that moment. Was he just tired and cold like all of us and in his frustration just wanted to get the situation resolved? Or did he enjoy being in the position of power, degrading another human being or maybe even feeling a sense of satisfaction by being really cruel? I feel that I remember this incident so clearly because I experienced the action of an oppressor and the oppressed so personally, so close by, and was powerless to do anything about it.

Aunt Hilde brought back the human aspect of violence. I believe people in general don't like to be cruel towards other human beings. So when we "civilized" people want to force other people to live or act in accordance with our wishes and demands, we organize our use of force as impersonally as possible. When we get some information about a "terrorist", located somewhere in a foreign country like Pakistan whom we want to kill, we don't even try to capture him. We get news that he plans to attend a family reunion at the occasion of a wedding. So we send a Drone carrying a television camera viewing the ground, flying out from a secret airport somewhere in Afghanistan. This Drone is totally controlled by a man or woman sitting in the cellar of the Pentagon in Washington or in an underground bunker somewhere under the mountains of Colorado, watching the start of the wedding. The officer in charge has orders to find and to kill this terrorist. He knows that there will be some "collateral damage" meaning some innocent bystanders in the wedding party will be killed too. His supervisor knows the rules. As long as only 35 civilians might get killed, the general in charge can give the go ahead. If the number gets above 35 then the president has to sign the order. At least this was the rule under former president George W. Bush. So once the officer in the Pentagon sees his target arriving at the tent he pushes the button, firing the rocket from the Drone, all the while comfortably seated in the Pentagon halfway around the globe.

After my first term as a student in Darmstadt, I returned from Hamburg. The news was bad: being so young, only 18 years old, I would not be admitted for the second term. There were few openings available and they would be reserved for those older students who had been soldiers since the early years of the war. What choice did I have? The American zone had been combined with the British Zone but the French Zone was still controlled separately. I had relatives in the French zone in Freiburg, in the southwest corner of Germany. I also knew that the local University had also re-opened. Why not try to get admitted there for studying physics and math. These were subjects I would need as an engineer. For the first time since the end of the war I would try to cross this border legally by getting a legal "card for exemption to travel" from Darmstadt to Freiburg from the Military Government of Germany Land Hessen-Nassau.

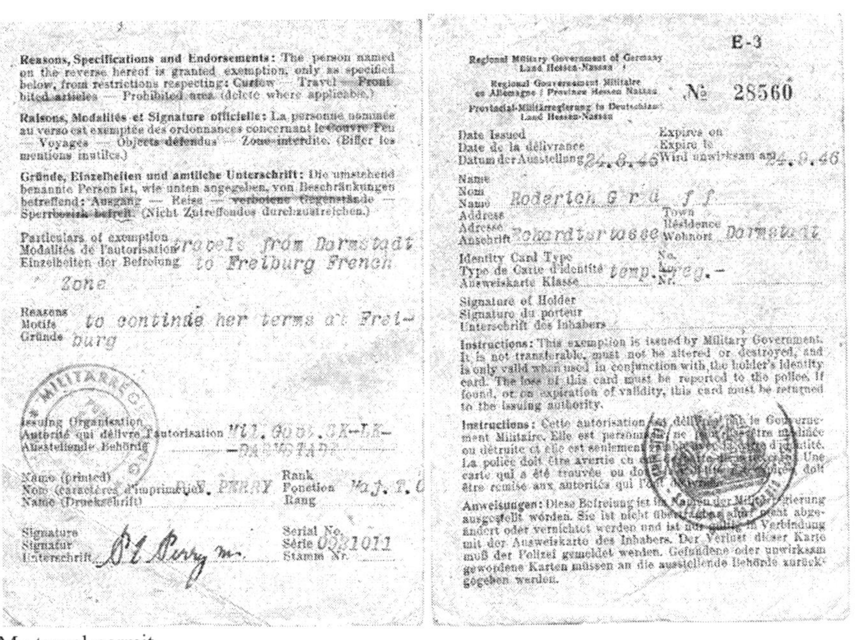

Reasons, Specifications and Endorsements: The person named on the reverse hereof is granted exemption, only as specified below, from restrictions respecting: Curfew — Travel — Prohibited articles — Prohibited area (delete where applicable.)

Raisons, Modalités et Signature officielle: La personne nommée au verso est exemptée des ordonnances concernant le Couvre Feu — Voyages — Objets défendus — Zone interdite. (Biffer les mentions inutiles.)

Gründe, Einzelheiten und amtliche Unterschrift: Die umstehend benannte Person ist, wie unten angegeben, von Beschränkungen betreffend: Ausgang — Reise — verbotene Gegenstände — Sperrbezirk befreit. (Nicht Zutreffendes durchzustreichen.)

Particulars of exemption / Modalités de l'autorisation / Einzelheiten der Befreiung: travels from Darmstadt to Freiburg French Zone

Reasons / Motifs / Gründe: to continue her terms at Freiburg

Issuing Organization / Autorité qui délivre l'autorisation / Ausstellende Behörde: Mil. Gov. SK-LK-Darmstadt

Name (printed) / Nom (caractères d'imprimerie) / Name (Druckschrift): D. L. PERRY

Rank / Fonction / Rang: Maj. T. C

Signature / Signatur / Unterschrift: D L Perry

Serial No. / Série / Stamm Nr.: 0031011

E-3

Regional Military Government of Germany Land Hessen-Nassau

Regional Gouvernement Militaire en Allemagne / Province Hessen Nassau

Provincial-Militärregierung in Deutschland Land Hessen-Nassau

№ 28560

Date issued / Date de la délivrance / Datum der Ausstellung: 24. 8. 46

Expires on / Expire le / Wird unwirksam am: 24. 9. 46

Name / Nom / Name: Roderich Graff

Address / Adresse / Anschrift: Eckhardtstrasse

Town / Résidence / Wohnort: Darmstadt

Identity Card Type / Type de Carte d'identité / Ausweiskarte Klasse: temp. reg. -

No. / No. / Nr.

Signature of Holder / Signature du porteur / Unterschrift des Inhabers

Instructions: This exemption is issued by Military Government. It is not transferable, must not be altered or destroyed, and is only valid when used in conjunction with the holder's identity card. The loss of this card must be reported to the police. If found, or on expiration of validity, this card must be returned to the issuing authority.

Instructions: Cette autorisation est délivrée par le Gouvernement Militaire. Elle est personnelle, ne [illegible] être modifiée ou détruite et elle est seulement [illegible] d'identité. La police doit être avertie en [illegible]. Une carte qui a été trouvée ou do[illegible] doit être remise aux autorités qui l'ont [illegible].

Anweisungen: Diese Befreiung ist im Rahmen der Militärregierung ausgestellt worden. Sie ist nicht übertragbar, darf nicht abgeändert oder vernichtet werden und ist nur gültig in Verbindung mit der Ausweiskarte des Inhabers. Der Verlust dieser Karte muß der Polizei gemeldet werden. Gefundene oder unwirksam gewordene Karten müssen an die ausstellende Behörde zurückgegeben werden.

My travel permit

CHAPTER 6

On June 22, 1946 I started out in the morning and was very lucky to get a ride on one of the few trucks traveling in those days for the first stretch to Karlsruhe, still in the American Zone. Here traffic was scarce as the border was only 30 kilometers away and hardly anybody was crossing from the American into the French zone. But early that afternoon a noisy, stinky truck stopped and allowed me to hop on. Like most trucks in those days, instead of using gasoline it had a gas generator mounted on the back of the truck which had to be reloaded with wood chips every hour or so. On a rural highway driving through a larger wooden area we approached the border. We were stopped by two French soldiers coming out of a wooden shack sitting amongst the trees 5 meters off the road. The driver was able to communicate with the soldiers in broken French. After a while the driver told me that the soldiers claimed that I had killed a French Guard during the night. He was not able to convince them that this was not possible, since I had come that day all the way from Darmstadt. The soldiers made me get off the truck and waved him to drive on.

Now I was alone with these two guys who led me into the wooden shack with my backpack. They studied my border crossing document and I started to worry. If they destroyed it and claimed I was a former soldier without proper papers, I could end up in France as a prisoner of war and join some of my class mates. There would be nothing I could do about it. To my great relief they finally returned the document to me. Now I had to unpack my meager belongings. They found my only watch, an alarm clock mounted in a leather case, a present from my godfather uncle Volker on the occasion of my confirmation. They seemed to like it and put it aside. Next was my knife which I used for cutting bread and preparing my sandwiches. Then I had to take off my high army leather boots, which had belonged to my father. Standing in my socks I could not help but smile at what happened next: They took out a postcard sized photo of my older sister and started to fight over it. Surely, my sister was very nice looking, I always thought so, but I had never foreseen her becoming a beauty queen! Well, these soldiers felt otherwise, and the photo disappeared.

I certainly felt mistreated. Having gotten back my papers I became braver and more outspoken: Officer! I demanded. I wanted to complain about these thieves. They looked at each other and nodded. Putting on a pair of spare boots and having put my belongings back into my backpack, one of the soldiers shouldered his machine pistol and waved me onto a path leading perpendicular to the road, straight into the woods. Well, this must be the way to the headquarters I thought. After about 300 meters the path got smaller and smaller and it became clear, this certainly would not lead us to an officer, but where to? Now the soldier stopped, took his weapon off his shoulders and into his hand and pointed straight into the wood towards he South saying "Freiburg!" This did not look good. Did he intend to shoot me here in a no-man's land, where it would take months or even years before anybody would find me?

But there was nothing I could do other than follow his command. I slowly and carefully headed out. Once in a while I would look back over my shoulder to the soldier who just kept standing there, following me with his eyes. The forest was not too dense here. No point to start running. I always tried to have at least one tree between him and me. Then I was separated from him by 10 meters, now by 15 meters. It was slowly becoming more difficult for him to hit me if he would try. Then I hardly could see him any more. My heart pounded No shots rang out. I walked a little bit faster. Would he follow me? He was out of sight now. I started into a slow run. I got out of breath. I slowed down again to a walk. I just kept walking. No shot, I was alive!

There is a German saying: "Himmelhoch jauchzend zu Tode betruebt!" which one can translate as 'Being happy sky high one minute and feeling deeply sad and close to death the next!' This is how I felt as I kept walking for half an hour through this forest until I dared to turn back to the road. Here I sat down in the early afternoon sun hoping for another car. None came. Finally, after another hour a small car approached and I frantically waved towards it. It turned out to be a jeep. It stopped, and who was in it? My two soldiers! They waved me in and off we went. Would we now go to the headquarters? After a few minutes they turned into a smaller side road and shortly thereafter stopped. One of them pulled me of the jeep and pummeled me with his fists without really hurting me too much and shoved me into the ditch next to the road, throwing my backpack after me. Off they went.

I collected myself and started to march back to the main road. After a while I saw a farmer working in a field. Still enraged, I told him my story. Where could I find the French headquarters to complain? "Young man "he answered," are you injured?" "No!" I responded. "Did they make you a prisoner of war?" No!" So what do you want! Get yourself to Freiburg and be happy about this very fortunate outcome.

Of course, he was right. I managed to get there by evening and poured out my story to my Aunt Annie, wife of uncle Fritz, a protestant minister. Yes, that had

been good advice, she told me, together with some stories of how they had experienced the occupation forces since the end of the war.
It happened to them a month ago! When the French arrived they needed some space. Contrary to the American soldiers, who would confiscate whole buildings by surrounding a house and forcing all inhabitants to evacuate within half an hour with just one suitcase, the French would more typically move in with a German family, inhabiting just 1 or 2 rooms of the home. Aunt Annie's single family house had received a big poster on the outside declaring if off limits to all French troops, as it was the house of a protestant minister. But one day a French officer came anyway and asked politely if he could have a room for just a few days so that his German wife could visit. "Sure," friendly Aunt Annie replied, "use our bedroom, we can move in with the children."

The visit was without incident and the officer left after a few days, leaving some food behind for the family. But while cleaning up after them Aunt Annie discovered the slippers of Uncle Fritz were missing. This was not acceptable to Aunt Annie. She had been a nurse during the First World War and had treated not only German soldiers but French prisoners as well. She felt she knew French mentality and could handle French officers even under these circumstances. She stormed off to the French headquarters, telling her story to the slightly amused French officer on duty. "Madam relax, I will look into it. Wait for a day or two." And within two days a French soldier appeared at the door and delivered Uncle Fritz's slippers. How little gestures like this can create so much good will!

This was the second encounter in her life with French soldiers. She was 17 years old when during the beginning of the First World War already living in Freiburg at that time, she volunteered to be a nurse in the local army hospital. Together with German soldiers a number of badly wounded French prisoners arrived. They appeared to be quite scared. Knowing French quite well she could not get these prisoners to talk to her. Finally, after a few weeks they started to loosen up. Yes, they expected to get killed. Their officers had told them that once they became prisoners the Germans would shoot them. Aunt Anni read to me out of her diary: "...after some weeks smiles appeared! One day the news made the round in the hospital: The French prisoners are back to normal, they even started to play Checkers!"

Yes, I had arrived in Freiburg, but would the university accept me and where could I stay?

A picture of the Freiburg MUENSTER surrounded by ruins in the center of the city, taken by me after my arrival there in 1946

The large home of my grandparents with its entrance shown in the picture on page… had survived the air raids intact, while across the street the school I had attended in 1942 had received a direct hit by a larger bomb. Until the end of the war they had used all the rooms in the house themselves with their daughter Eva, who managed a book binding operation in the rooms of the cellar. Now many rooms had been taking over by three other families, either bombed out in Freiburg or refugees from the East. Still, I could put up a bed in the attic filled with unused furniture and other stored goods. And I was very lucky: The University accepted me for the summer term in the department of mathematics and physics.

Like in Darmstadt during my first semester, all students had to help to clear the university grounds of rubble and assist in rebuilding the destroyed buildings. The following "Bescheinigung" – Confirmation note – states that I fulfilled my rebuilding duty during the summer term by working for 56 hours, of course without any pay. Did I inwardly or outwardly complain about this "loss" of time? No, Germany lay in ruins and who could start the rebuilding other than we the survivors, especially the young generation? No. I was happy to have survived, unhurt and alive, and to be allowed to study!

Bescheinigung

Herr / Frl. Graeff, Roderich

hat seiner / ihrer studentischen Arbeitspflicht beim Wiederaufbau der Universität im Sommer Semester 1946

mit 56 Stunden genügt.

My grandfather had been a Professor of Pathology and Anatomy at the University. And as luck would have it I was sent to the half destroyed Anatomy building to carry building blocks and mortar to the third level to rebuild an outer wall. Masons laid down the stones in a bed of mortar and it was not too long before I was allowed to try my hand at it. So I learned a new skill and from then on felt secure enough to perform similar tasks. Looking down from my third story work place into the courtyard, I could see medical students learning to dissect dead bodies. There were plenty of dead bodies in big tubs filled with a preserving liquid on the first floor, but the half destroyed building provided no place for the dissection tables. The students therefore placed the bodies on little wagons with a flat top and wheeled them into the sunshine in the courtyard to do their work. Well, it smelled accordingly! Death and life, so close together!

Dear reader, did you make it through this chapter? It got much longer than intended. Once I started to write it, more and more memories came flowing back. Maybe I had to get these stories out of my system. Maybe I had to write about them so that you, my children and grandchildren, you children of the world, will create a better world. Let me finish this chapter with the following picture:

Subway station in London during a German air raid [31]

It could have been that at the same time the bomber crews left Great Britain to start their long flight to Hamburg to deliver those two bombs which hit the subway station, that somewhere in northern Germany, German air force pilots started flying their bombers towards London where English women and children crowded into their subway stations, a nightly routine just as it was for the German families in Hamburg. Can you see the stupidity of all of this?

How stupid can one get! How is it that Governments can coerce or scare us into these situations where the killing of other human beings, the so called "enemy", seems to be the only logical solution to a problem? Children and grandchildren of the world, wake up, refuse to participate, refuse to kill under any circumstances, don't die for any grand cause. Live! Enjoy life and work for peace!

Chapter 7
The Second Law of Thermodynamics Can't Be Correct, Can it?

1948. After 2½ years of studying in Darmstadt I passed my half way examination. Under the German system used at that time, this was more or less the half-way point toward getting a Master Degree as a mechanical engineer. By this time we were supposed to have learned the basics, such as mathematics, physics and chemistry. Up to this point hardly any topics were concerned specifically with engineering. Unlike the American system where you have tests throughout each term, we had studied totally unchecked by professors or their assistants. In the half way examination you had to prove that you knew what you were supposed to have learned during the last two years. If you passed it, you could go on studying about subjects which were much closer to the field of mechanical engineering.

"Mass and Heat Transfer" was one of these subjects. We listened to Professor Krischer, an outstanding teacher and scientist, especially in the field of drying. His enthusiasm for his field showed in every lecture. I still remember when in one of his lectures I suddenly realized for the first time that mathematical equations could describe thermodynamic processes. By using differential equations one could calculate, for instance, how a wall of a building would change its temperature over time under the influence of air flowing over its surface and the exchange of heat through radiation with the environment. This was fascinating!

We still would write in our notebooks what we heard in the lectures. Though textbooks would slowly become available, up to then I did not own one. All I had were some pre-war books on physics and mathematics I had received from my uncle Volker. But one day one of my co-students Rudolph Wellnitz had an idea. He would provide a little place where students could bring two old textbooks they didn't need any more and exchange them for one other of greater significance to them. This way he would slowly accrue a sizable number of books enabling him eventually to open a real store. He started out with two wooden planks on sawhorses and placed them at noon in front of the Mensa door. Soon thereafter he fixed up a small newspaper kiosk, which had more or less survived the bombing and was conveniently located next to the main entrance of the university.

The little newspaper kiosk to the left standing before the destroyed physics department building in 1946
[25]

In the picture above opposite the little kiosk and on the very right one can see the corner of a destroyed building into which. Rudolph Wellnitz eventually moved and where he built up his little second hand store into the leading University Book Store of Darmstadt.

I went to his place and exchanged some old books which my father gave me from his library for my first technical textbook, *Ernst Schmidt: Thermodynamik, 3rd edition, 1945 [40]*. Proudly I carried it to my little room, and when I opened it, I was completely surprised. In this book I found many of the subjects which Prof.

Krischer had tried to teach us discussed in detail. This was a new insight for me that books existed which covered the same subjects as were taught in our classroom lectures. In high school I had become acquainted only with books filled with texts to be read or to be translated or others filled with math problems to be solved at home.

Einführung
in die
technische Thermodynamik

Von

Dr.-Ing. habil. Ernst Schmidt

Dritte, unveränderte Auflage

Berlin
Springer-Verlag

My first text book: Schmidt, Thermodynamik

Filled with enthusiasm I leafed through my new book. I stopped on page 314. It showed two bodies within a highly insulated enclosure, one covered with a black and the other with a highly reflective surface:

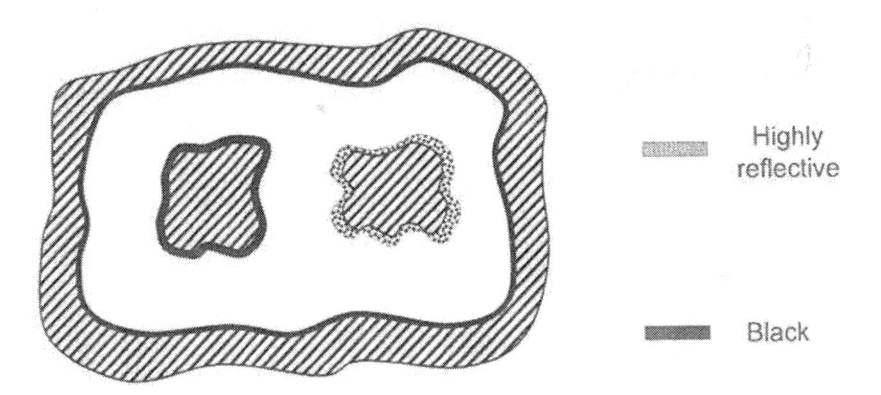

It discussed questions of the heat transfer by radiation between both bodies and the surrounding surface. It stated that all of these surfaces had to have the same temperature in accordance with the Second Law of Thermodynamics. Therefore, the emissivity of each surface had to equal its absorptivity. The amount of energy emitted by radiation had to be equal to the amount absorbed from the incoming radiation. Otherwise temperature differences on the various surfaces could arise. This would contradict the Second Law.

We had learned something about the First and Second Laws and I felt that I had some understanding of them. The First Law stated that energy could not be created or destroyed; it could only be transformed into another form like from the potential energy of an apple hanging on a tree to the kinetic energy of the same apple falling down. Heat was another form of energy as well as the mass of a body.

The Second Law of Thermodynamics, among other things, states, that heat can only flow from warm to cold, never in the other direction. It also confirms our every day experience that temperatures of bodies tend to equalize with the temperature of the surrounding air. A hot cup of tea will cool down until it has the same temperature as the room. A cup of tea would never suddenly start to get warm all by itself. In our classes we had learned that this flow of heat would take place via conduction within bodies, by convection within gases and liquids, through heat transfer between surfaces and their surrounding gas or liquids, or by heat exchange through radiation. The end result and conclusion of all of these flows of heat energy could only be: the temperature of all bodies within an insulated enclosure would ultimately become the same.

On page 320 Mr. Schmidt discussed the radiative heat transfer between the solid body F1 surrounded by a body with the surface F2.

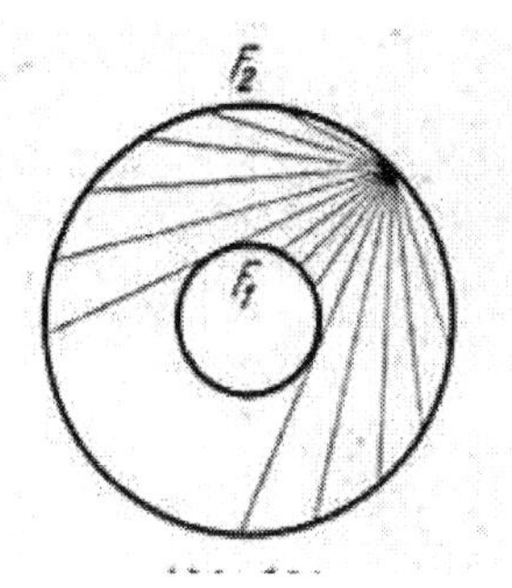

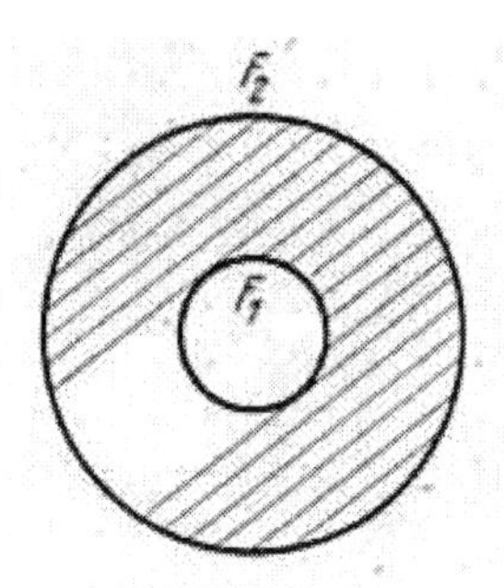

All radiation coming from the inner sphere F1 would reach the surrounding surface of body F2. The Second Law of Thermodynamics demands that only the same amount of radiation coming from the surface F2 should reach surface F1. That would mean that the direction of the radiation leaving surface F2 would have to follow Lambert's Cosine Law. This hypothesized that maximum radiation would radiate from the surface F2 perpendicular to the surface, and the radiation would decline in accordance with the cosine of the angle between the vertical and the perpendicular.

This made sense to me. But then I turned to page 317 where Mr. Schmidt switched from theoretical thoughts and black surfaces to discussing technical surfaces:

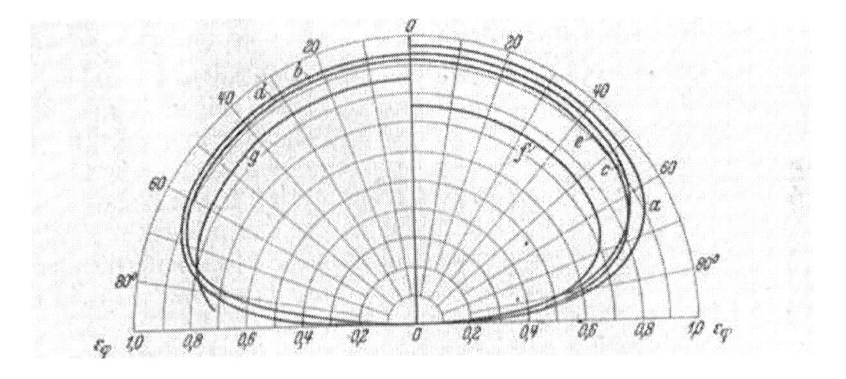

Apparently the radiation from surfaces of wood, glass, paper, or copper oxide closely followed that of black surfaces, but highly reflective surfaces such as chromium or aluminum showed a different behaviour. He even stated: The radiation of real surfaces is quite different from that of a black body. They radiate differently in their distribution by wavelength, and they do not follow the Lambert Cosine Law.

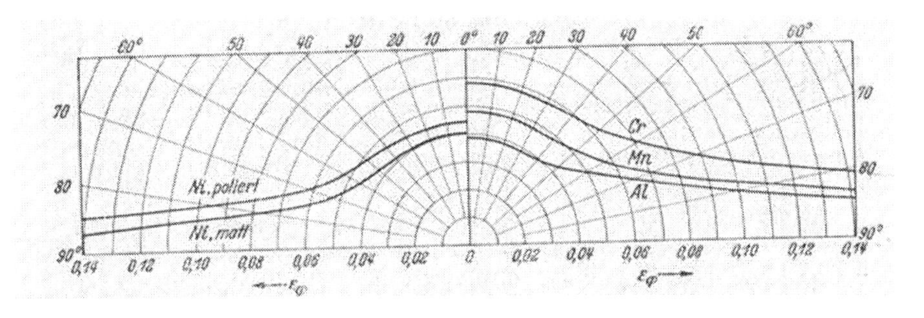

Radiation as a function of direction for materials with highly reflective surfaces.

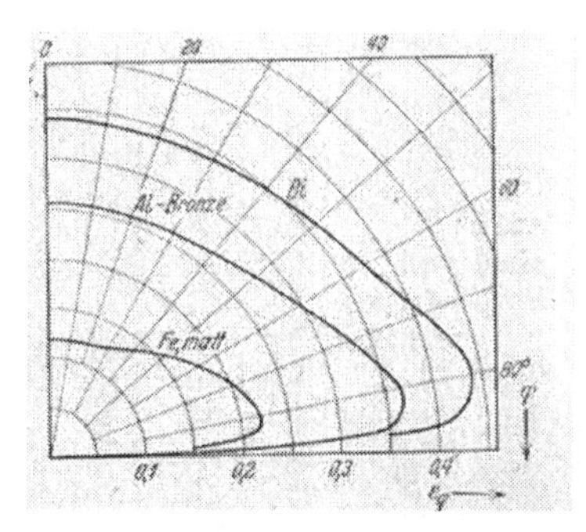

I stopped and wondered: On the preceding pages the Second Law of Thermodynamics necessitated radiation in accordance with the Lambert Cosine Law. And now, two pages later, I had to learn that real surfaces would not follow this law. How could this be? What would happen to the Second Law of Thermodynamics? Would it not follow that an inner sphere F1 with a black surface surrounded by an outer surface F2 made from highly reflective chromium would have to show a difference in temperature, since only part of the radiation it emitted to the outside would be returned? But if this really happened, would this mean that the Second Law of Thermodynamics was not correct in all instances?

I went to Professor Krischer and proudly showed him my new and first textbook. Then I told him about my problem. He looked at me speechless, just opened his mouth, threw up his hands in dismay and looked at me strangely. He hardly could speak. Finally I understood what he was trying to say. The Second Law of Thermodynamics was sacrosanct, a basic law of physics. To doubt it was unthinkable. He gasped for air. I realized I could not expect to get an answer from him, so I quickly left.

My basic question, which was what really happened to the radiative heat exchange between spheres with different surface characteristics, was left unanswered. Should I follow up this unresolved miracle? I was reminded of the answer I had received from my Uncle Volker after I had invented my first Perpetuum Mobile, better concern yourself with something more realistic.

Would I give up my quest for an answer to this question? If my professor or my co-students could not give me a clear detailed explanation, then I just would have to find out myself. Could I not, maybe, try to measure the anticipated effect, a temperature difference between the inner sphere and the surrounding surface?

How could I proceed? How could I set up an experiment with two spheres, one within the other? How could I measure a temperature difference between the inner sphere and the outer surface? In my engineering courses I had learned to use thermocouples to measure temperature differences. I went to the institute for material testing and I was lucky. I received one chromium wire and one chromium-nickel wire, each one meter long. Out of these I soldered a thermocouple. I found a bigger glass bottle, the inside of which I covered with aluminum foil. One end of my thermocouple I connected to the foil. The second point I blackened with a candle and hung it in the middle of the glass bottle. Now I only had to measure the voltage between these two points. If there was a temperature difference I should get a value. I realized that I could expect to find only a very small temperature difference resulting in a very low voltage value, if any at all. Normal micro voltmeters would not be adequate. My student friend Dieter Köttgen was able to help. His father was a professor in Giessen, at the Institute of Agriculture. When he came back from his next weekend trip home he brought along a mirror galvanometer from his father's institute.

This was really exciting. I would be able to perform a scientific experiment. So the next day my bottle was placed in the middle of the wooden floor in my little room and the thermocouples were connected to the galvanometer. It really worked! A small strip of light appeared on the wall of my apartment. Great excitement! Would it show a temperature difference in the right direction, warm in the middle and colder on the circumference of the bottle? Yes! The light beam moved in the right direction, 10 cm, 20 cm, and nearly reached one meter. I hardly dared to breathe. Could it be that the Second Law of Thermodynamics was not correct after all? But then, very slowly, the light beam stopped and slowly moved backwards. It went faster and faster, proceeded through its zero position and sped up, indicating minus values. What was going on? More than an hour had passed, and I had to leave for a lecture.

Coming back I just watched the light beam going back and forth between minus and plus values. Slowly it dawned on me. My measuring instrument was so sensitive that it reacted to any number of influences from the environment. My breathing, my moving around, the changing temperature in the room, all these factors had an overwhelming effect on my experiment. There was no way how I could isolate my experimental setup from the effects and influences of the surroundings. My setup was way too primitive. The next morning I sadly returned the instrument to Dieter Köttgen and delivered the glass bottle into a trash can.

Even today, I still have not received an answer to the basic question: How is it possible that with various surfaces we get changing directions of the radiative heat exchange and still keep supposedly equal temperatures on both surfaces? When I participated in a conference in San Diego 1n 2002 under the heading "Second Law 2002," I found out that there are some other people, genuine physicists, wondering about questions very close to this subject. They feel that the interaction of gas molecules with different surfaces in closed systems might create very small pressure and temperature differences, which seem to contradict the Second Law [24]. These are primarily theoretical thoughts, not proven yet by experiments, but they call it a "Paradox to the Second Law". Perhaps I am not the only one questioning the Second Law of Thermodynamics after all?

Chapter 8
America, You Have it Better, a Temperature Gradient in the Hilsch Tube?

1951. Six years after the end of the war, food coupons were pretty much gone. If you had money you could buy what you needed for your daily needs. Streets and sidewalks had been cleared of rubble, at least in those areas where houses had been left standing or had been repaired. Even Rudolph Wellnitz had moved from his little cubicle into a small store. Students could actually buy textbooks again.

I had received my degree in mechanical engineering in 1950 and worked as an auxiliary assistant when I got a job offer in a glass manufacturing company. My first monthly salary would be 380 German Marks, which translated into 90 US Dollars. Suddenly due to my membership and work in the German-American Club I received an unimaginable offer. The US State Department would fund a one year course in labour relations and democracy organized by the University of Wisconsin in Madison in 1951-52. It was an effort to re-educate a group of young Germans, primarily some activists in the labour union movement, together with some students, by exposing them to American democracy. Was I interested? This offer put me temporarily in a quandary. Should I endanger my first job as an engineer by going for one year to the United States? I did not have to deliberate long. I would go. My decision was made even easier when my future employer

agreed that this visit to the United States would give me new and very worthwhile experiences.
I said good-bye to Darmstadt in two parts. First I went around the city and tried to capture with some photos the atmosphere of some sights to be shared with students I would meet in America. Not far from my student quarters I found this little courtyard.

Courtyard in Darmstadt in 1950 surrounded by ruins

Curiously, a lone statue of a man or a woman in a courtyard had survived the bombing. Instead of facing houses and other buildings surrounding the courtyard, the statue now pointed only to weeds and bushes which had sprouted on the rubble of the bombed out buildings. I wondered what impression this statue had gotten from the violent behaviour of us human beings. It was hard to predict what response I might get when showing this picture to the American students I would meet in Wisconsin.

The second part was more pleasant. The evening before I had to leave I met with those three students with whom I had begun my studies in Darmstadt in 1946. By helping each other all four of us had managed not only to survive the difficult years after the war, but had succeeded in earning our degrees. The last year we all had secured positions as secondary or full time assistants in Professor Scheubel's Institute of Hydraulics. One of my friends provided two bottles of wine which we happily emptied together.

From left to right:
Rudi Przybilla-Singer, Guenther Dibelius, Alexander Haltmeier, Roderich Gräff.

Alexander, do you remember that when we parted we had to carry you home? I wonder why?

The next day our group met for the first time in Frankfurt. Our ages ranged from 18 to 25. 15 of us were active members in the labour movement while 3 had received their university degree. The small steamer Rhyndam from the Holland-America Line would take us to the United States. After 8 days of what seemed to me to be pure luxury on board this small ship with quite a few days of very rough weather we approached the coast of the United States. At least when travelling for the first time to the US this really is the way one should travel. It gave you the time to think what you left behind and what might lie ahead. You felt the distance which separated the Old World from the New. You could start to adjust to the coming changes. How different it is today when you arrive by air in New York , Chicago or Detroit after less than a 10 hour flight!

The fog horn sounded loudly, but where was New York City? Fog was all around us. And then suddenly above the fog and clouds the tops of skyscrapers appeared high above us. What an impressive view it was! This was America.

An overnight train with Pullman beds on both sides of a pass way through the middle of the railroad car took us to Chicago. I got up one hour too early, as I had not realized the one hour gain as we crossed into another time zone.

Madison, Wisconsin, the university, the town, my dormitory. What a wonderful place full of surprises! We received three meals a day, breakfast, lunch, and dinner in a cafeteria, and we could eat as much as we wanted. Everything was new and so different from my wartime and post-war experiences in Germany.

Bascom Hall, Madison, Wisconsin, USA

I took a freshman course in democracy. After six weeks there was a test, a multiple choice test, the first one I had ever taken in my life. The professor explained: There will be 80 questions, and you will mark on the form with your pencil either "yes" or "no." For each correct answer you will receive one point. Incorrect answers won't count. The total of all points is your score.

I heard this explanation, and I understood it. This would mean that I had 50% chance to get a point even in those cases where I did not know the answer, and just made a guess. But this could not be correct, I thought. Of course, a wrong answer surely must result in a negative point. Why had the Professor told us such an impossible story? Had I misunderstood him?

The questions were quite easy for me. But six times I wasn't quite sure about the correct answer, and following my own reasoning, I didn't answer with either "yes" or "no." I didn't want to get negative points. There also were some questions which I felt were purely questions of opinion like: "Communism is a cause for unrest, strife, and war." I knew that "yes" would be the expected answer, but I refused to mark it this way.

A week later I received my score, a 73. This was quite a good score, which I found out by talking to my friends. I wondered, where had I made mistakes? I went to the professor and asked for the correct answers. He looked at me and said: "Why do you want these? You got a 73, a very good score." It took quite a while to convince him that to know the correct answers for the ones I missed was more important to me than the score I received. Finally he understood, and within a week I received a listing of the correct answers.

By now, I realized that the professor had not lied to us when he explained the scoring system. I tell this story in such detail as it shows to me how distrustful I had become toward everybody after my experiences in the 3rd Reich, in Hitler's Germany. After this experience, nobody could be believed any more, no government or any person of authority. I was totally surprised to see how Americans trusted their government, at least in those years. Today, looking back, I wonder if this distrust was the deeper reason that I became interested and involved to such a degree in the Second Law of Thermodynamics, that I was able and moved to distrust even such a basic law.

In the afternoons we had quite a bit of free time. Having received my degree as a mechanical engineer and having worked on a problem of fluid mechanics in the

Department of Hydraulics in Darmstadt, I went to the same department in Madison, and asked if I could not perform some type of experimental work. Professor Faber was happy to help. He was interested in the Rankin-Hilsch tube.

How A Vortex Tube Works

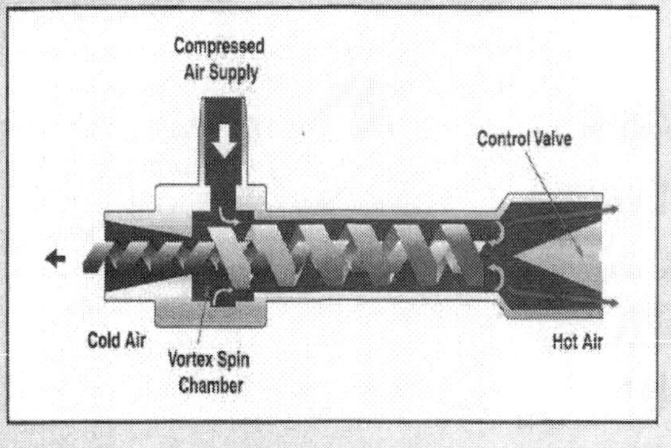

Compressed air, normally 80-100 PSIG (5.5 - 6.9 BAR), is ejected tangentially through a generator into the **vortex spin chamber.** At up to 1,000,000 RPM, this air stream revolves toward the hot end where some escapes through the **control valve.** The remaining air, still spinning, is forced back through the center of this outer vortex. The inner stream gives off kinetic energy in the form of heat to the outer stream and exits the vortex tube as **cold air.** The outer stream exits the opposite end as **hot air.** There is a detailed discussion of vortex tube history and theory later in this section.

It had been rediscovered recently that one could create a cold and a warm air stream by blowing compressed air tangentially into a tube. The company ~L~Corporation• 1250 Century Circle North• Cincinnati, OH, today.produces products based on this principle. I take the following excerpt out of their advertising leaflet:

The vortex tube was invented quite by accident in 1928. George Ranque, a French physics student, was experimenting with a vortex-type pump he had developed when he noticed warm air exhausting from one end, and cold air from the other. Ranque soon forgot about his pump and started a small firm to exploit the commercial potential for this strange device that produced hot and cold air with no moving parts. However, it soon failed and the vortex tube slipped into obscurity until 1945 when Rudolph Hilsch, a German physicist, published a widely read scientific paper on the device.

Much earlier, the great nineteenth century physicist, James Clerk Maxwell postulated that since heat involves the movement of molecules, we might someday be able to get hot and cold air from the same device with the help of a "friendly little demon" who would sort out and separate the hot and cold molecules of air.

Thus, the vortex tube has been variously known as the "Ranque Vortex Tube", the "Hilsch Tube", the "Ran que-Hilsch Tube", and "Maxwell's Demon". By any name, it has in recent years gained acceptance as a simple, reliable and low cost answer to a wide variety of industrial spot cooling problems.

A vortex tube uses compressed air as a power source, has no moving parts, and produces hot air from one end and cold air from the other. The volume and temperature of these two airstreams are adjustable with a valve built into the hot air exhaust. Temperatures as low as -50°F (-(60°C) and as high as +260°F (127°C) are possible. Theories abound regarding the dynamics of a vortex tube. Here is one widely accepted explanation of the phenomenon:

Compressed air is supplied to the vortex tube and passes through nozzles that are tangent to an internal counter bore. These nozzles set the air in a vortex motion. This spinning stream of air turns 90° and passes down the hot tube in the form of a spinning shell, similar to a tornado. A valve at one end of the tube allows some of the warmed air to escape. What does not escape, heads back down the tube as a second vortex inside the low pressure area of the larger vortex. This inner vortex loses heat and exhausts thru the other end as cold air.

While one air stream moves up the tube and the other down it, both rotate in the same direction at the same angular velocity. That is, a particle in the inner stream completes one rotation in the same amount of time as a particle in the outer stream. However, because of the principle of conservation of angular momentum, the rotational speed of the smaller vortex might be expected to increase. (The conservation principle is demonstrated by spinning skaters who can slow or speed up their spin by extending or drawing in their arms.) But in the vortex tube, the speed of the inner vortex remains the same. Angular momentum has been lost from the inner vortex. The energy that is lost shows up as heat in the outer vortex. Thus the outer vortex becomes warm, and the inner vortex is cooled.

I was supposed to measure the radial temperature distribution. So on my free afternoons I found myself in a huge laboratory filled with machines, gadgets, and various instruments, a paradise! As a first step I had to build a Hilsch tube. With a thermocouple I would try to measure the temperature distribution from the middle of the axle to the outside diameter. But I worried about the accuracy of the results. In the middle of the tube the air speed had to be close to zero while on the outside it would be at a maximum. Surely this difference in air speed would affect the reading of my thermocouple. The high speed on the outside diameter would falsely indicate too high a temperature. I had to find a way to measure the distribution of air speed from the axle to the outside. But how? I decided to build my Hilsch tube out of Plexiglas, a transparent material. I would then try to introduce some powder into the air stream and take a photographic picture of these particles lighted by a stroboscope. Depending on their speed and the duration in the stroboscope light I expected the powder particles would show up as short streaks of different length on the photograph. From their length I should be able to calculate their speed. But how could I get a stroboscope? We don't have one in our Department, I was told, but why don't you walk over across the street to the Department of Thermodynamics? Maybe they can give you one.

I crossed the street and introduced myself, told them what I needed, and was shown three different types. Which one do you want? I selected one and was asked: "Do you need anything else?" I said "No," and "Thank you," astonished that it had been so easy. "Good luck and good bye," were the parting words. But what about the paperwork? Wouldn't they have to ask me who I was? Put down my name and address, and have me sign a receipt? None of this was necessary. Coming from a totally destroyed country and university were we had learned to improvise even to get the smallest item it was unimaginable to me just a few minutes earlier that this could happen. I left with my stroboscope.

As part of the program we received some spending money which covered our essential needs. But I wanted to save some money to be able to travel through other parts of the States at the conclusion of our program. In an ad in the local paper the wholesale store for restaurant goods of Brothers Johnson was looking for some cleaning help to work evenings after closing hours. I applied and got the job. My wage would be 75 cents per hour, taxes and social security would be deducted. My task was to clean the store starting after 6 pm. I had to manoeuvre a huge rotating waxing machine around many tables displaying glasses and dishes for use in restaurants and wax the floor. This was fun. I learned it quickly, without bumping against the tables too often. As Mr. Johnson apparently liked my work he offered me more working hours on Saturdays and Sundays in his storage room mixing soap concentrate with water to be sold as cleaning agent. Soon thereafter I was painting the stairway. I felt I was recognized for the quality of my work so one day I confronted Mr. Johnson: “I believe my work is worth $ 1.00 per hour!”. He looked at me somewhat amazed and answered: what about 95 cents? No, I believe I am worth $ 1.00! Mr. Johnson agreed.

Soon I had saved up about 60 $. Then my square dancing partner Nancy told me that one of her friends was trying to sell his big old Chevrolet for $55. I was interested. I, the poor German student coming from a Germany where you could hardly own a bicycle, could own a car? When Nancy told her friend about me he took pity on me, he lowered his price to $ 49.00. So I became the owner of a 6 cylinder Chevrolet with a radio! From now on I, the cleaning lady, drove proudly in my own car to my evening job.

The friendliness which I encountered by practically all the people I met was very impressive. Was this natural in a democracy? But I also met with some contradictions. It was 1951, and even in this northern town of Madison, Wisconsin, black people encountered racism. I, a young German, growing up in a racist dictatorship was supposed to be re-educated. But when I asked the local barber why he wouldn't cut the hair of blacks, he replied, "I wouldn't mind cutting their hair. But if I served just one black person, I would lose most of my white customers." I wondered: How deeply felt and to what degree was it practiced, this democracy, which I was trying to grasp and learn about?

At the end of our course we went on a week-long bus tour through the south-eastern United States visiting factories and public installations. It was in a sugar factory near New Orleans, when I noticed in an assembly hall the usual drinking water fountains installed for the use by the workers. But what a surprise: there were two valves on opposing ends at each fountain. One indicated that the workers could get plain “white” water. The second, on the opposite side, said "Colored," and I thought, "How nice! They're offering fruit juice! This I had to try out. I went to the fountain marked "Colored" expecting a sweet taste. But I had hardly touched the handle when a white supervisor came over and waved me away saying: "This is for black people only! You have to use the other side marked 'White!" Separate but equal was the law of the land.

We visited a university for Blacks located near New Orleans. The students were practically all black whereas most of the professors were white. I asked one of them if he could find some students willing to meet with us for a beer sometime in the evening. He asked: How do you think this would be possible? Here in the South we have race separation. You could not walk with a group of blacks on any street, and you certainly could not visit a pub together. Finally, he found a student who invited us to his parent's home. As it turned out it was one of these magnificent big southern mansions extremely tastefully furnished. His father had made it, he was a medical doctor.

Back in Madison the student paper announced a play-writing contest. It had to be a one act play, the two winning entries would be performed. I decided quickly to participate. The topic for my play would be the problem of war and peace. Where there any methods other than war and violence to settle international problems? World War 2 had hardly finished and the United Nations had declared all types of war illegal. But already now the world was involved in another war, the Korean War. Just the night before I had watched a sketch in the Student Union. In the final scene a group of 17 male students dressed as women in short skirts danced a funny hilarious skit and ended up as a chorus line swinging their legs high into the air when, as a final picture they turned around, pulled up their skirts and exposed their underpants each with a letter spelling: GIVE BLOOD FOR KOREA. I cringed. I had experienced war as something terrible, not something where you could jump around in a funny dance.

So in some of the lectures which did not interest me that much, like the history of the American labour unions, I would write and write. Nancy tried to change my broken English into something more readable. My play described the main character, the radio announcer Paul and his somewhat simple girlfriend Ann. Paul was accustomed to announcing the arrival of some big Navy ships at some foreign port as a friendship visit.

IN THE PLAY Paul would predict to Ann what the next announcement would be like, as so many times before:

PAUL: (getting up and helping her rewind the tape on a wire recorder) I'll tell you. (in his announcer's voice) Attention, please! Here is a very important message from the government: On its peace voyage around the world, our fleet this morning entered the harbor of the capital city of Debterland. The town was blanketed with flags. Shortly after the arrival, King Giveme was welcomed by Admiral Sinkem on board the battleship Freedom.

"This friendship voyage by our fleet will show the entire world that we want peace" said our admiral. On the tour of the ship, King Giveme was much impressed by the new 21cm cannons. "We can destroy any ship within five miles of us with five salvos from these guns, and we hope to reduce that very shortly to three" said our admiral.

Ann: With only three salvos?

PAUL: Yes, with only three salvos. Highlight of the day was the signing of a friendship pact between the two countries. "We are proud to form an alliance with the most powerful nation in the world, with that nation which shows, not withmere words, but with action that it is for peace" said King Giveme at the conclusion of the negotiations. Officers from the fleet were entertained at a royal banquet given by His Majesty in the courtyard of the palace. Shortly before dusk the fleet left the harbour. The flags were put away again.

So it took Paul by surprise when he started to read the real announcement:

PAUL. Attention, attention please! I am going to read an important message from the government. Please listen carefully! This message vitally concerns every citizen!

"For several years the great powers of this earth have been arming themselves on a rapidly growing scale. Day by day more scientific ingenuity and productive capacity is being devoted to the development and building of weapons of war; and an ever-increasing part of life is being spent in learning to use these weapons. Each nation gives as its reason for this arms race the need for a strong defence force to repel threatened aggression.

We believe that modern warfare is so devastating that an army is no longer a protection for life and freedom. In a future war, a great part of the values which make life worthwhile will vanish. The few who would survive another war would be unable to think of themselves as victors in a world destroyed. We have viewed these developments for a long time with the deepest concern, but we believe that war is only one of many inhuman consequences of the basic attitude in life which overemphasizes the value of external forces. Our past attempts to find a solution were bound to fail because we weren't able to touch this fundamental cause. We came to the conclusion that the first step in the development of a new attitude will be the abolition of some of the external forces, and the attempt to help more and more people to develop their own inner values.

Therefore, we, the elected representatives of the people, with the fullest consciousness of our responsibility to our nation and to the world, have decided to make the following declaration:

All development and production of arms will cease immediately. Excluding those materials which can be used for peaceful purposes, all existing weapons and military accessories will be completely destroyed. To replace military training, a voluntary work force of young men and women will work in those countries which want their services. We will equip them with all necessary training and material to accomplish their work, In exchange, we will welcome all those from other countries who wish to live and work with us for a period of time.

For this program we will need only a small part of our present military budget. Another part will be used for converting our war factories to peace production. The goods manufactured in these factories will be at the disposal of all people who need them. In addition we will be able to increase our expenditure for cultural and educational purposes.

We are convinced that this program will have a great influence on human relationships throughout the world. We recognize what great individual sacrifices it will require, but we believe that our people have the moral strength to make this idea succeed!

(left side becomes dark)

The play continued, describing how Paul got all excited and wanted to participate in this new endeavour. At a later stage he, together with some other soldiers, approached the borders riding tractors instead of tanks, tractors they intended to bring as presents. Quickly a confrontation started as both sides insisted to be the "WE'S" and the opposing side had to be the "OTHER'S"! Before this question could be settled the tractors arrived to the surprise and happy response of the red soldiers, new modern tractors with new tires! The Polit Officer of the Reds smelled danger. He could not allow the acceptance of these presents to take place, so he demanded the tractors to be painted red with the lettering: "Our Army our Guarantor for Peace". This was too much for the soldiers who had brought the tractors, they refused to do that. The Polit Officer felt he had won when he suddenly saw Paul start painting the tractors red. This he could not let happen. He pulled his revolver and shot Paul. He felt remorse. He had killed a man. But in this moment Paul got up. He would not accept his death, it was not worthwhile to die this way for a doubtful better and more peaceful world. This was asking too much of any human. He would go on living as before, announcing the arrival of big Navy ships realizing, that this policy one day could end up only in a big conflagration, maybe killing everybody. He, Paul was not yet ready to try solving problems in an alternative, peaceful, non-violent way.

And that was the way I felt in 1952. The continuing violent confrontations and build up of arms could only lead to a major catastrophe. But to let go of all arms and weapons, for that I was not yet ready. The text of the complete play as written in 1952

I Am Not Ready Yet!

can be found in APPENDIX 1.

Well, I did not win, but I received two opinions from the judges in charge of selecting the winner of the competition and one from a history Professor:

Judge 1:

I'm Not Ready yet
Poorly disguised propaganda.
It does not say anything new
suggest any solutions that are not
already old and hackneyed.

(I'm not ready yet
Poorly disguised propaganda. It does not say anything new, suggest any solutions that are not already old and hackneyed.)

Judge 2:

This play has a significant message and is cast in an experimental form adaptable for Play Circle production-
There is a steady dramatic build up from the mid-point to the climax within the framework of fantasy; however, the early sections are dull and unimaginative, the play moves at a pedestrian pace and the language is uninspired, even trite-diction is then undistinguished. Concept of play resembles Brecht's even larger works in scope and intent; would like to see this staged, yet feel that in its present form, its characters fail to convey the emotional conviction and realizations that the words assigned to them connote.

BURR W. PHILLIPS
BOX 2042
MADISON 5, WISCONSIN

Dear Dr. Graeff:

Thank you for letting me read your one-act play. I think it has possibilities. If it is to be produced at the Union or elsewhere, please let me know.

At first I was a bit confused by the "cows in heaven", but after I had read to the end I went back and re-read the first part and it was very convincing.

I enjoyed having you here Sunday evening. I know you are very busy all the time (and I am too!), but I hope that we shall see each other again before your time in Madison comes to an end.

Sincerely yours,
Burr W. Phillips

Several months of free time were left before I had to return to Germany. How could I travel through other parts of the U.S. and earn some money on the way? I found a newspaper ad: "Come with your truck and pull a house trailer from northern Wisconsin to Georgia." Hundreds of these mobile homes were needed to house workers who were building a new plant.

My small truck with the mobile home

I don't remember today if I knew then what I was to find out later about this plant. This plant was being built for assembling the first H-bombs. This was the time of the Korean War and an intensifying Cold War. Now in 1952, only seven years after the devastation of World War II, was I going to help in the building of an H-bomb, a new monstrous atomic weapon, by pulling a house trailer to Georgia? At that time, I did not make the connection. I just saw the trip from Wisconsin to Georgia as a great opportunity to fulfil my personal wish to see other parts of the United States and earn some money. I did not consider the moral consequences of what I was doing. Was this not an example of how quickly we can overlook and compromise our basic beliefs and moral convictions as soon as our own personal interest becomes affected? Again, I don't remember clearly if I knew the mission of this plant. Most likely it was not even public knowledge at that time. But if I had known then the connection of my pulling a house trailer south to building a new terrible weapon, I doubt that I would have refused this opportunity for travel and earning a small amount of money.

In a final accounting I did not help the effort to build this plant as fast as possible too much. If the same thing had happened in Russia I probably would have been send to Siberia for sabotaging the armament goals.

I was just 150 miles away from my final destination when driving down a small inclining road suddenly my house trailer started to swing side ways. First a little and than more and more my car also began to sway left and right. My trying to slowly use the breaks and steady my car-house trailer combination did not work. Now everything happened very suddenly, but seconds seemed to last forever. My car was pushed across the road to the left, a gas pump of a filling station came closer and closer. Just before I hit it the back of my car was lifted up about two feet into the air, the sound of a big crash shook me up and my luggage fell all over me. I was pushed forward against the steering wheel - it was the time before seatbelts had arrived - and the car came to a stop, 3 feet before the gas pump. I realised, the house trailer had tumbled over, lying now on its side.

This was terrible; I had crashed the house trailer! Was I also hurt? I pushed the luggage back, I could move again, I touched my face, I did not feel any pain, apparently I was all right. The thought that I had not hit the gas pump, that there had been no car coming from the opposite direction raced through my mind, but then a very intensive feeling overwhelmed me, I just wanted to disappear and hide deep inside the earth!

Well, I didn't. There was really no time for it because the sound of the crash had attracted a big crowed coming from nowhere. And now something typical American happened: A man came running to my window:

"Are you hurt? Are you o.k.?"

And when I confirmed this, and still sitting in my car under some shock, came his second question:

"How much do you want for the trailer?"

This was one of these enterprising Americans who had created the wealth of this country: First full of compassion, then ready to act on any opportunity.

After one year, my visit to the United States came to an end. It had been a wonderful year. I had learned to love this country with so many friendly people. I had met a young woman, Nancy, my future wife. I had travelled north and south and across this vast country all the way to the west, visiting the most amazing National Parks along the way. This was more than most native Americans ever would be able to visit. I had not solved the secrets of the Hilsch tube. Even today, 58 years later, I believe the air flow and the functioning of this device has not been totally understood. With these experiments in Madison I was for the first time confronted with the question of the temperature distribution in a gas with a pressure gradient. At that time I did not see any connection to the Second Law of Thermodynamics. Forty-five years later this became a fascinating question for me. And today, in the summer of 2010, I start to ask myself, could the cause for the temperature gradient which I have been measuring during the last 10 years and which I have called T(Gr), could this be part of the explanation why inside the Hilsch tube great temperature gradients develop in a strong centrifugal field, similar to the effects of a gravitational field?

Chapter 9
From a Solar Stove to and a Solar Heated Plant to an Atomic Warhead

It is July 4, 1989. I am sitting in a window seat flying from Detroit, Michigan, to Kansas City, Missouri. In a few hours I will be 50 yards away from a source of very intensive energy, an atomic warhead. How will I feel?

Our airplane is still climbing. Below I can make out Ann Arbor, a town where my family has lived for the last 20 years. I also see a blue dot next to the interstate, a little lake just 20 miles west of Ann Arbor. In 1972 Nancy and I bought some land from an old farm around this lake. The property included an old farmhouse, a huge barn and some wooden outbuildings. We built a small summer cabin on the other side of the lake from the farm house, which we enjoyed with our children on the weekends for swimming, boating and hiking. Our small vegetable garden was not too successful. We discovered there wasn't much we could plant that rabbits and deer didn't also enjoy!

In the early seventies the cold war was in full bloom. Books like "The Population Bomb" by Paul Ehrlich and the "Environmental Handbook" edited by Garrett De Bell were being hotly debated. Too many people, not enough to eat, chemicals like DDT were destroying our fauna and flora. The future of the world seemed to be precarious. I wondered what was more likely to lead us to our own destruction: An atomic war with a quick death or the slow deterioration of the environment making human life miserable or impossible. Should we emigrate into the far north of Canada trying to escape this dark future? Buying the small farm just twenty miles away from our home in Ann Arbor seemed to be a good first step.

What could one do? As an engineer I could concern myself with the problems of our over-use of resources, energy, and especially atomic energy. I was going to design a solar cooking stove. But who would go outdoors in order to cook their lunch or dinner? My stove would have to be usable indoors, in the typical American kitchen. This was my idea: The cabin had a kitchen facing the south side. Outside, in the yard below the kitchen window would be my solar stove. The heat would come from a parabolic mirror located just below the kitchen window. It could be turned around an axle mounted in a north-south direction and so follow the sun. The parabolic mirror would reflect and concentrate the impeding sunrays to a focus point located just about a foot below the level of the working table inside the kitchen, which would be arranged just under the kitchen window. At the right location below the kitchen window there would be an opening in the outside wall about 50 by 50 cm or 1.6 by 1.6 feet. This would be sealed by a double glazed window. The concentrated rays would come from the parabolic mirror mounted on the outside of the house, enter through this vertical solar cooking window and would be absorbed by the black surface of the cooking pot. It would be hanging inside a round hole cut into the working table in the kitchen and surrounded on three sides by blackened solar ray absorbing panels. If I could construct a mirror with a surface of about 1 square meter, on a sunny day my stove should provide about 1 kW of heat.

I liked my idea. With some clay I made a form of the parabolic concentrator with a diameter of a little over a meter. From the Mother Earth Catalogue I ordered a flexible mirror. It came rolled up and consisted of burlap netting with a multitude of approximately 3 cm square mirrors glued onto it. A small space between the mirrors allowed sufficient flexibility to conform to my master. My form was covered with a sheet of cellophane. Around the edges I bent a metal tube defining the outside dimensions of my mirror. Next came a few layers of fiberglass, three pounds of polyester resin and catalyst, and on top of this my flexible mirror. Two hours of exciting waiting, and there it was: A beautiful, 1.2 square meter big rigid parabolic mirror.

Testing of my solar stove in our yard

The next weekend I invited some friends over to proudly demonstrate my new cooker. But you already know what happened. Of course it rained. So a few days later, while alone at the cabin, I watched half a gallon of water start to boil in the pot which I had stolen from Nancy's kitchen and blackened with some stove paint. This was the high point of my endeavor. Quickly I took out a saw and cut the hole below the kitchen window at the right location on the southern wall.

Southside of our cabin waiting for the solar stove

Surely Nancy wouldn't object now, not after I tell her about the boiling water. Well, a few minutes later she quietly asked "Wouldn't it be better if you somehow covered the hole you made in the wall so we won't find so many mice inside our cabin when we come back next weekend?" This is the end of the story of my indoor solar cooker. A year later we left Ann Arbor and sold our cabin on the lake. I really should return some day to find out whatever happened to the holes in the wall and the kitchen table!

My flight from Detroit to Kansas City continued. Typically during such a flight I try to talk to my neighbor a little bit. What was he or she thinking about the peace in the world, the threat of an atomic holocaust? How would the president respond if he were to be told right now that 10 rockets were approaching which would explode in 20 minutes, destroying 200 US cities? I would show my neighbors a questionnaire and most of my fellow travelers were quite interested and willing to fill it out. You, reader, might want to do it also. You will find the questionnaire on the next page. Fill it out before you check the tabulated answers at the end of this chapter.

My plane finally landed in Kansas City, where I met with Dr. Regina Birchem, a Biologist by profession and very actively involved in the work of WILPF, Woman's International League for Peace and Freedom. We were driving south from Kansas City about 50 miles. We repeatedly passed signs to rocket installations just a little off to the right and left of the road.

March 8, 1989

Dear Human Being:

Imagine you are Mr. Bush, Mr. Gorbachev or Mrs. Thatcher. During the night you are awakened with the news that ten intercontinental rockets, each with ten atomic warheads, are flying towards your country and could explode in 20 minutes. During the next 20 minutes you clarify this with your military advisors who believe that this not a false alarm, but a real happening. The opposite side contacted on the red telephone replies that it can only be a mistake. What are you going to do in the remaining 10 minutes before a possible explosion?

My reaction would be:			I believe that this would be the reaction of the present world leaders or their military advisors:	
Yes	No		Yes	No
O	O	1. Delay my decision and wait to see if the 100 explosions really take place.	O	O
O	O	2. Run in pajamas and robe to the "atomic-bomb-proof" bunker below my living quarters taking along my spouse and personal aide, but let the rest of my staff sleep.	O	O
O	O	3. Give orders to alert the civil defense system so that they can undertake, Hmm, what?	O	O
		4. Give orders to fire back right away with		
O	O	1 rocket) each with 10 warheads	O	O
O	O	10 rockets) each with 10 warheads	O	O
O	O	100 rockets) each with 10 warheads	O	O
O	O	1000 rockets) each with 10 warheads	O	O
O	O	5. Awaken my spouse, explain everything, and call our daughter, who is expecting our first grandchild, to say goodbye.	O	O
O	O	6. Give orders not to shoot back under any circumstance even if the atomic explosions should take place.	O	O

My reaction is: World leaders reactions:

Yes	No		Yes	No
		7. Give orders to wait, but once the warheads explode, to fire back		
O	O	- with the same number	O	O
O	O	- with 10 times their number	O	O
O	O	- with 100 times their number	O	O
		using our submarines, airplanes or land based missiles.		
O	O	8. Call my son who is in charge of a submarine carrying 20 rockets, each with with 10 atomic warheads, representing an explosive power 8 times the total used during World War II, explain the situation and give him the code word, and tell him that I leave it up to him to decide what to do, because if the alarm isn't based on a computer error, I most likely will die within the next 10 minutes.	O	O
O	O	9. I pray	O	O
O	O	10. I curse God	O	O
O	O	11. I blame myself for not having been more concerned and active to prevent this catastrophe.	O	O
		12. Remarks:..		
O	O	13. I believe that an atomic war will break out during my lifetime.		
O	O	14. I feel that the survival of myself and my family, and maybe all of humanity is today in danger of being destroyed by a nuclear holocaust.		
O	O	15. I believe that during my lifetime all nations will agree to abolish all atomic weapons.		
O	O	16. In thinking about these questions concerning atomic weapons I feel secure and believe that my government works effectively and is instituting the right solutions.		

I am a woman O
I am a man O
My home is(country):...............

Age:.........................
Student........Staff.........
Date:........................

Please fill out this questionnaire, discuss it with others and return to:
(Include a self-addressed envelope if you want to receive the results.)

It became apparent we were approaching a totally different form of energy. We would attend a work week with about another 50 peace activists camping around a missile silo near the towns of Montrose and Spruce. The missile we would discuss contained three atomic warheads in its tip, which each had the explosive power of 20 Hiroshima bombs.

In my luggage I carried two jars with white asparagus which had been harvested in the fields around the town of Weiterstadt, located 30 miles south of Frankfurt in Germany. Weiterstadt is the location of my small plant with about 100 employees. I intended to talk to the airmen guarding this intercontinental rocket, asking them to please not shoot this rocket at my small town in Germany.

I had announced my intention to visit this location with an open letter. The leading local newspaper, the Kansas City Star, published it as a paid advertisement. I also asked the local edition of Time Magazine covering Missouri to print my letter, but they refused. When I asked them for their reason they replied: It is our policy never to state our reason for a rejection. I was surprised and horrified. Even when you were willing to pay for it, the media of the press could that easily refuse to publish what you want to say. Was this part of a "free press"?

This was the text of our letter:

Open letter to the soldiers serving the Missile J2, between Montrose and Spruce, Route H, 17 miles east of Butler, Missouri

Dear Service Men and Women:

Starting on the Fourth of July, Independence Day, people from the United States and countries around the world will meet close to your missile silo to participate in the MISSOURI MISSILE SILO PEACE SCHOOL. We will talk about nuclear weapons and other topics related to peace and justice.

I have traveled here from my home in Germany to join this Peace School as a representative of the Foundation, "Gewaltfreies Leben," (Foundation for Violence-free Living). I also feel responsible for 102 employees and their families working in my factory in Weiterstadt, located 15 miles south of

Frankfurt. During my stay here I would like to meet with you, talk to you, get to know you.

Your great national Day of Independence! This was the starting point from which your young nation began to develop and live those many wonderful ideals later incorporated in your Bill of Rights, including the rights of freedom of speech and of assembly. Only these principles make it possible for me to join your co-patriots and to address you freely through this letter. I plan to similarly approach the soldiers of other countries, manning intercontinental ballistic missiles in France, Great Britain, Russia and China. But, as you might imagine, I will most likely encounter more difficulty in some of these countries whose citizens do not enjoy the same freedoms as you.

I'm sure that you, like soldiers all over the world do your duty as one way to guard and defend your democratic rights and the freedom of your people. You feel that you are an important link in keeping the peace in today's world.

Frankly, I'm very much concerned about your rocket. I worry that some day you might get an order to fire it, and that it might come down on my small town of Weiterstadt, eliminating it along with my family and the families of my 102 employees.

May I ask you not to aim your rocket at our town?

On the map next to this letter you can find Weiterstadt, located just 20 miles south of Frankfurt. We are a peaceful town, and I'm sure neither you, nor anyone else in this world feels threatened by our existence. We have no weapons and no weapon factories within our village. Actually we are famous primarily for our white asparagus, which grows below the ground in the fields around us.

The international airport at Frankfurt might be a more likely target for atomic rockets coming from either the East or the West. But please don't aim your rockets at this target either, because an atomic explosion there, 15 miles north of us, would have practically the same results as if Weiterstadt were hit directly.

Also on this map you will find the location of two atomic power plants, one 17 miles to the north/east at Kahl and the second, 15 miles to the south-west at Biblis. I needn't explain to you that an accidental hit there would have catastrophic results, not only for Germany but for the whole of Europe. The

accident at Chernobyl and its consequences pales in comparison.
I am really asking you never to fire your rockets. If one rocket were fired somewhere in this world, two rockets might be shot in retaliation from the country where the first one explodes. Four rockets might respond to these two in an increasingly-escalating exchange of destructive power. The whole of humanity would be threatened with annihilation, and no one would be able to pinpoint where this catastrophe had originated.

Like you, I feel responsible for peace in this world and try to do my part to preserve it for myself and for my children. Can we exchange our thoughts on this common goal sometime during the next few days, somewhere near your missile?

I can't help but see some strange connection between our asparagus growing beneath the ground and your underground rocket. May I invite you to come to our town of Weiterstadt to meet its people and the men and women working in my factory?

Join us in a meal of our best asparagus!

During the next few days you can find me next to your missile. You could also contact me or write to me through this newspaper.

With warm regards to you, and to you loved ones, sharing this world, working, hoping, and praying for peace

Roderich W. Gräff
Stiftung Gewaltfreies Leben
Egerländerstr.2-4, D-6108 Weiterstadt 2, West Germany

c/o 607 Church St., Ann Arbor, Michigan.

Have you ever stood next to an intercontinental missile ready to be fired? I tell you, it is impressive just to stand at the fence, 10 meters away from the one meter thick concrete block covering the missile, ready to be pushed sideways to enable the rocket to leave on a moment's notice in a terrific blast off. What stupidity of the human race to expose us, our children and families, the people all over the world to such a monstrosity!

All participants had agreed that this week-long meeting would be non-violent, no climbing into the enclosed area with the rocket or cutting the fence would take place. During the week nobody seemed to check on us, only once in a while a helicopter would appear for a few minutes encircling our camp site. The cameras installed on long posts overseeing the rocket sight where soon forgotten. Nobody hindered us when we hung a few peace messages onto the fence.

An intensive week together with beautiful people took place around the rocket J2. I reprint a report which I wrote at that time:

Demonstrating at a Missile Silo in USA

It is July 4, 1990, and for the first time in my life, I stand only 15 meters away from three nuclear warheads, each one having an explosive power twenty times greater than the atom bomb dropped over Hiroshima. These warheads, located in the top of the rocket 'Minuteman" J2, are aimed at three separate targets in faraway Europe or Asia.

One can see only the concrete cover platform, about 1.5 meters thick, with a total area of ten square meters. An explosive charge can move this platform on two tracks sideways opening the path for the flight of the rocket. It takes only 28 seconds from the arrival of the order to fire until the ignition of the rocket. This concrete cover plate is surrounded by a simple, three-meter high fence enclosing an area of the size of a basketball field.

Without local guards watching it these rockets installed 20 years ago are lying dormant, ready to go at a moments notice. On a little post an optical sensor is mounted which can detect a human being or even a rabbit moving within the fence. Soldiers stationed around ten kilometers away in an underground command post maintain these rockets around the clock.

Nearly 100 people have gathered for one week near this rocket facility to discuss questions of violence-free resistance against this threat to humanity. Some of the activists have spent time in prison, because they cut a fence, spilled blood onto the concrete platform, and then prayed. Phil Berrigan is here as well as Paul Kabat whose brother Carl is in prison serving his sixth year of an eighteen year prison sentence. His crime? He cut such a fence and for 10 minutes, using an air jack hammer, attempted to damage a similar concrete platform. We were writing and signing letters to all we knew who were presently serving time in prison for their peace activities.

Members of our group writing letters to the people who presently have to spend time in prison for their peace activities.

The people in this gathering around this rocket J2 feel that it is inhuman to accept rockets which can cause the end of humanity

But the farmers in whose fields these rockets are located, and most likely the majority of Americans and Germans, believe that rockets like these are needed to preserve world peace. Without a doubt I belong to the minority that sees no value in these rockets, but only a tremendous risk, and who is unwilling to accept them silently any longer.

I still had the letter addressed to the soldiers with our invitation to an evening meal with us. How could we deliver it? Where would we find the soldiers who supervised J2? We had seen a sign indicating a control center which we were told was in charge of nine rockets. Regina and I decided to visit it and try to talk to the soldiers.

The signs led us in our car about three miles away to a little dirt road, which ended in a lot covered by knee deep grass surrounded by a 4 meter high meshed wire fence. A huge gate constructed out of the same material was locked with a chain. Inside the fence we could see two small wooden barracks, but no people. Regina and I looked at each other - what could we do? There was no mail box, no bell to press. I took my camera out of my pocket trying to take a picture. Suddenly a voice sounded straight out of the sky. "No picture taking!" Well, it was not the voice of God but it came from a loud speaker mounted on a long pole about 5 meters above our heads. I pocketed my camera and talked to the loudspeaker: "Can we talk to somebody? We want to deliver a letter. Silence. After about one minute the voice returned: "Are you the people from J2?" "Yes", we replied. Again silence. After another minute or two of waiting the voice sounded: "Somebody will come to the fence!"

About three minutes later a soldier showed up on the inner side of the gate. "We want to deliver an invitation to a common meal tonight. We brought a jar with asparagus all the way from Germany." We rolled our envelope into a little tube and pushed it through the fence to the soldier. "Would you like to come tonight?" "I don't think our officer will allow that" he answered. "Couldn't you come when you are off duty?" "No, I don't think so." With these words he departed, taking our letter along with him. Our chance for a meeting was gone.

During my flight to Kansas City, during the intervening years and still today when the cold war apparently has given way to a so-called "war on terrorism", I feel deeply worried. How can my generation leave a world which still has thousands of rockets mounted in the plains of North America, Europe and Asia, rockets transported day and night in the submarines of various nations, ready to be launched at any minute towards us, our children and grandchildren?

Green energy from a solar cooker and an atomic warhead on an intercontinental ballistic missile! What greater contrast can there exist?

After my return to Germany I received eight letters as the result of our open letter published in the Kansas City Star. You can find them in Appendix… on page… They express a number of quite different views. Today, 20 years later, having these rockets and maybe even using them in defending our way of life is acceptable to the majority of the citizens of most countries. Yes, they feel, we need these rockets, as terrible as they are, to keep the peace!

I responded to the people who wrote to me with the following letter:

CHAPTER 9

Weiterstadt, August 6, 1990.

Liebe Mitmenschen,
Dear fellow human beings,

Thank you for writing to me in response to my Open Letter in the Kansas City Star, addressed to the soldiers manning the rocket J2. I am sorry to say, I did not receive any reply from them. I was therefore especially pleased to have received eight letters from you.

Six of these felt positive about my Open Letter, two were critical.
Here are three examples:

I read your ad in the K. C. Star of July 4, 1990 and was moved by it. Your concern for world peace and your anguish at the possibility of a nuclear holocaust is very apparent, and is shared by many in this country. Before swords are turned into plowshares there will have to be an individual decision to do so. Your call to everyone to sit down and discuss it is a first step to that end. I commend your courage and your disposition to bring the issue to the forefront of everyone's agenda. We are pleased to have you visit our country and participate in the celebration of this birthday.

and:

Mr. Gräff - what makes you think we need you to come here to Missouri, U.S.A. and tell us how to run our business? Who is paying you to do this? These missile silos are part of the reason for our peace. I suggest your people stay at home and try to help your own country. If we need you, we'll let you know.

and:

Why do you want to come here and cause trouble? Mind your own business!

Let me try to answer these questions:

I came not in order to cause trouble, but because I am troubled. I am troubled when I realize that we humans in East and West were willing to install thousands of these rockets with atomic war heads and to keep them operational and capable of taking off within 28 seconds after the command arrives, killing tens of thousands of people in one flash with each explosion, and thereby threatening the survival of the human race.

Yes, I am troubled, and if we want to effect any change of this terrible situation more and more people will have to become troubled.

Who is paying me to do this? I am not doing this for pay, but my conscience drives me to action. The expenses actually were paid by the foundation "Gewaltfreies Leben" (Foundation for Violence-free Living). I started this foundation one year ago funding it with royalties which I receive for inventions I have made.

I felt moved and impressed to stand 40 feet away from three atomic war heads, separated from them only by a wire mesh fence and a thick concrete plate. There the rocket J2 was sitting all by itself with no human beings around, ready and waiting to be fired.

You might recall that I invited the soldiers in charge of this rocket to a meal of white asparagus which I brought along from Germany. Nobody came. So one day I went to the command post J1 in charge of the rocket J2 and 9 other rockets. You can approach only as far as a chain link fence and talk through a loudspeaker system to some unseen human being. Finally, after ten minutes conversation with long interruptions a pleasant young soldier approached from the inside, and I was able to hand him several copies of our Open Letter through the fence. I repeated my invitation, but he was doubtful that the officers in charge would allow them to come.

How do I feel about my trip now? I am very happy I went. I had the privilege of meeting many of your co-patriots at the site of the missile, people deeply concerned about the peace in this world and willing to do something about it in a non-violent way. I feel sad that a discussion with the people directly concerned with this missile did not take place. We will have to keep trying.

A special thank to those of you who invited me to your home and to a meal. Sadly I could not contact you personally as I received your letters only after my return to Germany. Some of you indicated that you might come here some time in the future. Please, keep me informed. I would be happy to meet you at my plant and make your personal acquaintance.

Today, August 6, 1990, is the 45th anniversary of the dropping of the Atomic Bomb on Hiroshima. What can we do to prevent a reoccurrence anywhere in the world? Thank you again for writing me, which I take as an encouragement in my trying to live a life free of hurtful force.

With warm regards,
Roderich W. Graeff

So we failed in our attempt to have a discussion with these soldiers. But our open letter, published in the Kansas City Star had an interesting consequence. Half a year later we were sitting in the federal tax office in Villingen, Germany, discussing the tax preferential status of our Foundation, available to non-profit organizations. The official of the tax department said to us: We might have to revoke your special status because you asked American soldiers to disobey orders. You cannot do this under your tax preferential status. He showed us our open letter which we had reprinted in one of our newsletters. "Are you telling us that we can not ask soldiers not to shoot their rockets directed at my plant near Frankfurt? Would you give us the same advice if we were to ask Russian soldiers"? No, he replied, there is a difference between soldiers from an ally and those from other countries. The latter you can ask to disobey orders, but not the former ones.

"This is very interesting," I replied, "but whatever the rules might be: we will keep asking every soldier in the world, friend or foe, not to shoot at us! " Well, we kept our preferential tax status.

Today, January 1, 2010, how can we answer my question of August 6, 1990: "What can we do to prevent a recurrence anywhere in the world?"

For many this question is being overshadowed by the problems we encounter with what we call international terrorism. But what happened to the threat posed by this rocket J2? It might be interesting to look at the view stated not so long ago by General Lee Butler who became the commander of the nuclear forces of the United States in January 1991, just half a year after our visit to J2. He was the General in charge of all nuclear forces in the United States, working day and night for their readiness. But only 5 years later, 3 years after he retired, he had changed his views totally. Just read a few of his the remarks he made in 1996 at "General Lee Butler Reflects on Working toward Peace."

.....I became the commander of the nuclear forces of the United States in January 1991, almost thirty years later. Until the day I assumed those responsibilities, I had never been given access to the nuclear war plan of the United States in its entirety, even though in Washington I had policy responsibilities that directed the plan. I knew nothing about the submarine operations of the strategic nuclear forces of the United States, and I had an incomplete understanding of the process that would lead to a command from the president of the United States to unleash nuclear war in retaliation for a presumed strike. I just didn't know.

Beginning in early 1991, I went through a process that very quickly accelerated and confirmed my worst fears and my worst concerns. What we had done in this country, what I believe happened in the Soviet Union, and what I think will inevitably happen in any country that makes the fateful decision to become a nuclear power-to acquire the capability to build and employ nuclear weapons-is this: the creation of gargantuan agencies with mammoth appetites and a sense of infallibility, which consume infinite resources in messianic pursuit of a demonized enemy. When that happens, it quickly moves beyond the capacity for any single individual or small group of people, like the president, the National Security Council, the chairman of the Joint Chiefs, or the Joint Staff, to control them or to understand. Let me give you some illustrations of what I mean.

A Chilling Ballet

In those responsibilities of commander of the forces responsible for the day-to-day operational safety, security, and preparation to employ those weapons, I was increasingly appalled by the complexity of this ballet of hundreds of thousands of people managing, manipulating, controlling, and maintaining tens of thousands of warheads and extremely complex systems that flew through the air, were buried in the bowels of the land, or patrolled beneath the seas of the world. The capacity for human error, human failure, mechanical failure, or misunderstanding was virtually infinite. I have seen nuclear airplanes crash under the circumstances that were designed to replicate, but were inevitably far less stressful than, the actual condition of nuclear war. I have seen human error lead to the explosion of missiles in their silos.

I have read the circumstances of submarines going to the bottom of the ocean laden with nuclear missiles and warheads because of failures, mechanical flaws, and human error. I read the entire history, and when I came away from it-

because I was never given access to it before-I was chilled. I was chilled to the depth of my strategic soul.

Consider my responsibilities as a nuclear advisor. Every month of my life as a commander of the nuclear forces I went through an exercise called the Missile Threat Conference. It would come at any moment of the day or night. For three years I was required to be within three rings of my telephone so that I could answer a call from the White House to advise the president on how to respond to nuclear attack. The question that would be put to me in these conferences, and as it would be in the event, was "General Butler, I have been advised by the commander in chief of the North American Air Defense Command that the nation is under nuclear attack. It has been characterized thusly. What is your recommendation with regard to the nature of our reply?"

This situation comes close to the one described by my questionnaire of 1989 discussed above. So what did retired general Lee Butler say in 1996 about it?

That was my responsibility, and occasionally that call came in the middle of the night as my wife, Doreen, and I lay in our bedroom. I had to be prepared to advise the president to sign the death warrant of 250 million people living in the Soviet Union. I felt that responsibility to the depth of my soul, and I never learned to reconcile my belief systems with it. Never.

My third responsibility was to devise the nuclear war plan of the United States. When I became the director of Strategic Target Planning, another hat I wore as the commander of the Nuclear Forces, I went down to my targeting room, several floors underground. I told my planners that we were going to get to know each other very well because I wanted to understand the plan in its entirety. I think this story is the most graphic illustration of the evolution of my views and my concerns and, ultimately, my convictions. When I began to delve into that war plan, I was absolutely horrified to learn that it encompassed 12,500 targets.

I believe this was the time when it was public knowledge that in case of a nuclear war the city of Kiev in the Ukraine would be targeted not by 1, like Hiroshima, but by 32 atomic bombs, each 20 times more powerful!

Ending the Madness
It took me three years, but by three months I was absolutely convinced that it

was the most grotesque and irresponsible war plan that had ever been devised by man, with the possible exception of its counterpart in the Soviet Union, which in truth probably mirrored it exactly. Because what that plan implied was, among other things, in the event of nuclear war between two nations, in the space of about sixteen hours some twenty thousand thermonuclear warheads would be exploded on the face of our planet, signing the death warrant not just for 250 million Soviets, but likely for mankind in its entirety.

The second thing that I began to grasp was that neither in the Soviet Union nor the United States did any of us ever understand those consequences, because the calculation as to the military effectiveness of that attack was based on only one criterion, and that was blast damage. It did not take into account fire; it did not take into account radiation. Can you imagine that? We never understood, probably didn't care about, and certainly would not have been able to calculate with any precision, the holistic effects of twenty thousand nuclear weapons being exploded virtually simultaneously on the face of the earth.

Do we understand this now?

That was the straw that tilted my conviction with regard to the prospects of nuclear war, and ultimately to an unavoidable responsibility to end this. To end it! And by the grace of God I came to that awareness and I inherited my responsibilities at the very moment the Cold War was ending and, therefore, I had the opportunity to end the madness.

We shrank the nuclear warplanes of the United States by 75 percent. By the time I left my responsibilities, those 12,500 targets had been reduced to 3,000.

When I retired in 1994, I was persuaded that we were on a path that was miraculous, that was irreversible, and that gave us the opportunity to actually pursue a set of initiatives, acquire a new mind-set, and re-embrace a set of principles, premised on the sanctity of life and the miracle of existence, that would take us on the path to zero. I was dismayed, mortified, and ultimately radicalized by the fact that within a period of a year that momentum again was slowed. A process that I have called the creeping re-rationalization of nuclear weapons was introduced by the very people who stood to gain the most by the end of the nuclear era.
The French reinitiated nuclear testing at the worst possible moment, as the Comprehensive Test Ban Treaty hung in the balance. We have reinitiated the

process of deionization of "rogue nations"; what a horrible, pernicious misuse of language! What an anti-intellectual, dehumanizing process of reducing complex societies and human beings and histories and cultures to "rogue nations." Once you do that, you can justify the most extreme measure to include the reintroduction of nuclear weapons as legitimate and appropriate weapons of national security.

If we truly cling to the values that underlie our political system, if we truly believe in the dignity of the individual, and if we cherish freedom and the capacity to realize our potential as human beings on this planet, then we are absolutely obligated to pursue relentlessly our capacity to live together in harmony and according to the dictates of respect for that dignity, for that sanctity of life. It matters not that we continuously fall short of the mark. What matters is that we continue to strive. What is at stake here is our capacity to move ever higher to the bar of civilized behavior. As long as we sanctify nuclear weapons as the ultimate arbiter of conflict, we will have forever capped our capacity to live on this planet according to a set of ideals that value human life and eschew a solution that continues to hold acceptable the shearing away of entire societies. That simply is wrong. It is morally wrong, and it ultimately will be the death of humanity.

So these are the views of retired General Lee Butler. Maybe in addition to these excerpts you might like to read more about his views in another speech he gave in Washington:

General Lee Butler's Speech and His Joint Statement with General Goodpaster

December 4, 1996
General Lee Butler, ex-commander of the Strategic Air Command, called for the elimination of all nuclear weapons at a National Press Club luncheon on December 4, 1996. He also issued a Joint Statement with General Goodpaster

details in:
http://www.pbs.org/wgbh/amex/bomb/filmmore/reference/primary/leebutler.html

And finally, here come the promised answers of people who filled out the questionnaire. Practically everyone who answered felt that the government most likely would shoot back while they themselves were less likely to do so. In spite of the small numbers of people who filled out my questionnaire in 1988, it might be interesting to know that only 12 % of the West Germans and 22 % of the Japanese would shoot back compared to 69 % of the Americans. What could be the explanation for this difference? Could it lie in the fact that Germany and Japan had very recently experienced in their own countries the horrors of war?

In addition, 50 students from the University Penn State filled out the questionnaire. Surprising for me, these young people answered pretty much identically to the 73 other people from the United States, who of course had a much higher average age.

CHAPTER 9

October 31, 1988

ANSWERS TO THE QUESTIONNAIRE, "Dear Human Being"
(How do I or the leaders of the world respond to incoming rockets)

This is not a scientific study or poll. The main reason for the questionnaire is to get people to think about the meaning of having around 50,000 atomic warheads on this earth. Another reason is to evaluate the answers from various people all over the globe in order to find an answer to the question, "How can I help to eliminate this threat to the survival of humanity?"

Since spring of 1988 I have given the questionnaire to friends and relatives, work acquaintances, business associates, people sitting around me on airplanes, in barbershops, or at the next table in a restaurant. Up to October 31, 1988, the following are the answers to the two main questions:

1. Would I shoot back if ten intercontinental rockets towards my country or have exploded?
2. What do I believe the governments or their advisors, who control these weapons would do, would they shoot back?

	Answers	"I would shoot back"	"Government would shoot back"
West Germany	25	12%	78%
South America	5	20%	80%
Japan	9	22%	78%
Russia	3	33%	-
Israel	2	50%	100%
United States	73	69%	90%
Worldwide	114	46%	87%

Again, as this is not a scientific study or poll and the number of respondents up to now are quite small, one shouldn't draw too many conclusions from the results. Still, one result seems to be obvious: Independent if the respondent would shoot back or not in such a situation, with 87%, most everyone felt that surely one's own government would.

This is a sad result. If one, or a very few, rockets would fly across the ocean, either by design or by mistake, or if one government believes, based on a computer mistake, that some rockets are approaching, a command to shoot one's own rockets would go out. And as the other side would react the same way, within a few hours it would have resulted in a worldwide holocaust.

Are we willing to accept this situation? Am I? Are you?

Chapter 10
Good Bye to SOMOS, but I CRY FOR SARAJEVO

In the summer of 1992 I had reached an agreement with Mann&Hummel, a large company with 6000 employees, to purchase my company SOMOS. Ownership would be transferred on December 31 of that year. The basic sales Agreement had been signed but dozens of small and big questions still remained to be settled.

I was extremely busy when some amazing news arrived; A minister in Padua, Italy, Don Albino Bizotto, had received an invitation from the various churches in Sarajevo to come for a visit into the beleaguered city to show solidarity and remind the world about the terrible conditions of the 300,000 citizens living there under war conditions without the possibility of leaving. He hoped to get 200-300 people to join him on this dangerous trip. I was electrified. Was this not the peaceful, violent free method to counter the violence of military intervention? Was this not exactly the method Quakers would use, like the one I had experienced months after the last war when I met a small group of Quakers who came just to help the former enemy? Had I not described this method in my play: "I am not ready yet!" Was I ready now?

It was clear to me right away: Yes, I wanted to join. Of course, the timing for this venture early in December could not have been worse, considering my unfinished work in connection with the sale of SOMOS. But there was no question in my mind what was more important to me.

Nancy was understandably shocked. No, I should not go. What about her and the children if something were to happen to me. We had just watched a documentary filmed by a French television crew in Sarajevo. Under the heading "A day in the life of a 15 year old boy in Sarajevo "they followed him for one day with their cameras. After breakfast he had to deliver an order to his father, an officer fighting in the outskirts of Sarajevo against the Serbian encirclement. Coming back he joined a group of friends. Trying to avoid getting shot they ran across an open field and reached a cellar where a disco was installed. These contradictory impressions demonstrated his life in Sarajevo. In late afternoon he was driven back towards his home when they stopped for a man lying in the road bleeding profusely from a shot which had just hit him. Our 15 year old jumped out of the car to help.But he was also hit right away, he fell down. He was dead. The documentary stopped. The day and the life of this boy in Sarajevo were over. Terrible. I wondered how the television crew felt about their day.

Nancy's worries did not get smaller when a week before the departure date the organizers asked for my blood type. I had no idea and Nancy, being in our apartment in Florida tried to find out about it by calling all of my doctors. But none had ever determined it. Luckily, I was eventually accepted to participate without being able to report my type. Nancy realized how important my joining the group was for me and finally relented to my going.

This was the plan of the organizers: The Group would meet in Ancona, Italy, south of Bologna on the east coast of upper Italy. Staying over night there and after a planning meeting the next morning the night ferry would take us across the Adriatic Sea to Split in Croatia. From there on plans got somewhat hazy. Hopefully buses would take us from there via Mostar to Kiseljak, a small town 20 miles southeast from Sarajevo, still located in an area under the control of Bosnien-Herzogovina. Here we would stay in a school for one night. The next day buses would carry us through the Serbian occupying ring around Sarajevo to that city where we would stay 3 to 4 days with private families. The return was planned on a similar route, dependent on daily conditions.

Heavily packed with food and presents for the people in Sarajewo, leaving by train from Darmstadt, Germany, to Ancona, Italy

So I went. After my return I wrote the following report:

I CRY FOR SARAJEVO

December 5, 1992:

What should I buy to take to some of the 300,000 people, living in the besieged Sarajevo, being fired upon from the surrounding hills for the past 8 months, unable to leave, with food supplies dwindling, hardly any electricity or water, and winter approaching? I walk through the typical German supermarket, overstocked with goods, screaming, "Buy me, buy me, low calories, lean foods! " But I am looking for the opposite, food loaded with calories, like bacon with fat. But there are many Muslims in Sarajevo, so I look for fish, with little liquid or sauce, packaged as lightly as possible because I will have to carry it on my back in my backpack. As we don't want to use any local supplies, we will have to carry everything we need: a sleeping bag and a thermo-mat, as it might be below freezing and we will be sleeping on floors in unheated rooms, all the food and drinking water we need and little clothing beyond the clothes we wear. Most likely we won't be able to wash during the next 8 days, so there's no need for a change of clothing. We are allowed to bring 5.5 lbs of food for ourselves for 7 days, we are told. If we have more, it will be confiscated at one of the many checkpoints we will have to pass. So I try to hide much of the food I bought -- the chocolate, milk powder, nuts, raisins, butter, dried bread, tea and coffee -- in the many compartments of my backpack, squeezed in among my other supplies a ceramic filter with an attached hand-pump which can clean 1 gallon of water per hour, water disinfecting pills, some medicine, candles, matches and a first-aid kit. What do you choose for a child in Sarajevo, a light-weight toy, a piece of chocolate or something wholesome to eat?

In what kind of world do I live, that I have to make such decisions? As I walk among the early Christmas shoppers, tears well up in my eyes.

I cry for Sarajevo.

December 11, 1992:

After a night on the concrete floor, we pack up to get started

I am in bus no. 2, one of ten buses filled with nearly 500 men and women - mostly Italians, some Spaniards and a sprinkling of people from Germany, Austria, Great Britain and the U.S., all of whom believe in non-violence, in settling conflicts without weapons. Yesterday was Human Rights Day, the day we had planned to be in Sarajevo for an ecumenical religious service, following an invitation to us from peace groups in Sarajevo and the four major churches -- Roman Catholic, Greek Orthodox, Jewish and Muslim. But this morning we were turned back from the Serbian checkpoint which we would have to pass in order to enter the Serbian ring completely surrounding Sarajevo. Now, after driving through the no-man‘s land between the Bosnian and Serbian forces for the second time, we are stopped once again. This is our last chance.

A shushing sound goes through our bus. Be quiet, don‘t move, don‘t do anything to offend the soldiers, who are entering our bus to check us. A young soldier, probably in his early twenties, walks through the bus asking for our passports. His sub-machine gun bangs horizontally around his neck with the gun barrel pointed directly at the head of each of us in turn as he moves down the aisle. He appears to enjoy his position of power over us, slightly smiling, as we crouch in our seats. Open this bag! Now that one! The owner jumps, sending the pulled out items flying left and right. Nervously I touch my cotton bag with my foot. It is filled with a water bottle, some food, chocolate and my first-aid kit, into which I had stuffed a handful of letters which I try to smuggle into Sarajevo for a Bosnian lady. It would be too dangerous if they were found on her. We are quiet and scared, hearing only his voice and some muffled gun-fire outside. They pull some backpacks from the baggage compartment of the bus, go through them and put them back. They confiscate one of our boxes filled with medicine. The bus moves ahead 20 meters. New soldiers enter and check our passports once again. They bang their fists on the roof of the bus as though they were looking for hidden compartments. Finally, they wave us on, and we enter the Serbian ring around Sarajevo.

After all the disappointments of the day, we are now feeling great! We are on our way to Sarajevo itself; we are going to make it. We even feel somewhat protected, being in Serbian-controlled territory where most likely nobody will shoot at us, at least not until we enter the next no-man‘s land between the Serbian and Bosnian forces and then finally, Sarajevo itself.

But I was upset that we had compromised ourselves. Just to get closer to our destination, we had accepted the show of brutal force at the checkpoint without opposing it. Such little evils are the beginning of bigger evils, like the situation in Sarajevo, unsolvable now by any simple solution. And we create more and more Sarajevo's when we accept little evils in our daily lives. Why? Because we‘re too busy, or we don‘t want to get involved, or we‘re scared.

And I cry for Sarajevo.

CHAPTER 10

December 11, 1992, 6:00 p. m.:

We have crossed the Serbian ring keeping Sarajevo under siege, and another checkpoint. The Serbian police car which escorted us remains behind. Soldiers hunched down along the road follow us with their eyes. The no-man‘s Land between the Serbian besiegers and the Bosnian defence of Sarajevo lies before us. Our euphoria changes into anxiety. Will we reach Sarajevo safely? Once more we put our backpacks next to the windows and most of us lie down in the aisle between the seats in the middle of the bus, on top of one other. The only chance to get through without being shot at, we had heard, is during the early morning hours as the soldiers are still asleep after their nightly drinking bouts. The real war starts daily around eleven o‘clock in the morning. My watch shows 6:33 p.m. It is dark out there. The line of our ten busses with their headlights turned on provides a perfect target for the soldiers hiding in the mountains around us who are shooting into Sarajevo.
Our small road winds slowly down the mountainside. We can make out a few shot-up houses to the left and to the right. We reach a bigger highway. A road sign displays “Seneca 60 km, Sarajevo 10 km“. Our busses are speeding up. With the loud noise of the motors we seldom hear the cracking of rifles and canons. But the sky shows a continuous flashing of light, a reflection from the firing guns.

On this good straight road the busses drive faster and faster. Apparently this is the area which is known as “Snipers Alley“. Cars which try to cross this area of the no-man‘s land during the early morning hours drive as fast as possible to avoid the sharp shooters who watch this part of the road intensively.

Suddenly we stop. Our bus just stands -- a perfect target. What is the matter? Finally we move on again. The next morning are we informed that metal barriers had been placed across the road. This was the location close to the Sarajevo airport occupied by the United Nations. During the daytime United Nations soldiers are posted there. In spite of their knowing that we would be attempting to drive through, they had closed the road and had gone to bed.

We turn into a narrow dirt road running between fields and enter an industrial area with destroyed buildings on both sides of the road. Dark shadows appear between the ruins. Some homes are recognizable, and once in a while it looks like candle light shining behind darkened windows. Now we drive through a suburb of Sarajevo. It is 8:00 p.m. A few people, civilians and soldiers, walk along the totally darkened streets. One can recognize shot-up facades, burnt out streetcars, hanging electric wires, and once in a while a car with darkened head lights. A few people wave to us. Do they know who we are? The people lying in the middle of the bus slowly dare to return to their seats, the backpacks are handed to the back of the bus. We are close to the centre of the city and stop between two high-rise buildings, apparently out of sight of the surrounding mountains. Our driver opens the door of the bus and turns on the lights. A soldier with an automatic weapon hanging around his neck climbs in. He stands in the front of the bus and beaming broadly exclaims:

“I love you Italianos!
Gute Nacht!
Welcome to hell!”

We‘ve arrived at our goal, we are in Sarajevo.

December 12, 1992:

It is two o‘clock in the morning. For the past hour we have been driving through the dark, shot up city with a Bosnian soldier showing the way. Streets pointing toward the mountains surrounding Sarajevo, we have to cross fast with the lights turned off, as sharpshooters are waiting for their chance. Once between buildings on the perpendicular streets, we are comparatively safe, and the lights in the bus can be turned on. We will stay overnight in a high-rise gymnasium, which we approach in total darkness. The bus driver aims the bus toward an area between two flashlights, held by two soldiers. The bus stops next to the gym. Hungry, thirsty, dead-tired and keyed up at the same time, I climb the 3-story stairway in total darkness, using only my small flashlight covered with two fingers, making sure no light will shine through the large, broken, shot-out windows. We are high enough above the neighbouring buildings, so that these windows are clearly visible to the gunners on the surrounding mountains. I step over bodies, already stretched out on the floor, finally finding a spot, where I can bed down with my head partly beneath a leather covered vaulting horse, feeling somehow protected by it. It is stone cold, so I remove only my shoes and crawl into my sleeping bag. There is the continuous sound of small arms fire. Through some blown-out holes in the roof I see flashes of lights from the firing guns and the exploding shells, sometimes accompanied by loud noises. I munch a piece of dry bread, which I manage to find in the darkness. Am I dreaming? Is this reality? Was I not in my warm bed in Darmstadt just 5 nights ago? Is the whole world crazy?

And I cry for Sarajevo.

December 13, 1992:

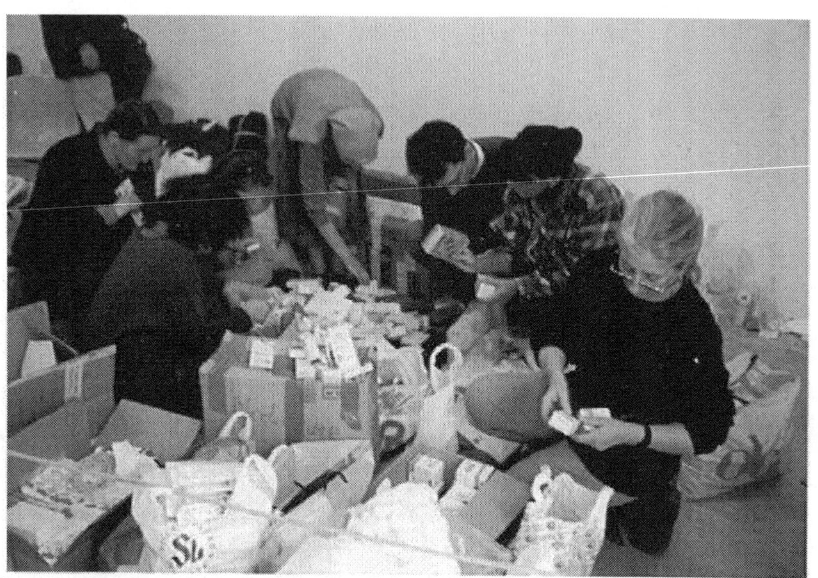

Participating doctor's sort collected medicines to be delivered to Sarajevo

It is Saturday morning, a few snow flakes are dancing in the sky, once in a while a sun ray pierces the clouds. Divided into four groups we walk through the inner city of Sarajevo. Some early risers, bundled up in their heavy winter coats, are walking along the streets, most of them pulling along a small wagon with large plastic containers, apparently searching for water. The houses on both sides of the street show shut up facades and once in a while signs of the impact of a shell. After searching for ten minutes, I finally see three unbroken windows. The others, destroyed by gun fire, are glued over with paper and tape, which is barely hanging together. Some are boarded up with wood. The rooms inside must be dark. Store windows are gone too, and there are no longer displays in the store windows. Once in a while a human being disappears into the back rooms. On the streets the garbage from the past month is piled up in huge mountains. What will happen to this garbage when warm summer days arrive? In front of a shop a man is sitting behind a small table. On the table are a few copies of a thin daily newspaper which is still being printed here in Sarajevo, along with three comic books and a book on gathering mushrooms!

Our group of about 120 people carries banners: “MIR-PACE-PEACE“
People look down to us from the three to five story buildings and applaud. Apparently they had heard about our arrival. We applaud in return. People in the street are crying. We are moved. We are really in Sarajevo.

To the continuous noise of gun shots we have slowly accustomed ourselves. We learn to watch out if we walk a street running parallel to the surrounding mountains, or if we attempt to cross an intersection where we‘re visible from the surrounding mountains. Quite often the dangerous intersections are relatively well protected here through vertical steel plates or large shipping containers. If not, then be prepared to run fast.
We meet people carrying broken tree limbs for fuel, and from some floors we see smoke filtering out through a stove pipe horizontally mounted through a boarded up window. Are there still any trees left in Sarajevo?

We reach a somewhat larger square, visible from the surrounding mountains. Luckily the mountain peaks are covered by morning mist. Maybe this protects us. Rifle shots ring out close by. We hear other guns being fired and exploding shells. We run.

We enter an orthodox church built in the l6th century for a service. The small congregation sings. The stained glass windows are protected from the outside by sand bags creating a mystical half-light in the church. We pray. Peace on earth?

Afterwards, we move along in the direction of a big movie house. In the midst of all this chaos and destruction, we pass a park where two city workers are sweeping the path with straw brooms. 1 can hardly believe my eyes.

The movie theatre appears ghostly, lighted only by a few candles. Ministers of the various religions, politicians and members of the local peace movement speak to us. We sing.

“We shall overcome some day“ and
“Peace in Sarajevo some day.“
When?

I leave feeling dizzy. My pulse races, measuring 130 beats per minute. Do I need fresh air?

In the courtyard of the theatre, the citizens of Sarajevo who wait to talk to us, come and go. Snatches of Serbian, Italian, English and German hit my ear. "We all will be slaughtered; we need weapons; could you please call this number in Konstanz in Germany, and tell them we are fine; Here's a letter, will you take it along?"

An elderly man approaches me. With trembling hands he shows me some papers. "I am an architect, a professor. I have taught in Germany and in Finland where I own a house. I was visiting Sarajevo 8 months ago when suddenly I couldn't leave anymore. I am Yugoslavian by birth and have a Bosnian passport. Can you help me get out of here? I need a visa and an invitation from Germany. This is my only hope. "

I promise to try, but how can I contact him? He presses a small note into my hand with the United Nations address in Split. Sometimes, he thinks, they carry mail to Sarajevo.

He walks with us, showing us some of the historical sights of Sarajevo. Suddenly exploding shells sound noticeably closer. He stops. "I would go on, "he says, "but I don't want something to happen to you. " You had better return." Deeply moved, we say goodbye, see you again. See you again?

5 members of our international group waiting for the return ferry in Zadora, Croatia on the Adriatic Sea.

December 14, 1992:

Only two nights later I am sitting in a night train, taking me from Ancona through northern Italy and Switzerland to Darmstadt in Germany. Day breaks and the rising sun is reflected on spectacular mountains and peaceful villages. But emotionally, I am still in Sarajevo, with the people I left behind, people who want to leave but can't. Soon I'll be home. I'll take a bath, have a warm meal in a heated house with glass in the windows and no gun-fire in the background.

And I am crying for Sarajevo.

December 18, 1992 8:00 A.M.:

People are waiting in the halls of the Immigration Service in Darmstadt. I had made an appointment, can enter right away, and am seated across the desk from the official. "I would like to invite Professor A from Sarajevo, and I need a visa for him. "

My friendly official tries to help. "Maybe you can get one. Are you willing to pay for all costs while this person stays in Germany? You have to prove that you can provide living quarters. What is your yearly income? Do you have your last tax return with you? This form you have to sign. "

I skim the form with glazed eyes... livelihood, living quarters ... doctor and hospital costs.. ... nursing care if needed ... all must be paid by me in total, even if the person asks for asylum and doesn't leave again. If you don't fulfil your obligations, you will be prosecuted to the full extent of the law, including possible seizure of your property and your bank accounts. I become agitated. The guns of Sarajevo still ring in my ears. "This man lives in Sarajevo, gets shot at daily, runs like a rabbit across streets, and you want proof that I can provide a vacant apartment for him? Nobody can say how Professor A could even get out of Sarajevo. The visa, you tell me, can be sent only to the German Embassy in Zagreb, 200 kilometres away from Sarajevo. We don't even know if the news about this visa will ever reach him. "

The sympathetic official understands, but he has to follow the rules. Finally he is satisfied that Professor A will find space in my apartment. I pay 15 marks. In three days they could send a fax with his visa to Zagreb, provided the Immigration Service in Cologne does not object.

I thank him, and I cry for Sarajevo.

Three months later, March 13, 1993:

Within two days 1 receive two air mail letters postmarked in Romania. One has a stamp, "Humanitarna Organizacija Adria. " On the back of the envelope is a sticker explaining that a private person, Mr. Rod Jones, with an address in England, carried 7000 letters out of Sarajevo. One of the letters I received is dated January 29, the other, February 23. Both are from Professor A. The news about the visa has indeed reached him. He writes:

"Dearest Mr. Graeff, - Many thanks for your letter with the visa and your invitation to Darmstadt. Today 1 received your letters dated December 21, 1992 and January 4, 1993. As you can see, letters need a long time, but what can you do. That is the way it is here.

I really don't know when I can leave here. Everything is closed by the aggressors. We are terribly imprisoned, like in a big concentration camp. All roads around Sarajevo are closed. The only exit follows the Ilidja-Korridor, the way you came to Sarajevo and left again. This road is deadly for us, because all of us in Sarajevo are sentenced to death by the evil ones, once they catch us.

We have no water, heat, lights, or electricity, and we survive only from the spoon of humanitarian help from afar. It's not easy to be here.

If I ever get the chance to leave, I will try to visit with you for a short time, with God's help.

Yours, E.A.

And I am in Darmstadt, hearing gun shots in my ears, and thinking of Professor A and the citizens of Sarajevo, and I cry for the Sarajevo's all over the world.

Note: Finally, in July 1993, Professor A. was able to leave Sarajevo with a convoy of Moslem children. He recuperated with us in Darmstadt for several weeks before returning to Finland.

This was my report which I wrote for the News Letter of my STIFTUNG GEWALTFREIES LEBEN – Foundation for Violence-free Living -, after my return. Was this trip of 250 people worth anything at all, and what? One participant later claimed that during our stay only one civilian died by Serbian gun shots while the long-term average was 2. Nancy was a nervous wreck during those days, listening to the TV News all day long in our Florida apartment. There was only one short announcement of our visit. The world was not interested, it hardly noticed it. Professor A surely thought our visit was worth it as he could leave Sarajevo as a consequence of it. The same is true for the parents of my worker in my plant. They eventually received my letter and I was able to secure a German visa for them. And what was the effect on the people in Sarajevo, who cried when they saw us walking through their city? So the greatest effect certainly was the effect the trip had on the participants. We had experienced very emotional days together within a group of great people. My feeling for justice was certainly sharpened.

On the bus trip from Sarajevo to the coast we all felt a great relief. We had succeeded to get to Sarajevo and out again and nobody had gotten hurt, I was wondering: When in Sarajevo, why did we not give our passports to some persons there enabling them to leave on our buses instead of us? On the way out these passports most likely would not be checked very thoroughly. I felt I would have participated in this. But it was now too late to propose and to try it. Alternatively I thought, what would happen, if all catholic or protestant ministers in Germany organized a meeting in Sarajevo, would the world take notice?

It was a very emotional arrival in Ancona with hundreds of people awaiting us and singing peace songs when our boat slowly docked. As I had to get back to Darmstadt as soon as possible I rushed to the railroad station and caught a night train going north. It was overcrowded with people without finding a seat, all standing in the corridors. I was very tired, so I tried to form a seat out of my back pack when I noticed a compartment with curtains drawn nearly closed where only two men, may be 30 years old, were lying sleeping on the 2 benches. I asked the by-standers more with my hands than with the few words I knew of Italian why nobody entered this compartment. The people standing in the corridor just shook their heads, indicating, there you better not enter. Were these some drug dealers, some tough men with whom you'd better not mess? But I just came from Sarajevo where people got shot at every day, these men could not threaten me. I went into the compartment. One of the men just looked at me, sat up, and I could sit down. I waved to the other people standing outside in the corridor to come in, but nobody followed me.

Soon the train stopped at a station. It was night now but I could look into a brightly lightened waiting room next to the platform. I saw policemen moving through the room and checking something stopping at each person sitting there. Then an elderly couple was apparently asked to get up and leave the room, maybe a couple without a ticket? I was enraged, how could the police be so insensitive, did they not know that it was cold outside? I felt like getting off the train to interfere, but right then – maybe luckily - the train started up again.

Did this visit by 300 peace activist change the way the world deals with violence? No. Did it change the participants? I believe, yes. Did it affect me? It certainly did. It was not long afterwards that I thought of a symbol for NON VIOLENCE, the BLUE ROSE. Since that time I wear daily a T-shirt with the BLUE ROSE symbol, every day, reminding me to think and act free of violence, violence-free.

And it did affect other people as I found out just now when I leafed through this book:

Excerpt from [28]

<u>LETTERS FROM SARAJEVO</u>

Voices of a besieged city

Anna Cataldi

Translated by
Avril Bardoni
(Published in the USA in 1994)

Sarajevo, 14 December 1992

My darlings,

It was exactly eight months ago that you set off, very apprehensively, on your "excursion". You disappeared round the corner at the top of the stairs.

I can still see that so clearly in my mind's eye that it seems only a few days ago. And yet the time has been long indeed, for time, these days, strikes cruel blows.

I am sitting in the living room that has, over the last two days, undergone the latest in a long line of changes. As I probably told you in a previous letter, we borrowed an ancient oil-burning stove and, since it is now almost impossible to find any oil, we have converted it to burn wood, which though scarce is still available. So the stove has become – as in most apartments - the focal point around which everything revolves. We have, indeed, given it pride of place in the middle of the room.

We took the old furniture from Keenan's study, the armchair from my room and the old white rug and blankets from the bedroom, and put them all in here. So we now have a place that looks really nice and cosy, especially when the stove is burning brightly.

Your letter was delivered last night by the Italian pacifists. Their arrival was quite a moving event, reminding us for a whole day what it feels like to be in touch with the rest of the world. When we first saw them, in such numbers, we couldn't believe our eyes. When they started to produce letters from their rucksacks it never crossed my mind that there would be one for us too. We were so delighted, more than you can imagine. To us these pacifists seemed like supernatural beings that had descended from another planet to our incredibly gloomy, devastated city. People regarded them certainly with gratitude for their good will and enthusiasm, but also with amazement, like adults watching children perform heroic deeds.

Perhaps we have grown tired of speeches of solidarity and support in a world where nothing changes.

Semir met some of them and gave them the letter for you together with your phone number. How are we living? Well, I have to admit that we have gradually adapted ourselves and now live this strange life as if it had always been so. We don't think very much about what is happening around us, but just

try to solve our daily problems. We read. Every now and then I translate something, invariably on a subject totally unrelated to the war, and wait for the time when my translations will once again be of interest to someone. I should have liked to do more, but it would have meant taking an enormous risk that, in my opinion, would not have been warranted by the result. I believe it is better to be patient and to save one's strength for what may happen later.
I hope, however, that the worst is now over .Even this dreadful year is on its way out!

When one of us leaves the house the other two worry very much. But to worry over that with everything else that is happening is a superfluous emotion.......

Today, looking back, this is what WIKIPEDIA writes about the Siege of Sarajevo (From Wikipedia, the free encyclopaedia)

Bosnian parliament building burns after being hit by Serbian tank fire

The Siege of Sarajevo is the longest siege of a capital city in the history of modern warfare. Serb forces of the Republika Srpska and the Yugoslav People's Army besieged Sarajevo, the capital city of Bosnia and Herzegovina, from April 5, 1992 to February 29, 1996 during the Bosnian War.

After Bosnia and Herzegovina had declared independence from Yugoslavia, the Serbs, whose strategic goal was to create a new Serbian State of Republika Srpska (RS) that would include part of the territory of Bosnia and Herzegovina, encircled Sarajevo with a siege force of 18,000 stationed in the surrounding hills, from which they assaulted the city with weapons that included artillery, mortars, tanks, anti-aircraft guns, heavy machine-guns, multiple rocket launchers, rocket-launched aircraft bombs, and sniper rifles. From May 2, 1992, the Serbs blockaded the city. The Bosnian government defence forces inside the besieged city were poorly equipped unable to and break the siege.

It is estimated that nearly 10,000 people were killed or went missing in the city, including over 1,500 children. An additional 56,000 people were wounded, including nearly 15,000 children. The 1991 census indicates that before the siege the city and its surrounding areas had a population of 525,980.

After the war, the International Criminal Tribunal for the former Yugoslavia (ICTY) convicted two Serb generals of numerous crimes against humanity in their conduct of the siege. Stanislav Galić and Dragomir Milošević, were sentenced to life imprisonment and to 33 years imprisonment, respectively. One of the 11 indictments against former president of Republika Srpska Radovan Karadžić is for the siege. The prosecution alleged in an opening statement that:

"The siege of Sarajevo, as it came to be popularly known, was an episode of such notoriety in the conflict in the former Yugoslavia that one must go back to World War II to find a parallel in European history. Not since then had a professional army conducted a campaign of unrelenting violence against the inhabitants of a European city so as to reduce them to a state of medieval deprivation in which they were in constant fear of death. In the period covered in this Indictment, there was nowhere safe for a Sarajevan, not at home, at school, in a hospital, from deliberate attack."

Typical street scene in a neighborhood of Sarajevo during the siege

STUPIDITY, STUPIDITY, STUPIDITY!

Criminal stupidity! This happened 20 years ago in the heart of Europe! And right now, 20 years later, our world is still involved in wars, today, right on this very day! People get killed!!!

WHY? WHY??? WHY?????

Why do we accept this?

Can't we learn???

Chapter 11
70 Years Old; It's High Time to Clear Up my Student Questions

What do you plan to do for your 70th birthday? In Germany, as you get older, a round birthday requires a reunion of family and close friends. The "birthday child" has to answer: What will be your plans for the future during these, your ongoing retirement years? A voyage to Antarctica, a trip to Las Vegas, more time to spend in your garden, your chess club or your grandchildren, community work or church activities?

From left to right: Sister in law Edith, my sister Sophie, my wife Nancy, theauthor, Dr.Alexander Haltmeier, My sister Hilde and my son Michael, October 1997.

Did I have any plans on that day, thirteen years ago? Just about five years earlier, when I was 65, I had sold my small company where with 100 employees I had designed, built and sold dryers for the plastics industry for nearly 30 years. These were dehumidifying dryers, dryers working with dry air in order to accelerate the drying process of plastic granulates. I had always enjoyed the creative part of this work, trying to come up with improved designs. To lower the energy consumption of these dryers was an ongoing concern for me. A lower energy consumption not only made our products more competitive but it gave me a good feeling. I was working for a reduction of the negative environmental impact caused by any energy consumption. I helped to reduce the energy demand, which on the one side was a sign of our living standard, but on the other side had been so often the underlying cause for worldwide conflicts.

I had enjoyed all aspects of my work, including the most important but also the most difficult one, interacting and dealing with all questions connected with the well-being of the employees.

But then there was always this pressure in the background to come up with the funds to pay for the materials you needed, for the energy, taxes, insurance, fees of all sorts, and of course for the wages and salaries payable during the last days of each month. When these responsibilities were transferred to the new owner, I felt a great sense of relief.

The pleasure I derived from tinkering with new ideas I did not have to give up right away. To pursue this, I needed only a piece of paper and a pencil. For a while, ideas kept coming, which resulted in some new patents. But once you stop working daily in your field of expertise, you slowly lose touch with it. It was time for something new.

Something new or something old? My questions since my student years about some aspects of The Second Law of Thermodynamics had never left me. In the intervening years whenever I met a scientist willing to listen, I approached this subject. Concerning my very specific question, namely the irradiative exchange between an inner sphere with a black surface surrounded by an enclosure with a highly reflective surface, they could not provide an answer. We are working in a totally different field, they told me. And the Second Law of Thermodynamics? I received the same answers which I had gotten 50 years ago from my Uncle Volker and from Professor Krischer: Don't question The Second Law! Spend your time with something more sensible. But the contradictions I had seen in arguments around The Second Law remained. I wondered: With better methods and the experience I had gained in the intervening 50 years, may be I should retry my amateurish first measurement I performed in 1947 and try to find an answer to my basic question? That would be meaningful and enjoyable!

But did I have at my age enough energy and time left to undertake such a difficult task? Was I young enough to come up with any kind of result? I felt healthy, hiked for hours through the beautiful woods of the Black Forest, played tennis with much younger people at least once a week and swam whenever a pool was available. My involvement with the foundation "Gewaltfreies Leben" (Violence-Free Living) and other work still connected with my former activities would leave me enough time. The financial burden of experimental work should be manageable. And my newly added office in my home in Koenigsfeld would provide enough space for some experimental work.

And I had a plan how to proceed. I would start basically with the same question I posed in 1947: Should there be a temperature difference between two surfaces, the black surface of an inner sphere and the highly reflective surface of a surrounding surface, highly insulated towards the environment, under equilibrium conditions?

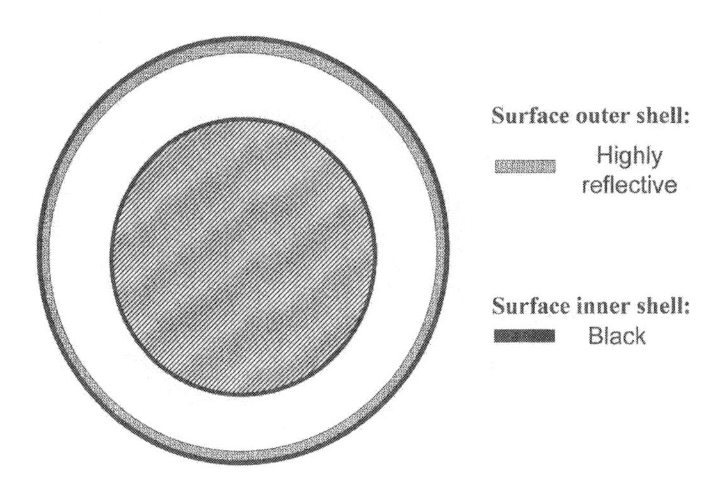

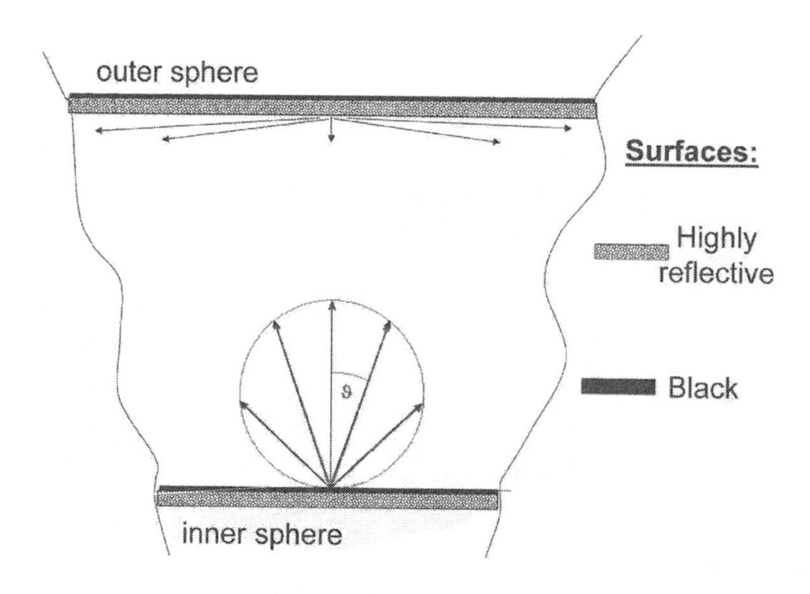

All the radiation emitted from the inner black surface would hit the outer sphere but only a small portion of the radiation reflected or emitted there would directly hit the inner sphere.

Of course, this time around I would try it quite differently. It was clear that if there was any temperature difference at all it would be extremely small. Otherwise it would have been discovered long ago. My whole set-up would have to be much more sensitive. Could I increase the sensitivity maybe by a factor of 10 or may be 100?

As a measuring method to measure a possible temperature difference I would stay with thermocouples. This is a passive device needing no energy input. There was no danger that unwittingly it could introduce some energy through its use into the experiment influencing and falsifying the results. This could happen with a thermistor, another type of temperature measuring sensor, which needed to be heated to make its temperature measurement.

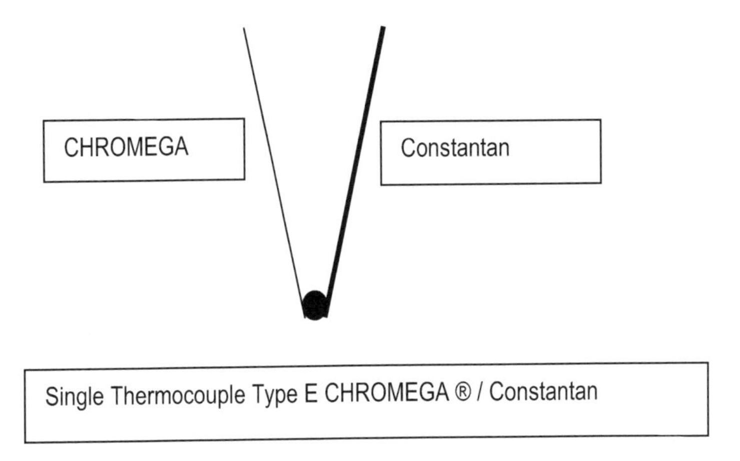

Single Thermocouple Type E CHROMEGA ® / Constantan

Instead of using just single thermocouples I could arrange a number of them as thermopiles, a number of single thermocouples connected in sequence. Maybe I could connect 100 of them, multiplying the signal of just one thermocouple by this factor of 100.

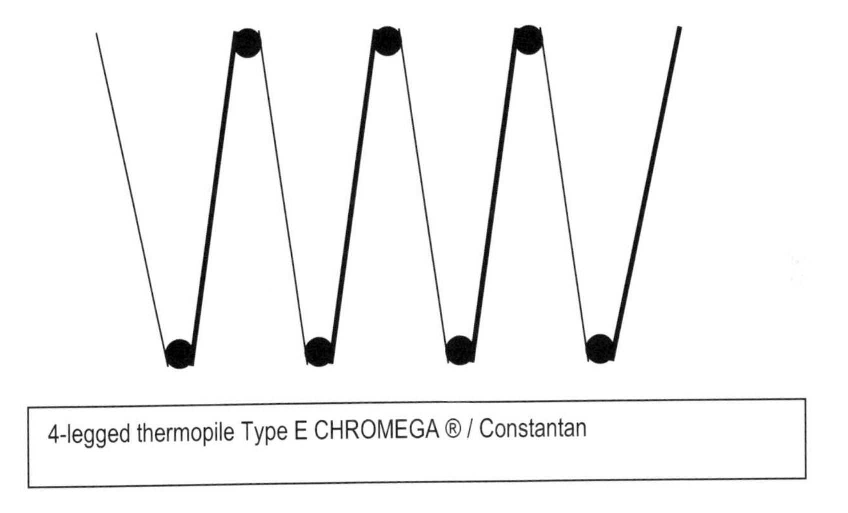

4-legged thermopile Type E CHROMEGA ® / Constantan

Were there other ways to increase the temperature difference I tried to measure? Yes. I would start with an inner sphere, surrounded by one outer sphere. And then I would use the exterior surface of the outer sphere as the next inner sphere by arranging another sphere around it. And so on and so on. Was it possible to do this with 100 spheres gaining another factor of 100?

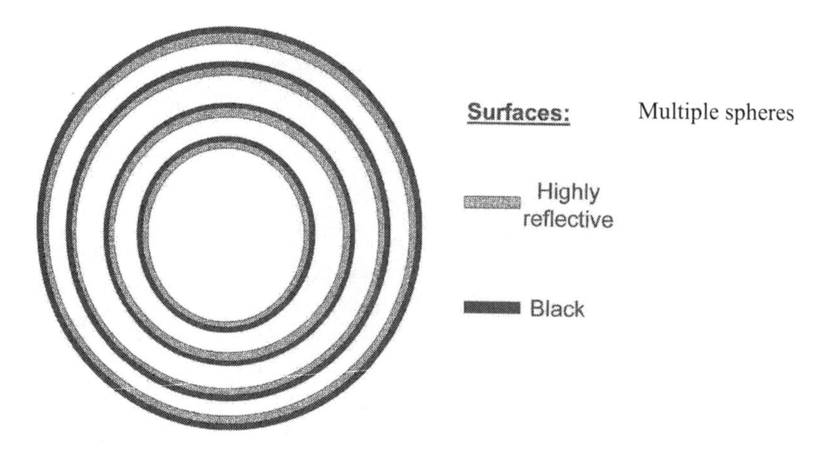

In 1947 the mirror galvanometer I used would have been able to measure a temperature difference of not much less than 1/10 of a degree Celsius. I soon found out that nowadays I could buy a Multimeter at an affordable price which would make it possible to measure a temperature difference of down to 1/1000 degree. This would be another factor of 100.

Three factors of 100 each meant a total increase of the sensitivity of a possible experiment by 100 x 100 x 100 = 1.000.000! This represented a reasonable chance to increase my signal by a factor of 1 million! This sounded pretty good to me.

I told Nancy of this plan, as well as my greater family and my friends. They looked at me politely but sceptically. How long would this endeavour last? Maybe 3 years, I declared. By that time it should have become clear if there was a chance for success or if I was chasing after an illusion.

My friend Alexander Haltmeier announced a birthday present: He would give me a standard Physics book: Gerthsen Physik, 19. Auflage, Springer. Should I, an engineer, learn more about some basic laws of physics before starting to measure in an area I really didn't know much about? I had to agree; I was an engineer, not a physicist. But exactly this fact gave me an advantage, I felt. Through his knowledge of the laws of physics, a physicist would be totally convinced that

The Second Law of Thermodynamics, one of the very few basic laws of physics, was correct, now, in the past and in the future. A physicist so indoctrinated and convinced would never try to measure something possibly contradicting this belief. And if he would in his work somehow stumble on an experimental result indicating a possible contradiction to The Second Law, he would not doubt The Second Law but the correctness of his measurements. Only someone like me, an engineer, hardly touched by any detailed knowledge of physics, would try such an outrageous undertaking.

It reminded me about the method I used when I tried to come up with a new idea, a new invention. When you try to invent something, many advisers would propose, look up all the patents already published in the field of your interest. This might give you new ideas and you will not reinvent what has been already invented. I used to do just the opposite. I never looked at existing patents, because I felt that they would channel my thoughts in the direction others had already tried. This would stifle my imagination, I hardly would come up with something radically different and new. So I always followed my own thoughts and ideas without looking at the existing patents in that field. There was plenty of time to do that once I believed I had invented a new worthwhile product or process. This method had worked well for me. I would continue to also use it in this new endeavour.

I started to get exited about this new task. My birthday guests departed. My work began!

Chapter 12
Searching in Junkyards, my First Experiments: A Total Change in my Research Direction!

How Should I report my experiments in physics which I started 13 years ago? How do I describe them if I want to reach interested lay people but equally real physicists? How can I convey the feelings and thoughts I had throughout these thirteen years which led me to progress two or three steps forwards, then to a path which led me nowhere, sometimes two steps back and sometimes to real new insights? "Himmel Hoch Jauchzend. Zu Tode Betruebt" is a German Proverb, roughly translated: Feeling happy sky high one moment and totally hopeless the next! Yes, that characterize very well how I felt quite often. But exactly that was it and is it, what makes my work so interesting and exiting: To work on a project which after 10 years convinced me: Yes**, heat can travel from cold to warm**, and yes, this does **not** contradict the Second Law of Thermodynamics, and yes, until today nobody believes me! So I am the only one in the whole wide word who knows this truth! Yes it was, and still is, very exiting!

So this is my plan: I will describe in the following chapters the sequence of events, what did I do, what kind of experiments were performed, what did I learn step by step, who helped me along and who did not even wanted to listen. But I will describe this in terms understandable by the interested layman or laywomen and people interested in technical ideas and progress, like a technician or an engineer. This means that I will leave out most all of the mathematics or details of the laws of physics which I encountered and which had an influence of my plans and experimental decisions.

But you Physicist, Chemist, Engineer or Mathematician, I want to reach you too. Therefore you will find in the second part of this book, in the Appendixs, the detailed description and the scientific background of my work and where you can read through a number of papers which I wrote throughout the years and which nobody wanted to print. So all of you, I hope, will find in the next chapters whatever you are looking for, more or less in chronological order.

Would it be possible to perform the planned experiments as a private person without the support of a staff and the possibilities a university institute could provide?

To be able to start with my experiments, I needed some space which I could use as a little laboratory. A natural choice would be our house in Burgberg/Germany which Nancy and I had built in 1959 and which we used through our life as residence whenever we were in Germany. After I sold my company SOMOS in 1992 I concentrated all my professional European activities in this location. To get more office space I added a subterranean addition to the house connected to the library already existing in its cellar.

Our German home with the laboratory annex

Another important aspect was that this location was the working place for my 20 years long employee and co-worker Mrs. Gisela Hofmann who took care of all aspects of the office work. With her knowledge of Physics and computers she was very well suited to supervise my future experiments in my absence.

I intended to perform some parallel testing in the United States. As soon as I got results in Germany I wanted to find out: Would experiments in the U.S. provide similar outcomes? In 1998 we moved from Florida to Ithaca into a senior home into a 1 bedroom apartment, reducing our living space considerable. For my office work I had rented an outside apartment big enough for an office and also to provide living space for Snjesana and Davor Bakula the refugees from the Balkan which we had sponsored to help them to immigrate to the United States. My office room was small but big enough to make small experiments. Having Davor in the apartment had the advantage that he as an engineer could supervise my experiments while I was visiting my German office.

Experiments like those I planned are normally being performed in university laboratories. Here you have the necessary scientific staff, handy men of various trades, tools and machinery, and most of the instruments and data collection systems you might need, not to speak of the multitude of different materials to construct your test setups. Naturally, I hade none of these items and could provide only limited amounts of funds to purchase them. Also what I might need was still pretty unclear at this early stage.

I did know that I would have to have lots of insulation material to protect my experiment from the influences from the temperature swings in this surrounding room. Very well suited for this would be polystyrene foam in the form of big panels or small beats as often used in packaging. Useable for surrounding the inner space of the experiments would be Dewar glasses. I started to visit scrap places. This worked very well, especially in Germany. Every big village or small town has one, organized for the use by the public who can bring their scrap material and leave it there free of charge to be recycled. You also can come to collect some of those pieces for your own use. So my office in Germany started to get filled with lots of insulation material. Here you can see Gisela Hoffmann working on her desk, slowly getting surrounded by piles of it.

Is this office becoming a junkyard?

Dewar glasses are used in thermo bottles which could be bought quite inexpensively in supermarkets in the States or in Germany. Slowly I had quite a collection.

Collecting thermos bottles and Dewar glasses

It was somewhat harder to find short pieces of thick walled aluminum tubing which I wanted to use to reduce the temperature gradient existing in the laboratory to values as small as possible.

In my search for these materials I found a wonderful commercial scrap yard in the small town of Villingen, just 10 km away from Königsfeld. Here Mr. Hans-Peter Jessen was in charge. He excepted scrap materials primarily from companies which would deliver it in bigger lots. I have never seen a more orderly and clean place for scrap materials.

Here I purchased various parts of tubing, wiring and insulation material. One day I found Mr. Jessen standing in front of a huge pile of scrap copper wiring,

Hans-Peter Jessen

compressed into small but heavy packets. Wouldn't it be wonderful to have a sufficient big number of these to build out of them a housing for my experiments? Wouldn't these thousands of copper wires not create a space of nearly equal temperatures over height? But one packet would weigh about 15 kg and as copper is quite expensive would cost about € 100 each. As I would need at least about 20 of these, that would amount to more than € 2000. This was getting to expensive. I informed Mr. Jessen about my dilemma. And he said: "Take how many you need, you can return them later!" He didn't ask for money, he didn't ask for a cheque for security, he didn't even count the packets I took or had me sign a paper. So soon thereafter, my heavily loaded car arrived next to my office-laboratory.

Bundles of compressed copper wire arriving at my office.

And from then on, whenever I needed a piece of his scrap empire, Hans-Peter Jessen just said: “Take it, you can return it later!” “Thank you, Hans-Peter Jessen!”

More difficult became another question. Very soon I wanted to install experiments with a great height, greater than my office allowed. In one corner of our large office room a spiral staircase led upstairs to my bedroom. Here it would be possible to install tubes in an upright position up to a length of 3 meters and still have enough space at the bottom and at the top to load them with an experiment.

It was quite an endeavor to get these 3 meter long tubes into the office and to move them into an upright position into the space on the staircase. Luckily the wooden stairs were held by screws on the supporting steel structure so it was easy to take some of them out to make space for the long tubes. Slowly the stairway filled up with these tubes and their surrounding insulation. The day came when the stairway become too crowded to remain usable. If I wanted to visit the laboratory in the cellar coming down from my living quarters I had to leave the apartment through the main entrance door and step down on the concrete stairway leading into the cellar, pass through the library and finally arrive in the laboratory. This was a long, cold way to come down to check about the progress of an experiment, especially at night! But what would I not do to check on the progress of a curve of a new experiment moving slowly in the right or wrong direction!

To measure the temperature gradient I had decided to use thermocouples. These are passive sensors delivering a voltage in proportion to the temperature difference to be measured to the meter without delivering any heat to the point of measurement.

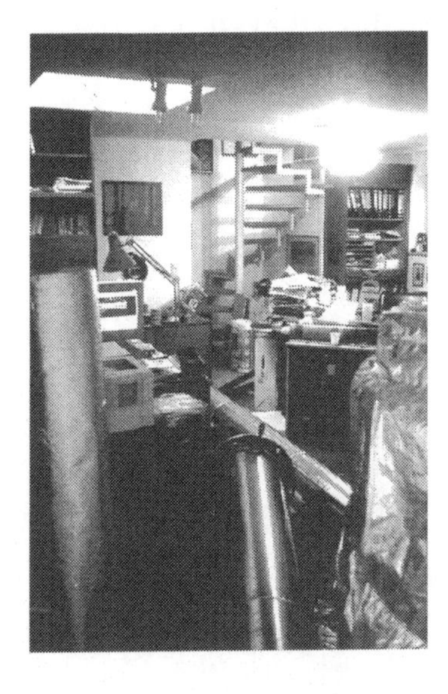

A thermocouple consists of two wires of different materials, typically connected at one end by being welded or soldered together. They would run the length of the vertical column in which I would perform the experiments. If there was a temperature difference between the top and the bottom, the one we wanted to find, based on their heat conductivity the thermocouples would tend to reduce the temperature difference we wanted to measure considerably. Therefore it was important that these wires had a diameter as small as possible. Rather than to buy finished thermocouples with such a small diameter which were quite expensive, I purchased the thermocouple wires separately wound on plastic spools which one could purchase in first class quality from the company Omega:

Thermocouple wires on spools together with some Dewar glasses

On the left and the right of the picture, you see some of these spools. In the middle are lying 4 Dewar glasses. The 3 with the narrow neck are taken out of commercial thermos bottles while the biggest one on the left was purchased from a company which produces these Dewar glasses in a large variety of sizes, primarily for the use in laboratories for instance for the storage of liquid nitrogen.

The tools I needed are the typical tools you can find in nearly any household. The picture shows a scissor for cutting the thermocouple wires and pliers for taking off the Teflon insulation electrically insulating the wires. A normal hand wood saw comes in handy to cut the polystyrene foam panels into correct sizes.

A very sensitive Multimeter I purchased from the company Keithley. It was capable to measure Voltages down to 10^-7 Volt. By using thermocouples type E this would mean that I would get a resolution of 1mK when using a single two legged thermo couple and down to 0.1 mK when using a thermopile with 2 times ten legs. The Multimeter could be easily connected to a data collecting computer with the software coming with the multimeter. We only had to purchase additional special software so we could watch the graph of our slowly developing data as coloured lines on the computer screen.

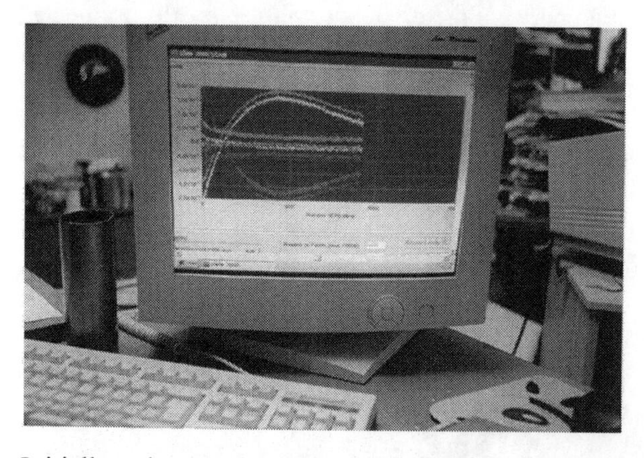

Computer for data collection

Initially as long as we were not able to print out these slowly developing curves, I would take Polaroid pictures in order to know on the next day what had happened in my experiments the day before.

Soon our two laboratory rooms were filled to such a degree that no space was left for preparing and putting together the first experiments. So to the slight disapproval by Nancy I invaded more and more often our living room.

It looked very similar in our home in the US, at that time in Naples, Florida, on a 6. Floor apartment of a condominium near the beach. The balcony was the perfect place to spray plastic drinking cups and plastic champagne glasses black on the one and silver on the other side. Soon I also had to move partly assembled experiments into our living room.

Nancy recuperating on the couch while I take over part of the living room

Bad news arrived here: Testing showed that Nancy had breast cancer. We decided for an operation but against radiation or chemo therapy. In addition Nancy switched to a diet based on natural foods free of colouring, chemical additions or sugar. She followed it very consciously and it worked: two years later she was declared free of any cancer!

After a few month collecting materials, tools, sensors and meters I was finally getting ready for my first experiments. I wanted to start by repeating my very first experiment which I performed in my student flat in Darmstadt in 1947. The original aim would be the same, measuring a temperature difference developing between black and highly reflective surfaces, but I would try to eliminate the temperature influences of the environment and to perform the experiments with much more sensitive instruments.

I planned to begin with a sequence of spheres, one inside the next, again and again, something like Russian nesting dolls, at least 10, maybe 100 nestling within each other, having a black surface to the outside and a highly reflective one towards the inside:

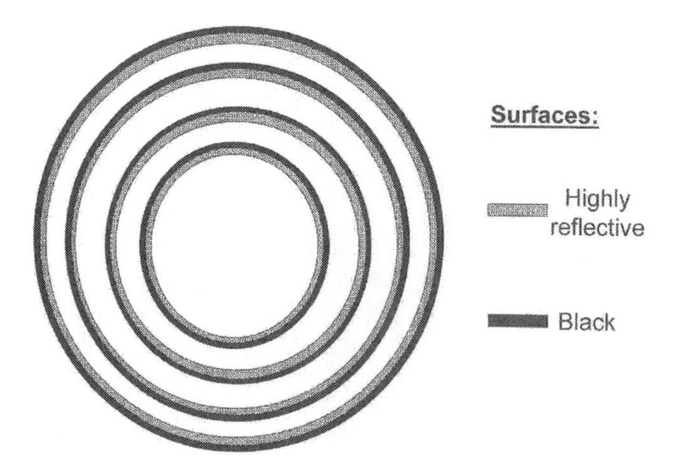

But how could I build such a structure? It would be very difficult, time consuming and expensive. Perhaps I could assemble something close to it. Could I wind a long but narrow strip of a thin film black on one side and highly reflective on the other into a ball? Soon I was standing in the garden spraying a metal foil I had found in the kitchen with a matte black lacquer coming from a spray can used in automobile repair work.

Spraying a metal foil with black lacquer in my yard outside of the laboratory

I cut the foil into small 2 cm wide strips. As I needed some distance between overlying film layers for radiation to be able to take place between the black and the highly reflective surfaces I intended to wind the foil strips very loosely into a ball after first crumbling the metal foil, thereby creating hills and valleys. Of course, overlapping layers would touch somewhere, reducing any temperature difference which I hoped would build up between them and which I tried to measure. My hope was that those areas which did not touch would create enough of an effect adding up from the inside out to become measurable.

In order to protect my ball from the temperature influences from the environment I could put them in one of my aluminum tubes with a thermocouple measuring potential temperature differences between the centre and the outer layers of such a ball.

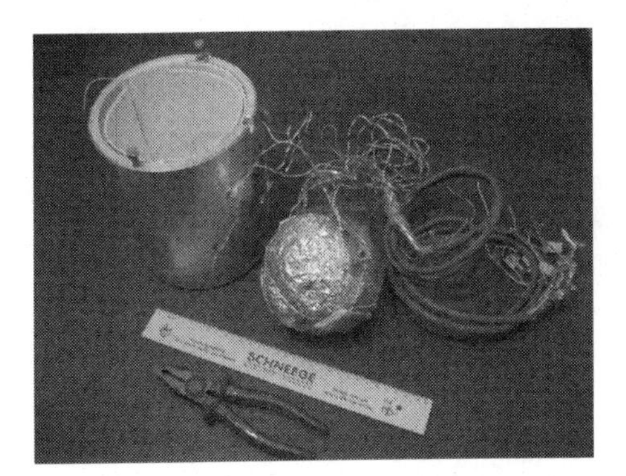

Then I wondered: If it was too difficult to create dozens of nesting spheres what would happen if I lined up a large number of sheets in sequence, black on one side, highly reflective on the other side, again with a short distance from each other to allow a radiative heat transfer from one to the next? This way I should be able to arrange my sheets spaced by a distance of one millimetre without too much difficulty. I could assemble 100 sheets per 10 cm or even 1000 per meter, multiplying the looked-for effect by these factors. But actually to build such a system was not a simple as it looked on paper. How can you create 1000 pieces of a very thin material, black on one side and highly reflective on the other? How do you actually make the 1 mm thick ring on the circumference which would create the distance between each sheet?

Well, why not use drinking cups! One could purchase 200 for about $3.00. They kept their shape when you sprayed one side with a matte black lacquer and the other side with a metallic silver spray. And when you stacked them up, you would get automatically a distance between each of them of about 1 mm.

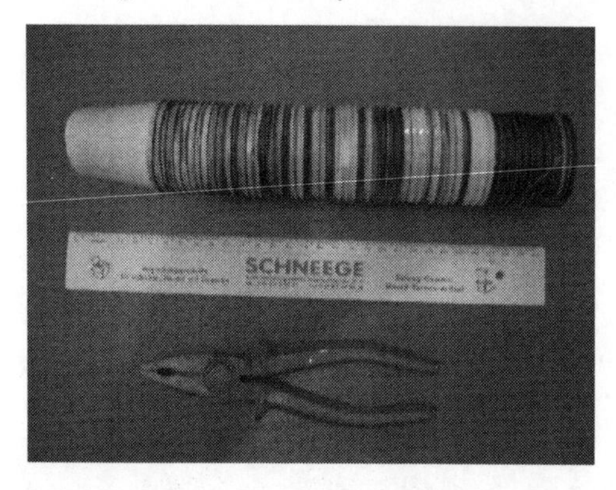

Drinking cups sprayed black and with reflective paint

I also bought very inexpensive plastic champagne glasses. Without the bottom stands they also would stack up nicely with a distance between them of 1-4 mm, dependent on the distance from the axle. 4 groups of 20 each I arranged vertically stacked onto each other.

Spraying cups and Champaign glasses on our balcony in Florida

Each group of twenty I sprayed differently, e.g. inside black, outside metallic, inside metallic and outside black, inside no lacquer and outside metallic, and many other combinations. Thermocouples allowed me to measure the temperature difference between the first and the last champagne glass of each group

A stack of plastic champagne glasses with varied surfaces stacked up to get insulated

I took this stack of plastic champagne glasses and laid it n the floor. In a first test I tried to insulate it against the temperature variations of the room by winding it up in a thick carpet. Then I measured the voltage generated by the temperature of the champagne glasses in each section. Theoretically one should expect that after an initial time all temperatures should become equal. But they were not. How come? What could be the reason? Was it based on the effect I tried to find or could the voltages be influenced by some electric field, maybe from the wiring in the house or from a nearby television station? I surrounded my lay up with some metal film which I then grounded. I could not detect any change of my measured values. Could it be influenced through a magnetic field? I turned the whole lay up from an east-west to a north-south direction, and then back again. Yes, I got some changes. But each time I turned my lay up, I got different changes.

Finally I realized it:

My measured values had nothing to do with electric or magnetic fields. They just reflected the changing temperature fluctuations taking place in my laboratory room. I had to improve the insulation. Maybe I could control these better if I arranged my experiments within a number of aluminum tubes and put them in a vertical direction? And around these aluminum tubes I would put several layers of polystyrene foam for an increased insulation.

Still, the temperature of the whole setup fluctuated too much, following the changing temperatures in my room. I had to create more mass within my experiments. So I got 2 aluminum tubes with diameters of 16 and 22 cm, 3 meters long, with a space of 2 cm between them. This space I could fill with water which hopefully would help to equalize the temperatures over height and reduce the temperature fluctuation of the room to a much smaller fluctuation within the centre of the tubes where my experiments in the form of these champagne glasses were located.

Very slowly I learned how to improve my insulation for my experiments mounted in a vertical position. So one day inside my 3 meters long water-filled aluminum tube, I arranged of a few hundred plastic drinking cups. And with this improved insulation I got more and more reproducible results, similar temperature differences today and on the next day.

And my multimeter showed values of a few thousand degrees C, a few hundred or even a few tenths. But slowly I had to recognize a very strange phenomenon: More often than not I would measure a negative temperature gradient, meaning that the temperatures at the top in the innermost setup of my experiments were lower than at the bottom! This was very strange because in the room, in the area around my experiments, the temperature gradient was always positive, the temperatures higher at the top than at the bottom! This outside temperature gradient I believed as being understandable and correct, I knew, this always happened in heated or cooled rooms. Cold air is heavier that warm air, the molecules closer together, and assemblies of these colder molecules would move towards the floor in a room, displacing the warmer molecules there and pushing them up to the top of the room. If one did not use ventilators hanging under the roof and recirculating the air, one would typically measure a gradient of about +1C per meter of height, warm at the top, cold at the bottom.

But if I accepted this positive gradient as being present, the inner gradient could not show the opposite direction, cold at the top and warm at the bottom. Yes, may be it could show up for a short time on the way to reach equilibrium conditions, but not over longer periods or as average values over time! What did I do wrong, where did I make a mistakes?

As long as I did not solve this puzzle, there was no point to continue with my task trying to study the radiative heat transfer between surfaces within spheres. Now, after one year of starting my experiments I had to clarify this point first: How could I get equal temperatures around my experimental setups, and specifically, how could there be a negative gradient within a positive one?

The direction of my studies had totally changed. “Goodbye” to the radiative heat transfer between surfaces, “Good Day” to the temperature gradients in columns of gases!

Chapter 13
Does Gravity Affect the Temperature Gradient? Measuring T(Gr), Was it that Easy?

My first experiments had not been for naught, I had learned a lot. I was able now to make thermocouples out of very thin wires. I knew I had to improve my insulation around my experiments. The area around the actual inner axle should have very good heat conductivity in order to get an equal temperature over height, as close as possible. It should further have some weight so any temperature variation in the room would be tempered and influence my inner axle as little as possible. The data collection worked, I was able to follow the experiments on the computer screen and save the data, minute by minute.

One of my first new experiments trying to fulfil these conditions was mounted inside a metal can where in its middle a thick walled aluminum tube was fixed. Within this tube one vertical column was inserted in the form of a Dewar glass filled with some metal powder or more often with a gas, like air. In the very middle, the inner axis, was the thermocouple, most often a 5-legged thermopile for measuring the temperature gradient.

Metal can with heavy aluminum tube

In the scrap yard of Hans Peter Jessen I found some heavy cast iron rings. They would be arranged within a box made out of polystyrene foam panels for an outside insulation.

.

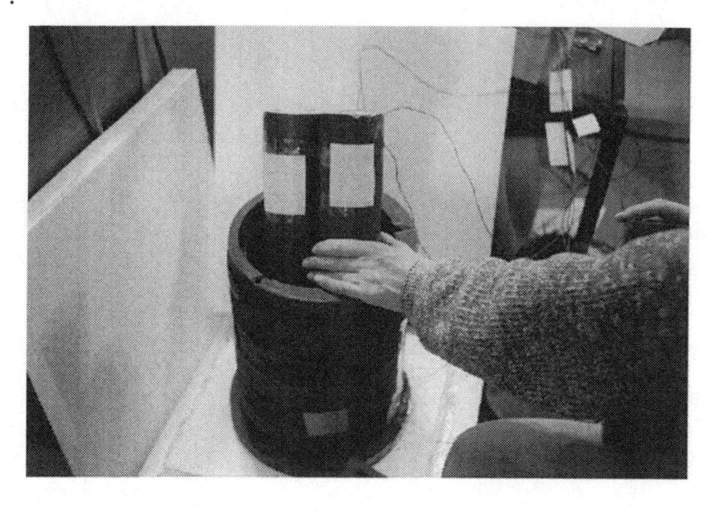

Heavy iron rings surrounding my experiment

Inside the iron rings was enough space for 5 different experiments, mounted in a vertical position next to each other.

Heavy iron rings with 5 experiments

The inner 5 tubes might contain sand, metal powders like lead and copper, or fibres made from different plastic materials. All empty spaces towards the outside I filled with insulation material, like small polystyrene foam beads or PET fibres.

By now I had two Keithley Multi meters, each with 10 channels allowing connection of 20 sensors. They collected values every 5 minutes. For the printouts I averaged 24 consecutive values over 2 hours which resulted in pretty smooth curves.

During these first few months I really was working in the dark. What materials should I measure, what heights should I use, what combination of insulating materials would work best, how many legs should the most inner thermocouple have to give meaningful results, not too close to zero?

Trying and trying, learning and learning, no clear cut results. But then very suddenly I tried a combination which caused me to get up through out the night 2 or 3 times. In the middle of the experimental setup I had arranged a Dewar glass filled with air and a very fine glass powder. The powder was supposed to suppress any convection currents of the air which might otherwise develop within the Dewar glass negating any temperature gradient which might develop.

I got results which were amazing. The innermost thermocouple showed a negative gradient while the outside tubes showed a positive one! I had defined for all my graphs that a higher temperature at the top compared to the bottom would mean a positive gradient. As the inner most values measuring the gradient inside the Dewar glass filled with air and a very fine glass powder was fluctuating between -0,003 and -0,013 K. The higher values appeared during the day and the lower ones during the night hours. This minus sign meant that all measured values would be located below the zero line. The Dewar glass was mounted inside a thick walled aluminum tube and was showing a gradient which hardly fluctuated with values of practically zero. This aluminum tube was mounted in an aluminum barrel with a diameter of 63 cm filled with fine sand. The sand showed a gradient of +0.1 K!

I got very excited! Was this already showing me, for the first time over a long time span, an inner negative temperature gradient surrounded by a positive gradient in the sand? Could it be that gravity was the cause of this? Did I hit the jackpot so quickly getting such sensational results?

During the day the gradient sneaked closer to zero, but during the night it fell and fell: I could not wait until the next morning, I had to descend into my laboratory cellar every 3 or 4 hours to check, not woken up by the sound of an air raid siren but by my inner clock. Was the gradient still falling, showing bigger negative values, maybe setting a new record? Or was it moving up towards the zero line, maybe even crossing it?

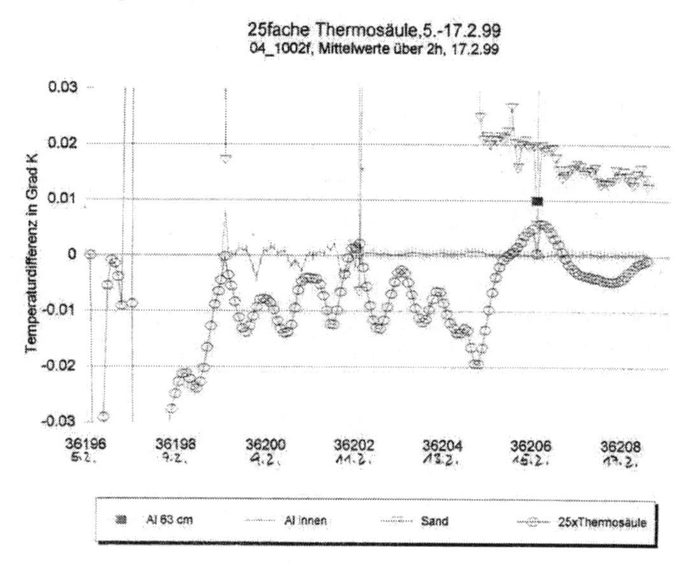

Six to seven days and nights passed with very similar results. Was it that easy to prove my theories which started to form in my mind? Would this not mean that heat could travel from cold to warm?

This next picture shows the same graph on a bigger scale.

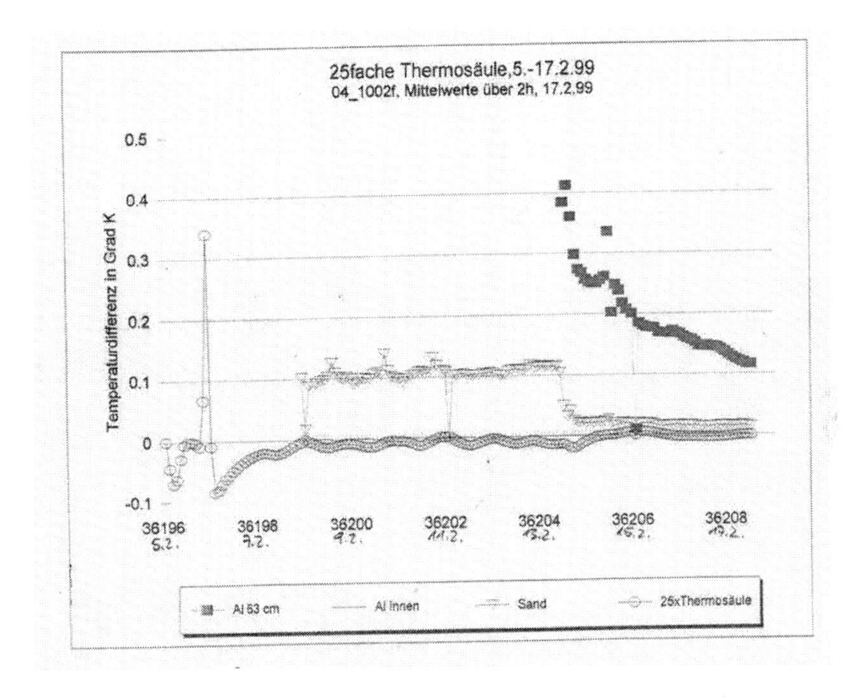

The values of the inner axle just below the zero line, meaning that the top is colder than the bottom, and the value of the surrounding aluminum tube and the sand with positive values, indicating the opposite, the top is warmer than the bottom.

Well, after 5 nights of broken sleep I could sleep through the nights once more. The curve of the inner axle had broken upwards through the zero line, indicating higher values for the top, lower ones for the bottom. Clearly something had happened, the cause of which I could not detect. It probably had to do with the fluctuating temperature gradient in the room. But as I did not use thermistors at that time, I had no way to establish this. Still, it had been very exciting if only for a few days. For any confirmation or disproof of this phenomenon I needed more follow-up experiments, as it turned out, lots of them, over many years to come.

How can I remember the progress I made years ago with the hundreds of experiments which I performed over the last 13 years, looking back from today? When did I follow a wrong path, when did I get a real new insight? Maybe the following will help:

What do you give a long time co-worker as a birthday present? A box with liquor filled with chocolates? A tie? A pretty scarf? Well, no, I prefer something connected with our common work or experiences during the past year. So in December of 1999, 2 years after we had started with our experiments, I gave Gisela Hoffmann in a picture frame a printout of my test performed in Ithaca, New York, the USA experiment no. 13. Now, looking at it after so many years it gives a good picture of how far we had come with my experiments at that time, how far or how not so far. The presented graph showed for the period from September 17th to November 26th 1999 the temperature gradients measured at four different locations, within a copper tube filled with lead powder, on the surrounding Dewar glass, on the outside glass container and finally on the outside plastic barrel.

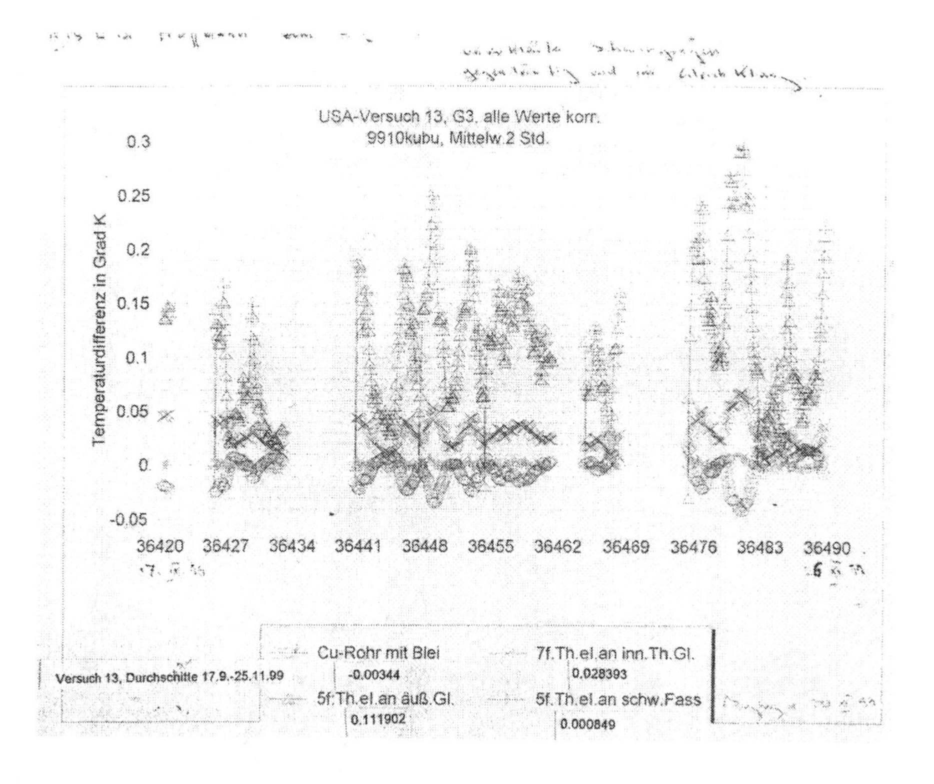

The hand written heading says:

"Gisela Hoffmann on December 2nd 1999,
Lived, experienced and formed together
Unexplained oscillations
Opposite each other and in harmony."
Test 13, averages September 9 - November 11, 1999

And underneath it gave the average values of the temperature gradients in Kelvin K for the total test period as follows:

Inside the copper tube filled with lead powder: - 0.00344 K
On the outside of the enclosing Dewar glass: + 0.028393 K
On the outside of the surrounding glass container: + 0.111902 K
On the outside enclosing black plastic barrel: + 0.000849 K

These curves represented one of the very first results showing a negative gradient in the inner axle of the experimental set-up over a longer time span, meaning that the top was cold and the bottom was warmer. But the amazing thing was that the surrounding containers showed a positive gradient, warm at the top and cold at the bottom, and this not just for a few days but for a longer period of 10 weeks. So already 2 years into my experiments I had found some very surprising and inexplicable results as shown on this birthday present. It created more questions than answers, but all very exciting to follow up. For instance: Why did the gradient of the inner copper tube oscillate just opposite to the values of the surrounding Dewar glass?

Let me jump to the present I gave Gisela Hoffmann 2 years later on December 2nd 2001.

1. 75 Innen Alu
2. 65 Alu-Rohr
3. 71 äusser Thermo 3x
4. 74 Thermo 5x
5. 74 Thermos 3x
6. 74 [illegible] 10x
7.
8. 70 Thermos aussen 3x
9. Trommel [illegible] innen 3x
10. 77 Dewar aussen 5x
11. 77 Dewar innen 5x
12. 681 Alu-Rohr [illegible]
13.
14. 75 Metallfolie + 5x
15. 72 Alu[illegible] aussen 5x
16. 75 Thermos [illegible] 5x
17. Trommel innen 5x
18. 76 Thermos aussen 5x
19. 76 Thermos innen 5x
20. 76 Glashohlkugeln 5x
21. 681 Alustab 3x
22. 75 Metallfolie Inn– 5x
23. 05 1x in Zelt [illegible]
24. 682 Glashohlk. = Cola 682 [illegible]
25. C-Drähte 24x
26. 682 3A unten 10x
27. 682 Cola Flasche [illegible] 3x
28. 811 PVC aussen 5x
29.

gisEla Hoffmann zum 2 XI 2001:

B 74: Gemeinsam aufgeschrieben zeigt: (Kanal 4)

– 474 / 14 / 3 / 6.1 × .17 = –0.0314 V/m !

Wo liegt die Wahrheit 22

Raumtemperatur °F

Kanal
1. 63 Dewar Al + Hohlk. gelb
2. 682 Glashohlk. = Cola 10x
3. 811 Thermos aussen grün 5x
4. 74 Glashohlkugeln blau 3x
5. 75 Metallfolie [illegible] lila 5
6. 75 " " " h lila 5
7. 70 3 Stahlzylinder weiss 3x
8. 77 Glas Kugeln braun 5x
9. 721 Metallfolie gelb 5x
10. 76 Glashohlk. rot 5x

As we were short on voltmeters and computer memory for storing the experimental results, twice a day I filled out paper forms with values read from a voltmeter which I turned to 40 different positions recording the respective voltage values of a number of different experiments. For this birthday I gave her a copy of one of these sheets covering the time period from October 19th to October 25th 2001. The sketch in the middle shows the experiment B74. Within an inner Dewar glass filled with hollow glass spheres and air, with a height of 180 mm, the temperature gradient is measured over a height of 170 mm. This inner Dewar glass is further insulated by a couple of larger sized Dewar glasses. The text on this present reads:

Gisela Hoffmann on December 2, 2001:
B 74:
Together written down
Channel 4 shows:
*= 474/14/3/6,1*0,17= -0,0314 K/m !*
Where does the truth lie?

So this result of October 2001, 3 years after the beginning of my experiments seemed to indicate again the surprising result: there is a negative temperature gradient in the middle axle within a positive gradient on the outside. This experiment, B 74, is very similar to the results of B 76 about which I report in the summer of 2002 at the conference in San Diego. This again confirms very nicely the very basic result: heat can travel from cold to warm! Already at that time, December 2001, I could see this result but I didn't dare to believe it to be true. At that time I felt that there must be another explanation correctly fitting within the accepted rules and laws of physics. So during the following 9 years I measured hundreds of other set-ups concentrating more and more on the question: could heat really move from cold to warm, could gravity be the cause of this result?

Still, my tests at that time where shots in the dark. I had no idea, no theory how to select optimal materials of solids or liquids, or the kind of gas to be tried in a meaningful, systematic way. Which type would give me the greatest gradient? How would this be influenced by the height of the set-up, by the type and thickness of the insulation, by the diameter of my thermocouples? Questions upon questions, and hardly any answers.

But this was changed dramatically at an unplanned stopover in Pittsburgh during a "lost", or "gained", New Years Day!

Chapter 14
Pittsburgh; Do Stones Have Feelings? Calculating T(Gr)

Do you like to fly? When I was 16 years old I started out like the Wright brothers with a very short flight, flying only about 10 meters or 30 feet. I had volunteered to become an air force fighter pilot as I wanted to shoot down those bombers which, after burning my home town Hamburg to the ground, continued to do so with one German city after another.

I was still a member of the anti-aircraft auxiliary force when I was sent to learn to fly a glider as the first step before being put into a real airplane with a motor. It was somewhat scary. The novice was strapped to a wooden seat in the front of the glider which was sitting on top of a sand dune 50 miles northwest of Hamburg. But around your seat there were no walls which would give you a sense of security, there was just air and a view of the ground. Behind you were some wooden beams, a wooden lightweight wing covered with thin cloth and in the back two rudders, one for steering left or right, one for up and down. These you could control with the stick in your hand, moving it left or right or forwards and backwards. Eight of my co-trainees manned a long rubber rope, two others held onto the end of my glider. On the command of our instructor the eight started running down the sandy hill on the top of which we had pulled the glider in the hot summer sun. As the rope stretched, the pull on the glider became greater and greater until upon a command the two guys in the back let go. I was pressed onto my seat as the glider was pulled ahead and cleared the ground. I was really flying! Everything went very fast, there was no time to think. I was only elevated about 30 feet above the ground, and it took only five or ten seconds to cover the 30 meters until I landed. But I had kept the glider pretty much on an even level, steady and straight; during this short flight, I had passed my first test.

In the 1950's, when I started to fly in commercial airplanes, I enjoyed flying, at least as long as the weather was fine and nobody had to use those paper bags in the pockets in front of your seat. Before the jet age arrived propeller planes, although noisy, flew quite low and through the big windows you really could enjoy watching the landscape passing slowly below you. Airports were small; you could arrive for your plane at the last minute. Once, in 1962, when I was a few minutes late for my flight to the States, I was put into a little Volkswagen and driven onto the tarmac up to my plane, which was ready to take off. This happened at Frankfurt airport, which is hard to believe nowadays.

Are you scared of flying? Living with my family half the time in Germany and half in the United States, crossing the Atlantic became a normal routine for me every few months. When the jets arrived I initially did not enjoy their high speeds when taking off or landing, but I got used to it. However, I did start to take one precaution: I would cut my toe and finger nails before a flight. This is not a task I especially enjoy so I delay it as long possible. But it can't be avoided from time to time, so why not do it just before a flight? This might sound strange, but let me tell you the reason for this. As you know, I am an engineer. If I think about problems I try to imagine the theoretical extremes or border lines. For instance when a plane takes off and has a hard time gaining altitude because it is too heavy, having ice on the wings or for some other reason, it might hit the fence at the end of the runway or just clear it. How could a slightly too heavy load result in a terrible crash instead of an uneventful beginning of a long flight? What weight could make a difference? What about 10 kg, 1 kg or 1 gram? Would an even smaller weight difference decide the fate of the crew and the passengers? I have pondered this question quite often. The answer: Cutting my nails could just be the weight difference needed to save my airplane, right?

After cutting my finger nails on December 31, 1998 I took off from Ithaca, New York to Frankfurt, Germany via Pittsburgh. In Pittsburgh our non-stop flight to Frankfurt was delayed. An engine had a minor problem which they hoped to be able to fix quickly. Well, it took longer than expected. So the airline announced: "If you want to be sure to cross the Atlantic tonight, quickly board a flight to

Philadelphia where you still can catch another flight to Frankfurt or take your chance that we will be successful with our repair and wait here." Most people left but I decided to wait. Rather than sit in the overcrowded airplane starting out from Philadelphia filled to the last seat I saw myself lying down across four middle seats for a good nights sleep in a nearly empty plane.

Four hours later they told us the repairs could not be done. Sorry, you will arrive in Frankfurt 24 hours later than planned. We have a motel room for you tonight, come back tomorrow. What a way to spend the evening. It was 11 P.M. By now the New Year was approaching and I had to try to get to my motel room. I waited for the limousine in a practically empty arrival building. Indoors it was nice and warm, outside it was snowing hard and a bitter cold wind was blowing. There I stood near the entrance, getting colder and colder, waiting for the car to take me to my motel. What a way to spend New Years Eve. Then I noticed a decorative flower bed next to the entrance filled with of a number of fairly good-sized stones:

Stones at the airport in Pittsburgh

I wondered: How do these stones feel during this cold night one hour before the year 1999 begins? Surely they consisted of billions and billions of atoms and molecules, all held in place by the electrical forces of their neighbours, but still rapidly swinging back and forth in accordance with their temperature. Those molecules, swinging up and down, would they be affected by gravity I asked myself, being slowed down on their upward swing and accelerated on their way down, somewhat similar to freely flying molecules in a gas? Would the stones therefore be somewhat colder at the top and warmer at the bottom, as I believed it happened in a gas? My thoughts were interrupted by an approaching limousine which delivered me to my motel. Fifteen minutes later I was asleep; the New Year had arrived.

I had planned quite a few activities for my first day in Europe. Now I would arrive a day late. Had I now lost a day, arriving 24 hours later than planned or had I received the gift of a day, totally free of any tasks or commitments? That's how I felt when I awoke on New Years Day, 1999. I could think about those cold stones at the airport, uninterrupted by daily demands. At this point in my experiments, the acceleration and deceleration of gas molecules on their way up and down seemed to be clear to me. I had wondered if molecules in solids might be similarly affected by gravity. I had started a test with an aluminum block, 10x10x30 cm, which was actually running in Germany right now.

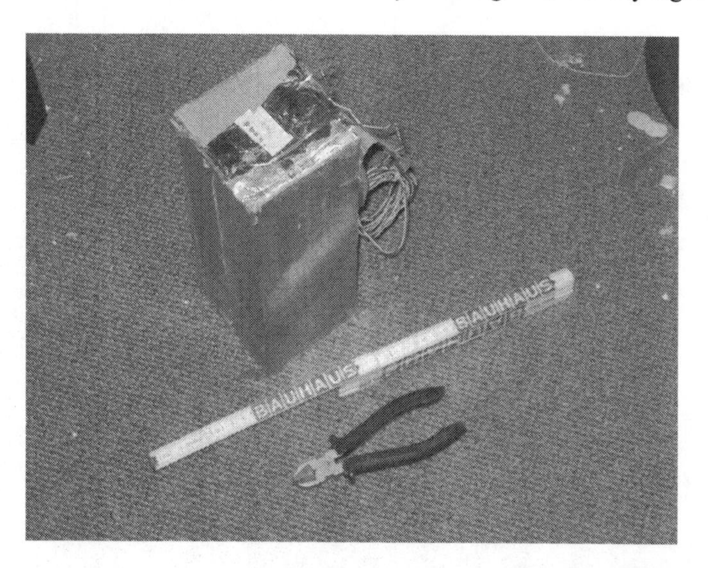

I wondered if it really should be similar in solids as in gases, or even in liquids. If a temperature difference developed, would its value be different in different materials? Could it be calculated? Could I possibly calculate these values which I tried to measure?

Those cold stones made me think. If I imagined a molecule in the stone at the moment it was swinging downwards, or similarly a molecule in a gas, would not its potential energy, which it had at its highest position at the top, be converted into kinetic energy during its downward path? This would mean an increase of its downward speed until it reached its lowest point. But the speed of a molecule was a measure of its temperature. Could I not simply calculate the potential energy at the top, deduct its potential energy which it would have when it reached the bottom, when it was stopped there, before being turned around for its upwards swing, and convert this energy difference into a temperature increase of the molecules involved?

After a quick breakfast I sat down in my room with an empty piece of paper and a pen. Of course I did not have any physics book with me but I knew:

The potential energy Ep of a body was equal to its weight G multiplied with its height difference H, so I wrote down

$$Ep = G \times H$$

and as weight G equals mass times gravity factor g, I added as the next line

$$Ep = G \times H = M \times g \times H$$

Wonderful! Ep would have a dimension of energy like Joule. This energy would have to be converted into a temperature increase of the molecules involved, because where else could it go. This increase which I would call T(Gr), or the temperature increase due to the effect of gravity. To calculate the temperature increase of a body due to an amount of energy added is a very normal calculation for an engineer, as the temperature difference equals the energy introduced divided by the mass of the body M and its specific heat c. So I could write down

$$\text{Delta } T = T(Gr) = Ep / (M \times c)$$

Combining the two equations I got

$$T(Gr) = M \times g \times H / M \times c = g \times H / c$$

or $$T(Gr) = - g \times H / c$$

A "-" sign made sense because gravity is directed opposite to H. This also would mean that the temperature at the top should be lower than on the bottom.

Now for the first time I had a formula for the calculation of an actual temperature difference. If this formula was correct then the temperature difference T(Gr) that I tried to measure depended only on the gravity factor g, which is practically a constant all over the earth, the vertical height difference H and the specific heat c of the molecule involved.

Now it became quite interesting. Maybe I could calculate this value, for instance for air, right here in my motel room. Then for the first time I would find out, if this value would be high enough for me to be able to measure it. Alternately I might find out that the value was way too low for any hope of success. But for calculating it, I needed the correct value of the specific heat of air in the correct dimension. How could I find it without access to any physics book?

I knew that for gases there were two values of c. One, cv, had to be used if you heated or cooled a gas while keeping the volume of the gas constant. The other, cp, was used for conditions with a constant pressure. I remembered the value of cp for air as being 1000 J/kg,K (joule per kilogram and degree Kelvin or C) as I had to use this value often when calculating the size of an electric heater to heat the air in the dryers I had designed and built for many years. So for now, without a book to look up the correct values, I would use cp and would worry later if cv should be used instead.

Remembering that the value for g was 9.81, it was easy to calculate for the first time T(Gr) for a height of 1 meter and I got quite excited when I wrote down

$$T(Gr) \text{ for air} = - g \times H / c = -10 \times 1 / 1000 = -.01 \text{ K/m } !$$

I was happy about this value because I knew that my instruments in my laboratory in Burgberg were capable of measuring a temperature difference of .01 C, or better written as .01 K, or one hundredth of one degree C. But was this a reasonable value? Would not his value have to show up in the atmosphere around the globe? During a flight the pilot sometimes announces the outside temperature. Was this not typically a value around negative 40 degrees Celsius when flying at a height of about 10,000 meters? Well, my newly calculated value of -.01 K/meter would mean a temperature difference between the surface of the earth and this elevation of 10,000 meters of -100 degrees Celsius. This is somewhat higher than what the pilot tells us, but still in the same ballpark. In the atmosphere there are also convection currents, air movement which would tend to reduce the temperature gradient created through the effect of gravity. The value which I had just calculated could exist only in an air mass without any movement. It sounded quite reasonable.

The stones at the airport had been the trigger for this calculation. Solids do not have different values of cp and cv but only one value c, and I thought the value for stones was something like 800 Joules/kg,K. Calculating with this value for stones in the flower bed at the airport with a height of 10 cm (0.1 meter) I found a value for my stones as

T(Gr) = -10 x .1 / 800 = -.0012 K

Well, this was a rather small value, probably too small to be measured.

Then I started to think about the earth. If gravity had an influence on the temperature distribution in solids, then it should influence the temperature of the core of the earth. With my new formula I should be able to calculate it. But what was the radius of the earth, the value I needed for H? I didn't have those numbers in my head, but I remembered the frequent flyer miles I got for one transatlantic flight: about 4000. Assuming that such a distance was about 1/5 of the earth circumference, the radius of the earth should be about

5 x 4000 miles /2 pi = 7,000 miles or about 11,000 kilometers.

Assuming that the specific heat might be close to the value of rocks which I knew was around 800 joule I could calculate for the interior of our earth

T(Gr)= 10 x 11000000 /800 = 130000 K !

This sounded much too high as I seemed to remember that the temperature in the interior of the earth was supposed to be more like 10,000 degrees. Would the high temperature I just calculated for the earth not become noticeable at the surface of the earth? No, I thought, it would not, at least if I assumed that the earth's interior would still be a solid under the very high pressures existing there. T(Gr) represented a temperature gradient under equilibrium conditions, constant over time, kept at that level by the influence of gravity. And no heat flow by conduction should take place over this gradient. Therefore the surface of the earth would not feel the extremely high temperatures of the interior at all. And certainly nobody had ever visited the interior of the earth to measure the temperature there. So why could the high temperature I just calculated not be the correct one instead of those printed in the text books? Sitting in my motel room I did not know, nor could I find out, why geologists argued for a temperature of around 10,000 degrees. I wanted to find out as soon as possible. Maybe the temperature was not quite as high as I had just calculated it. If the earth interior was liquid, then at those high temperatures the material there would be quite fluid. The temperature gradient would make the material flow, the colder material being heavier than the hotter material would create vertical convection currents. Material from deeper levels would rise while the colder material would fall. This movement would tend to equalize the temperatures somewhat.

But I felt great. Just by using my head, a pencil and a piece of paper, I had calculated the temperature of the earth's interior and found a value, which maybe nobody had ever calculated before! Correct or incorrect, I felt a real high on this January 1, 1999. Thanks to a broken-down airplane engine, I gained this special, unplanned, free day in my life.

Yes, airplane disasters are rare, and I had cut my toe nails. But one never could know for sure. At least my new insight in calculating T(Gr) should not get lost if my flight were to end in a disaster. Therefore I wrote out my new formula for T(Gr) and the steps which let to it on a piece of paper and faxed it to Nancy. I felt better. At least my formula would survive my death.

And more important: if my new formula

$$T(Gr) = -g \times H / c$$

was correct, I would find different temperature differences for different gases depending on their specific heat. The smaller their specific heat the higher their temperature gradient should be and therefore easier to measure! Suddenly I had a method to make sensible selections of the gases I was going to use in my future experiments.

My flight to Frankfurt on January 1, 1999, proceeded uneventfully. After a good night's sleep after my arrival in my own bed the next morning I calculated in a spreadsheet the T(Gr) values for a number of gases:

T(Gr)calc2009	Gases			T(Gr) Calculation			
2.Jan 1999		air	Argon	Freon 12	Xenon	Radon	Jod
molecular weight		29			131	222	
Density	kg/m3	01. Feb	Jan 63	05. Apr	05. Apr	Sep 73	
sp. Heat cv	J/kg,K	1003	518	611	158	92	
sp. Heat cp		715	314	542	95	56	
speed of sound	m/sec	340			600	200	
heat conductivity	W/m,K	0.024	0.018	0.01	0.006	0.003	
T(Gr)	K/m	-0,013	-0,019	-0,016	-0,063		

For air the table shown I got a value of -0.013 K/m, the same value I had written down in my motel room, but xenon had a much higher value with-0.06 K/m, six times as high. Would I be able to make experiments with such a rare and probably expensive gas?

If my calculations were basically correct, this presented a great advance. From now on I would be able to select those gases for my next experiments which promised the highest temperature gradient. The logical next step should be: do my present test results come close to these theoretical, calculated values?

Looking back today to that strange New Year's Day 1999, I still get a very good feeling. I, the poor engineer, developed the first formula, just with a pen and a piece of paper, a formula allowing the calculation of the temperature gradient I tried to measure. What I did not know then: In a good 3 years time, exactly on March 5, 2002, another insight would increase my theoretically calculated T(Gr) values by a factor of 5!

Chapter 15
Wow: An Improved Theoretical Explanation for my Measurements?
My Web Page www.firstgravitymachine.com

2001, 4 years after I had started with my experiments I felt quite excited with the results. They seemed to show clearly that a temperature gradient existed in my test setups being cold at the top and warm at the bottom, and this in an environment with a positive gradient, warm at the top and cold at the bottom. I went to the wonderful physics library of Cornell in Ithaca trying to find out if anybody else had ever before tried to measure this gradient. Looking through some 40 different current magazines issues displayed in the magazine room I suddenly found a paper in an offbeat magazine by an Andreas Trupp describing the history of a dispute between the physicists Boltzmann and Loschmidt in 1876 in Vienna, discussing this very question [4]. Apparently this had been a hot topic during those years. I really should try to contact this Andreas Trupp. Surprise, surprise: He had an address in Germany.

During my next visit to Königsfeld, Germany, I gave Andreas Trupp a call. He was actually a PhD of law, teaching at a Police Academy in the eastern part of Germany. He sounded very nice and confirmed his interest not only in my experiments but also in the work of my foundation "Gewaltfreies Leben", (Violence Free Living). He accepted my invitation to come to Königsfeld to our next seminar in December 2001 which we would combine with a one day discussion about my experiments.

At the arrival time of his train in Sankt Georgen, our closest train station, nobody from Berlin arrived. When I returned and checked the next train one hour later, I was a few minutes late, and it had apparently already passed. In front of the station on a big stone around a flower bed a man was sitting looking like a homeless person, maybe 40 years old, and showing a three- week old beard. He had a big knife in his hand cutting off thick slices of bread and pieces of a large sausage, enjoying the sunshine. Normally I don't approach homeless persons with a knife in their hand, but bravely and somewhat timidly I tried: "Hello, have you seen the 12 o'clock train passing by and did anybody get off the train?"

He looked at me and said yes, the train left already, only a few people had arrived. "I'm looking for a guest from Berlin, maybe 40 years old, do you remember having seen him? "

"Well, I came from Berlin!" It was Andreas Trupp.

This started a friendship based on our common interest in physics, history, non-violence and politics. He had studied the historical dispute between Boltzmann and Loschmidt and felt that Loschmidt had a good argument for the existence of a temperature gradient. He published his thoughts in his web page.

After his 2 day visit he wrote in our guest book:

From the 8th to the 9th of December, 2001, for 24 hours (with a short interruption for a little sleep) the Second Law of Thermodynamics, the police, the violence in society and other topics were covered (really a work for titans). Furthermore an action plan was agreed upon to finally show the world its path towards the future. A highly interesting visit lies behind me.

Andreas Trupp.

Andreas was a wonderful participant in our discussions. "Why don't you extend your measurements to the use of centrifuges to create a much greater temperature gradient, much easier to measure?" And, echoed by Alexander, he added:"

Roderich, you are a nice experimenter, but leave the theoretical questions to the specialists, the theoretical physicists."
"Fine", I answered, "but up to now I have not heard a peep from anybody trying to think of a theoretical explanation for my experimental findings. Without it, how can I plan new meaningful tests?"

My recent experiment B74 measuring air within a Dewar glass filled also with glass powder had produced very nice results, showing a long term average negative gradient T(Gr) of -0,07 K/meter. On the one side, to get such a high value for the gradient was very nice. It brought the reassurance that the cause for this gradient could not be convection currents of the air between the glass powder creating an adiabatic gradient like it existed in the earth atmosphere. This adiabatic gradient or laps rate could reach a maximum of only -0,01K/meter. It was just slightly lower than the value which I had calculated and written down for the first time during my New Year's Day 1999 experience in Pittsburgh based on the formula

$$\mathbf{T(Gr) = -g \times H / c}$$

or in words: The temperature gradient in vertical columns of gases equals the gravitational constant g, multiplied by the height H and divided by the specific heat of the gas. Calculating with this formula one got a value of -0,014 K/m for air. So this new measured value of -0,07 K/meter created the problem of being 5 times too high for my formula, the only formula I knew.

New ideas or insights I seem to be getting very typically at night, during my sleep or while dreaming. I fall asleep with a thought or a problem and the brain seems to be working on it developing a solution or answer during the night. In the morning any result has to be verbalized in order for me to become aware of it and this happens with me best during breakfast with another person with whom I can discuss such a nightly thought. During recent years this takes place quite often with Gisela Hoffmann as we have breakfast together when she is in town working on the documentation of my experiments. So on March 5, 2002, I suddenly declared out loudly:

Königsfeld,
5. März 2002

Erkenntnis des Tages.

Endlich geklärt im Frühstücksgespräch mit Gisela Hoffmann:

Zweiatomige Moleküle

fallen $\sqrt{5}$ Mal so schnell vom Baum

wie ein Apfel.

Roderich Gräff

This, for me, was a tremendous statement. We typed it up and sent it right away to Alexander. It reads in English:

Koenigsfeld, March 5, 2002.

Insight of the day.

Finally clarified in a breakfast discussion with Gisela Hoffmann:
Diatomic Molecules
fall $\sqrt{5}$ times faster from a tree
than an apple.

Roderich Gräff.

Correct?

Gisela Hoffmann protested right away: All masses, including single molecules and apples, fall with the same speed (in a vacuum). Well, maybe she was right, but at the moment I was not interested in the speed of an apple but of a single molecule.

I felt this was a very important insight as finally it gave an explanation why my measurement results of T(Gr) were 5 times higher than my old formula allowed. The former formula from now on would have to be multiplied by the number of freedoms n of the molecules in question, so from now on it would read:

$$\mathbf{T(Gr) = -g \times H / (c / n)}$$

As air has 5 degrees of freedom, this changed the result of my old formula for air from -0.14 K/meter to –0.07 K/meter, which was very close to the long term average value I kept measuring. In addition, this seemed to be a nice independent confirmation of the basic correctness of my experiment B 74. It is somewhat difficult to explain it clearly. But let me try:

The underlying thought of my new insight was this: A molecule in a free fall gets accelerated by the force of gravity, but all energy induced and showing up as kinetic energy, affects only one degree of freedom. In a diatomic molecule like oxygen or nitrogen constituting air, this would be only one degree of freedom out of the five these molecules have. And as the speed of the molecule would be proportional to the root of the energy added, the speed of the molecule would be the root of 5 times faster than that of an apple. Why faster than that of an apple? Because in an apple the added energy would be distributed during the fall to all five degrees of freedom as the molecules within the apple are in constant contact which each other. Bouncing continuously against neighbouring molecules would result in a thermalisation of energy which is another way of saying, that the added energy is distributed to all degrees of freedom. (Note: As it turned out later, the part with the apple was incorrect, but this did not influence my reasoning of the speed of a molecule.)

Let me demonstrate this with the following two pictures:

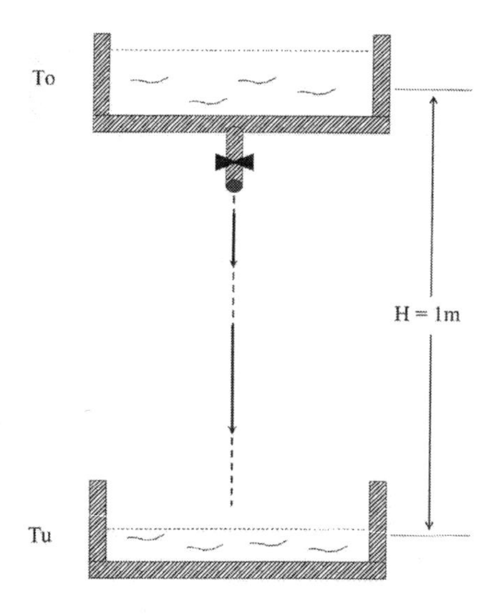

This first picture represents my insight from my "Pittsburgh" experience leading me to the formula

$$T(Gr) = -g \times H / c$$

Here water drops from the upper container into the lower container through a height of 1 meter. In the lower container the water comes to rest. The potential energy of the water in the upper container over the height of one meter has been converted to heat, heating the water by

$$T(Gr) = -g \times H / c = -9.81 \times 1 / 717 = 0.0137 \text{ K}$$

assuming that all heat stays in the water and none entered into the lower wall of the container. This example represents quite well the case with an apple, as an apple consists mostly of water.

In the second picture I assume that instead of water, falling down and staying down, a diatomic molecule like oxygen or nitrogen similar to a billiard ball falls through the height of one meter. It does not stay down when hitting the lower wall but is reflected from its surface and returns to the upper wall.

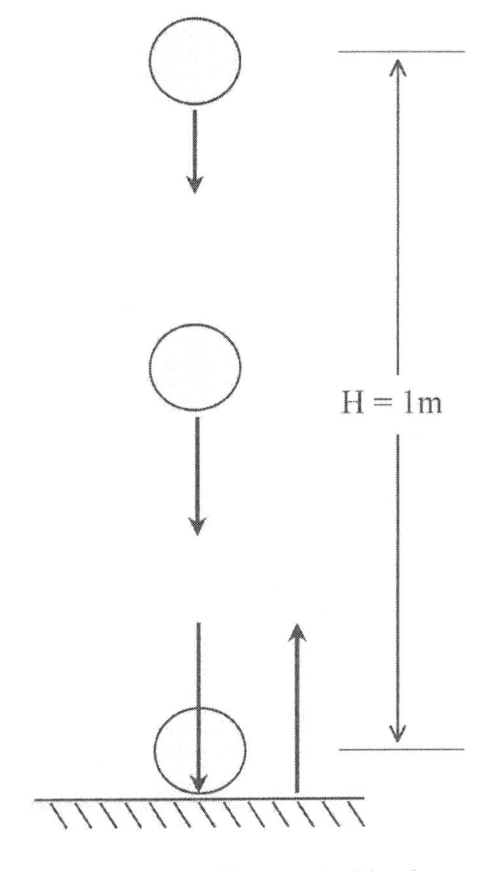

At the moment of impact the kinetic energy of the molecule is transformed into a momentum impacting onto the surface of the lower wall. The size of the impact is proportional to the square root of the vertical speed component of the falling molecule, which is the square root of 5 bigger than in the case of the falling water, where the energy of the water remaining at the bottom is distributed equally between all degrees of freedoms. So our T(Gr) formula has to be changed by multiplying it by the number of freedoms n, so it changes to

T(Gr) = -g x H /(c/n)., as stated above.

Herewith I have tried to explain in a few words what my insight on the morning of May 5, 2002 consisted of, I am sure not very well and too short to really make it clear. Well, during the morning of that day, during breakfast, it was not totally clear to me either. It took quite a while until I felt that I understood it well enough to be able to really explain it clearly. If you, reader, are interested in a more concise explanation please check Chapter 24. Still the result of my May 5 insight was that my formula stood up to my experimental results and, up to today, no real specialist in the field of thermodynamics has been able to point out to me any inconsistencies or incorrectness in my argument.

When I decided on my 70th Birthday in October 1997 to start with physics experiments about a question which had plagued me since my student years, I thought that in 3 years I would have reached an answer. Either I would have found some interesting results confirming my suspicion that the energy exchange by radiation between a blank, highly reflective surface and a black one did not quite follow the Second Law of thermodynamics, or it had become clear that everything took place in accordance with our physics books. These three years had passed late in 2000. I had to admit that my original question had totally changed to another one. Not the heat exchange by radiation was the aim of my experiments anymore but the question, if a vertical column of molecules in an isolated system would show a temperature gradient or not. So I kept working and after another 2 years in early 2002 I felt I had something to tell the world. Yes, opposite to today's thinking I had found a negative temperature gradient, cold at the top and warm at the bottom. I had sent a paper covering my experiments with air to the physics magazine Annals of Physics but they declined to publish it. So I felt it was a good alternative to publish my findings myself on my own Web Page.

I had never worked with publishing something on the internet but my friend Davor Bakula offered right away to perform the necessary work. Davor and his wife Snjezana were recent immigrants to the United States coming originally from Bosnia and Serbia. Having fled to Germany there in the late 1990ies, while trying to escape the civil war going on in the Balkans, Snjezana had shown up one day at my little plant in Weiterstadt, Germany, asking if we could not employ her. We really did not need any help at the time, especially from a young student who could hardly speak German. But she was a refugee and I sympathised with her remembering my first job in the US as a visiting student working at nights cleaning a store in order to make some money. So I asked my long time secretary, Mrs Ursula Matthes in charge of all personnel to offer her any job we could find, maybe as a cleaning lady. Was she interested? Oh yes, she was. And soon thereafter, when we needed help in our production department assembling loaders, and with her German fast improving, she was promoted. A year later she again advanced, this time to a white collar job, quickly learning how to become a purchasing agent.

A problem arose in 2000. The fighting in the Balkans had stopped and the time to return to their homeland had come. But, being married by now to a husband coming also from the former Yugoslavia, what was the home land of this young couple? Growing up they were Yugoslavs, but now, both of them coming from Croatian/Serbian mixed families in the middle of the Muslim part of Bosnia, they had no place to go. The Balkans were still in the aftermath of a civil war. As a couple they would not be welcome in either place.

Snjezana and Davor Bakula

Luckily the United States started a program for 10000 of these mixed marriages by allowing them to immigrate to the US. Nancy and I decided to sponsor them, meaning that we would guarantee their upkeep in case they would not be able to provide for themselves and needed financial assistance. We found an apartment in Ithaca which was big enough for them but which also left one room for me to move my office there and which was big enough for performing some experiments. Davor being an engineer easily grasped what I was trying to do and assisted me in my work whenever I needed help. And having grown up with computers he designed in no time my webpage and filled it with the texts I provided.

The time was right to create a webpage as I felt that I could report solid experimental results. And in the following I will reproduce those pages published in early 2002, as they document quite clearly how far I had come four years after

I had started with my work. Without having produced this webpage in those years it would be difficult to recreate today the situation of early 2002. How far had I come, what did I know then, where was I right, where was I incorrect at that time?

The complete original webpage with later additions can be viewed in Appendix 2. Here I will just describe the early pages published up to 2002.

I started out with a short introduction so that my plans and intentions could be quite easily understood by anybody, not necessarily an engineer or physicist. Here comes my introduction:

From deep in the Black Forest, January 6, 2002

Welcome! What led you to this web-page? Was it the idea of "Peaceful Energy", or "the Blue Rose", or "Gravity Machine" which struck a chord in your imagination?

Maybe one day you can visit me in the Black Forest area in the southwest corner of Germany and we will take a hike through the pine woods surrounding us with clean air and stillness. The Blue Rose began its life here in this peaceful area as a sign for non-violence.

And here you can find the first Gravity Machine, at least as far as I know, it is the first. Sure, there are power installations, which create energy out of gravity, like the generation station in La Rance in France, using the change in the water levels between high and low tides to generate electrical energy. The potential energy present in the moon circling the earth is thereby converted. Each tide slows down the moon, taking it further away from earth, a process, which in the far distant future will come to a sad end. Lovers beware!

No, my Gravity Machine is different. It also creates electrical energy, but it does this by using the chaotic, kinetic energy of millions of molecules in a gas. It is not Maxwell's demon who sorts the fast molecules from the slow ones, but a gravity fairy who creates temperature differences in a vertical column of gas. This is the starting point for creating a PERPETUUM MOBILE OF THE SECOND KIND. Since two years my Gravity Machine uses these temperature differences to generate electricity without the introduction of any energy from the outside! Not enough yet to heat my meals, but the principle counts, and time will tell! Interested? Keep reading. In the coming months this website will develop step by step. It will give you an overview over my new BOOK. Some time in the future:

GRAVITY MACHINE
My Search for Peaceful Energy

will be available and tell you the story of the First Gravity Machine in much more detail. In case you are interested let me know your address and I will inform you when the book is available. The BOOK will contain all the information you need to build your own gravity machine. Will you be among the very first to demonstrate a PERPETUUM MOBILE OF THE SECOND KIND?

April 27, 2002
Updated January 26, 2003

CONTENTS

This website describes experiments I started 5 years ago. They are still ongoing, because the results up to now seem to indicate a big surprise, even a sensation: Warmth can travel from cold to warm!

I describe the experiments, some results, and the theory behind them. You can browse through the topics listed on the left side of each page. You will find plenty of links in the text.

I like to whet your appetite enough to let me know if maybe you want to order my book - whenever it is ready -

BOOK GRAVITY MACHINE
My Search for Peaceful Energy

Here you will find all information necessary to build your own GRAVITY MACHINE. The text is embedded in my experience with non- violence, a topic closely related to our ever-growing need for energy. The headings of the book chapters are these:

Well, I don't have to list them here, you are holding the book, finally finished 8 years later, in your hands. I then continue:

April 23, 2002

INTRODUCTION

Warmth can flow only from warm to cold. Every body knows that. Nobody doubts it. Physicists teach it. The SECOND LAW OF THERMODYNAMICS confirms and states it:

"Heat cannot pass by itself from a colder to a warmer body"
Correct?

Well, not quite. Because you have to exclude me. And once you study this web page and, once it is ready, my BOOK

GRAVITY MACHINE
My Search for Peaceful Energy

you might start to wonder. It might even entice you to build your own gravity machine, a PERPETUUM MOBILE OF THE SECOND KIND. Wouldn't it be nice to sit in your bathtub and watch a boat swimming around you on its own power with the energy to drive the screw of your boat coming solely from the latent heat of the bath water?

Are you one of the few energy conservation conscious people, who, in winter, don't drain the warm water after your bath but let it sit until the next morning in order to use its inherent heat to warm the house? Over night the bath water will cool down to room temperature, in accordance with the SECOND LAW OF THERMODYNAMICS. If you heated the bath water with heating oil or gas, the same energy source you are using to heat your house, then you got your warm water for your bath free of charge, because by releasing its heat into the bath room air, you are saving an equal amount of energy heating your house.

When you enter your bath room the next morning, you will see your dirty bath water still sitting in your tub, but your face will brighten. Your little boat is still running around in circles! Where does it get its energy? Could it be possible that its movement demonstrates a perpetuum mobile of the second kind?

During the second half of the 19teenth century a spirited discussion went on in the circles of the leading physicists of the time. And not every one of them agreed with the conclusions. L.Boltzmann (1) and J.C.Maxwell (2) argued, that a vertical column of gas under equilibrium conditions would have to show equal temperatures over height. Otherwise the construction of a perpetuum mobile of the second kind would be possible.

But Loschmidt (3) disagreed. He felt that gravity would cause a stratification of temperatures, cold at the top, warmer at the bottom.

Throughout the years theoretical arguments were put forth; and they changed over time. As long as the results agreed with the SECOND LAW OF THERMODYNAMICS, they were eventually accepted nearly universally as correct. A good summary is given by A. Trupp [4].

Who was right? Why had nobody, at least to my knowledge, ever published experimental results disproving the ideas of Loschmidt? His opinion should have had some weight because he was the first who counted the gas molecules in a cubic centimeter, a pretty high number. Here everybody agreed and we all learn this number as the Loschmidt Number even today.

Six years ago, I asked myself, was it possible to measure the temperature distribution in vertical columns of gas in a meaningful way with the limited resources available to me? Was this possible for me, an engineer, not even trained as a physicist? Or could this work be done only by an outsider like myself, as a physicist would not even try, as he knew the result already before starting out: the temperature at the top had to be the same as on the bottom. The SECOND LAW OF THERMODYNAMICS demanded it. And that is sacrosanct.

Well, I dared to try! The results are surprising. Yes, warmth can travel from cold to warm!

The Second Law

New statement of the Second Law

The Second Law of Thermodynamics can be stated in a great variety of ways. One of the first one comes from Clausius in 1854 [11]. No process is possible for which the sole effect is that heat flows from a reservoir at a given temperature to a reservoir at a higher temperature. This statement implies that the process takes place within a closed system with no exchange of matter and energy across its bounders. It also implies, like any other presently used statement, that the closed system might be exposed to a force field, like gravity, and that in spite of this, the statements remain valid. Contrary to the statement by Clausius, the reported results show that in a closed system under the influence of a force field like gravity heat can flow from a reservoir at a given temperature to a reservoir at a higher temperature. In order to keep the validity of the Clausius statement and many similar ones one has to exclude the influence of force fields. This leads to a new general statement of the Second Law:

In closed systems, with no exchange of matter and energy across its bounders AND WITH NO EXPOSURE TO FORCE FIELDS, initial differences of temperature, densities and concentrations in assemblies of molecules will disappear over time resulting in an increase of entropy. And in order to state the difference of the new statement of the Second Law of Thermodynamicsto the old statements explicitly: In closed systems, with no exchange of matter and energy across its borders, FORCE FIELDS LIKE GRAVITY generate in macroscopic assemblies of molecules temperature, density and concentration gradients where the temperature differences can be used to generate work resulting in a decrease of Entropy.

This text, written in early 2002 shows nicely, that already at that time my description of the meaning and the results of my experiments are very close to the wording which I am using today, 8 years later, as described later in this book.

April 24, 2002

The Dancing Molecules

In the late 19th century a heated discussion went on about the temperature distribution in vertical columns of gas. L.Boltzmann [1] tried to prove with his calculations that a vertical column of gas under equilibrium conditions would have to show equal temperatures over height. Otherwise the construction of A PERPETUUM MOBILE OF THE SECOND KIND would be possible.

J.C.Maxwell [2] agreed. Based on the same assumption he calculated the speed distribution of molecules. He found that it would be identical for different gases and depend only on their temperature.

But there was a lonely dissenting voice. J. Loschmidt [3] disagreed. He felt, that gravity would cause a stratification of temperatures, cold at the top, warmer at the bottom. But Loschmidt died in 1895.

Throughout the following years theoretical arguments were put forth, but they changed over time. Only an equal temperature over height was consistent with the SECOND LAW OF THERMODYNAMICS and as long as the theories showed such a result, they were accepted nearly universally as correct. A good summary is given by A.Trupp [4].

Did anybody ever measure this temperature distribution? I am not aware of any published results. If you are, please let me know.

When I started with measurements, I took the following approach: I look at a container filled with a gas with such a low pressure that the free path is longer than the vertical distance of the walls:

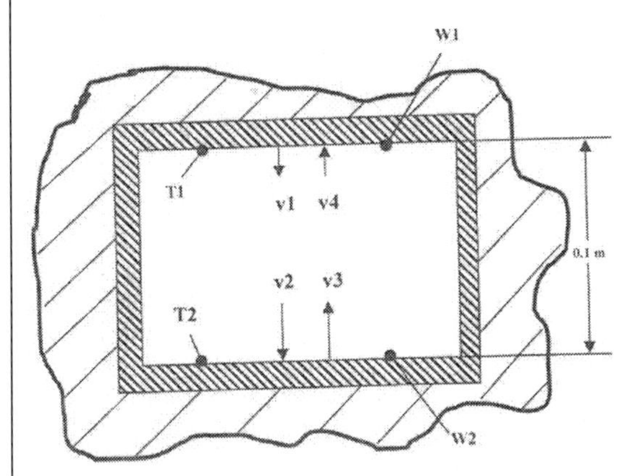

With an assumed inner pressure of .0001 mbar, the free length of the molecules at room temperature is about .6 meter. If the vertical height of our container is .1 meter, a molecule flying between the upper and lower walls would bounce back and forth between these walls with hardly ever hitting another molecule. On its downward path it would be accelerated by the influence of gravity, increasing its speed. On the upward path the opposite happens; the speed of the molecule is reduced.

But the speed of a molecule corresponds to its temperature. We also know that a molecule bouncing off a wall leaves the wall, on the average, with the temperature of this wall. If initially the walls had the same temperature, then the upper wall would lose energy as it accelerated the arriving "slow" molecules while the lower wall would receive an equivalent amount. The upper wall would get colder, the lower one warmer as energy would be transferred from the top wall to the bottom one.

This thought experiment seemed to indicate that a temperature gradient would slowly build up. If we wait long enough until we reach equilibrium conditions, then the upper wall would have to show a lower temperature than the lower wall, as it gets hit by molecules with a slower speed than the lower wall.

My BOOK will describe this happening in more detail. It points out that this process would take place not only in a rarefied gas but also in a dense one. Here a molecule would bounce from molecule to molecule and would hit a wall only rarely. Still, the stratification of temperatures, cold at the top and warm at the bottom would take place just the same. Our main interest is of course, to find out, what size of temperature difference we can expect to get. This I will discuss under TEMPERATURE DIFFERENCE IN THE GRAVITATIONAL FORCE FIELD.

May 1, 2002

Temperature Differences in the Gravitational Force Field

In THE DANCING MOLECULES I described that gas molecules, arranged in a vertical column should show a temperature difference between the temperature at the top T1 and the one at the bottom T2. I call this difference T1-T2 = T(Gr), the temperature difference due to gravity.

In trying to measure it, it is important to know the course of it and its size. Once we know its course we can plan meaningful test setups, once we know its size we can judge if the measurement of this difference, if it exists, appears possible or not. These questions are covered in the following paper:

Temperature Differences in the Gravitational Force Field

In THE DANCING MOLECULES I described that gas molecules, arranged in a vertical column should show a temperature difference between the temperature at the top T1 and the one at the bottom T2. I call this difference T1-T2 = T(Gr), the temperature difference due to gravity.

In trying to measure it, it is important to know its size. Once we know it we can judge if the measurement of this difference, if it exists, appears possible or not. In order to calculate it we will initially look at molecules inside an enclosure with a low pressure. This would represent a rarified gas where the free path of the molecules is greater than the distance between the top and the lower walls. A molecule will fly between walls with hardly ever bouncing into another molecule.

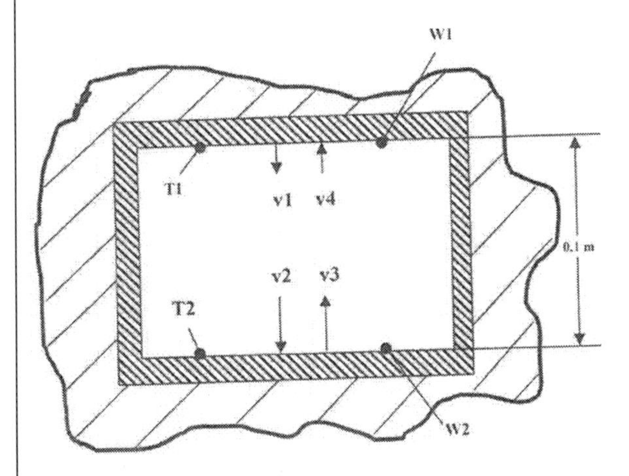

Basically, the potential energy which a molecule has at the top of a container is converted into an increase of its speed on its way down to the bottom. Speed is a measurement for temperature. So the potential energy at the top is converted into a temperature increase on the way downward. So T(Gr) = T1-T2 can be calculated as it is proportional to the potential energy of the molecule and inversely proportional to its mass and specific heat, or

T(Gr) = - M x g x H / (M x c) or
T(Gr) = - g x H / c ("-" because gravity is directed opposite to H)

So T(Gr) depends on the gravity factor g, which is practically a constant, the vertical height H and the specific heat c of the molecule involved.

Calculating T(Gr) for air with cv = 718 J and taking H = 1 meter
T(Gr) for air and a height of 1 meter = -.014 K or
T(Gr) per meter of height = -.014 K/m

In this calculation cv has been used. But the use of cv or cp makes sense only if you have a system where an amount of energy is added or taken out. In our case we have an isolated system where no matter or energy is introduced. We have to use another specific heat which I call c(Gr). This specific heat has to be used if a molecule is accelerated or decelerated by gravity. In this case only one degree of freedom of the molecule is involved. In a monatomic gas c(Gr) would then be only 1/3 of cv. In a diatomic gas like nitrogen or oxygen with 5 degrees of freedom, c(Gr) would be only 1/5 of cv.

If n is the number of degree of freedom we can write

$$T(Gr) = - g \times H / c(Gr) \text{ with } c(Gr) = cv/n$$

When calculating for air, T(Gr) would be

T(Gr) per meter of height = -.07 K/m.

Of course, in an enclosure with a rarified gas only some molecules travel directly from the upper wall to the lower wall. Depending on the dimensions of the enclosure many will hit the vertical walls on their way up or down before hitting the upper or lower walls. But if we assume that the vertical walls show the same linear temperature distribution as the column of gas, cold at the top and warm at the bottom, then a molecule bouncing off a vertical wall before hitting the lower wall would leave the vertical wall on the average with the temperature of that wall at that height and again only one degree of freedom would be involved on its way to the lower wall.
The temperature distribution in dense gases follows the same argument. I discuss this in detail in my BOOK.

To actually measure this calculated value for T(Gr) of -.07K per meter of height, cold at the top and warm at the bottom, seems to be feasible, as this value is high enough to be measurable. But we have to realize that the actual value will be lower as the effects of radiation, convection and heat conductance in the vertical walls will all have the effect of lowering this difference. Most of my test setups have also a height not of 1 meter but only of about .25 meters, as it is very difficult to provide an environment over a height of 1 meter which shows a sufficiently equal temperature distribution over height. The only positive effect comes from the use of multiple thermocouples arranged as a thermopile.
All these effects make it necessary to measure temperature differences

corresponding to about .005K, a difficult, but not impossible task. Check out the measured values under TEST RESULTS.

So now, in early 2002, I describe on my webpage the results of my experiment B 74. As I will discuss this experiment in the next chapter, I won't do it here but show the last pages of my webpage as I stated them in 2002:

GRAVITATIONAL ENERGY

Where does the energy comes from which is being produced by my gravity machines? This is not a question easily answered.

It does not come out of the gravity field. My gravity machine is in reality a heat pump which transports energy from the upper part of the closed system to the lower part. If you don't take energy out of this closed system then the maximum delta T will develop between the upper and the lower level. The upper part represents the heat reservoir of the heat pump. I you start to take energy out of the system, for instance through a thermocouple or a heat exchanger, you will have to replenish this energy at the upper level. Just imagine that the upper level is connected to a vast heat reservoir with a constant temperature. Through the effect of gravity my gravity machine creates a temperature difference by creating a higher temperature at the lower level without the introduction of any work. .

A PERPETUUM MOBILE OF THE SECOND KIND

If a vertical temperature gradient develops under the influence of gravity, would this allow the building of gravity machines generating work out of a heat reservoir? If you answer with yes, would this constitute a perpetuum mobile of the second kind? You might have to buy my BOOK to find out.

DESIGNING A GRAVITY MACHINE

I will need quite a few pages in my BOOK to describe this in enough detail, to enable you to build your own gravity machine. If you don't want to wait, come to one of my SEMINARS

SEMINARS

We offer two types of seminars.

Type 1: Duration 1 day.
Introduction to the theory of gravity machines; demonstrating a gravity machine at work; discussing test insulation, data collection and test results of a number of gravity machines; showing the construction of a gravity machine in all details; possible applications.

Type 2: Duration 2 days.
Same as Type 1. In addition, you participate in the building of a gravity machine from start to finish. If you so desire, you can purchase all parts of a gravity machine and build your own, taking it home with you at the end of the seminar.

Locations: **Germany:**
Koenigsfeld near Villingen, located in the south-west corner of Germany.

USA: Ithaca, NY.

Languages: German and English.

Then, after informing about seminars I offered in Ithaca and Koenigsfeld, I asked:
Who will repeat my experiments, will it be you? I ended my webpage:

OUTLOOK

My BOOK describes „My Search for Peaceful Energy." For gravity machines to produce useful energy in an economical way many more steps in testing and designing have to be taken. The most important questions:

Who will pick up the baton and confirm or disprove my test results, showing a TEMERATURE GRADIENT in isolated, vertical columns of gas or liquid?

Who will be the first one to build a GRAVITY MACHINE which produces useful amounts of energy?

Will it be you ?

Read my BOOK; send me your commentary.

And most important: THINK, SPEAK and ACT in NON-VIOLENT WAYS!

Once my webpage appeared on the web I was astonished how quickly people found out about it. In about one year's time about 3,000 interested parties visited the pages every month! Some of them ordered my book, which had not even been published yet. But hardly anybody wrote to me with some commentary or questions. That surprised me.

Chapter 16
I Report a Negative Temperature Gradient in Air 200 Doubters in San Diego, Is it just all Theory?

It was not long after this "night-dream-insight" experience when Andreas Trupp called me: One professor Daniel P. Sheehan planned a first initial conference under the title

QUANTUM LIMITS TO THE SECOND LAW

at his University of San Diego, California. It would take place in 2 months' time, in July 2002. He planned to attend, would I be interested?

Of course I would be. My recent experiment B74 measuring air within a Dewar glass containing glass powder had produced very nice results, showing a long term average negative gradient T(Gr) of -0,07 K/meter. If I could write a paper about this experiment maybe it would be accepted for the conference in San Diego..

A week later I wrote to professor Sheehan attaching the paper. He answered right away: Time for proposing papers had run out but he felt that these experimental results were very interesting and he invited me to demonstrate them within the poster session.

Six weeks later I checked in at out little airport in Ithaca to fly to San Diego, California, with a flight change in Philadelphia. How would the participants of the conference react to my paper claiming that work could be produced out of a heat bath under the influence of gravity? As this contradicted what we all had learned in our physics lectures especially about the Second Law of Thermodynamics, it should create a little sensation, I wondered. I just overheard near the check-in counter that another young man, maybe 35 years old, would also be flying to San Diego. I approached him: “Are you planning to participate in the conference in San Diego?” “Yes” he said “are you going too?” We started a lively conversation. Did he know San Diego? How many participants would we expect? Would he give a paper? How many foreign participants would come? No, he would expect only a small group without any foreigners, especially at it lasted only 2 days. Only two days? Three days was my information.

It turned out that he was an insurance agent joining a conference by his employer, a big insurance company. For nearly 10 minutes we had discussed our conference, finally realizing that we had talked about 2 completely different meetings. It is amazing to me how often this happens: I discuss a subject very lively with persons I know quite well and after 2 or 5 or even 10 minutes we realize we had talked about quite different subjects without noticing it. I wonder how often this happens when, after this exchange of thoughts and feelings you part without the realization you had talked about totally different items.

The town of Ithaca is located about 50 miles south of Syracuse in upper state New York. It is a small town with approximately 50,000 people; most of them are connected somehow to the University of Cornell, one of the top universities in the United States. Being a small town it has only a small airport which is serviced by US Air. So flying from Ithaca to San Diego meant that I had to fly first in the opposite direction, all the way to the East Coast, to Philadelphia. Here I would catch another US Air flight all the way across the country to San Diego. Luckily, I lost my insurance agent here so sitting in a window seat on my next plane, I had 3 hours to think about what to expect at the conference.

The topic was:

Quantum Limits to the Second Law

What did this mean: “Quantum Limits”? I really didn’t know much about Quantum Physics. I was educated as an engineer and I don’t remember ever having heard about Quantum Physics during my education. And what I was bringing to the conference had not much to do with Quantum Limits, I thought. I was bringing results of macroscopic measurements. And I claimed in my paper in the summary:

Measuring the temperature distribution in isolated spaces filled with a gas and a powder a vertical temperature gradient was found, cold at the top and warm at the bottom, as argued by J. Loschmidt. [3].

This result seems to be a contradiction to the Second Law of Thermodynamics. If correct, it would make possible the creation of work out of a heat bath.

Was I, an engineer, not even a physicist, claiming that the results of my measurements, if correct, were a contradiction to the Second Law allowing the production of work out of a heat bath? This was really a tremendous statement. If I would have read this sometime before I started with my own experiments, I probably wouldn't have believed it as I knew, it could not be possible. If correct, it would mean that you could put a ship in the ocean and the ship could propel itself, without having to burn a fuel like coal or oil to drive its engines. It could move just by using the heat out of the ocean's water.

I had sent my paper to the organizers of the conference about 2 months before. Somebody must have read it. The organizers must realize what sensation the results of my measurements represented, provided of course, that they were correct. If correct my experimental setups represented a Perpetuum Mobile of the Second Kind. And that could not exist, as every body knew.

My plane was passing Pittsburgh when I began to read my paper starting at the very beginning: (If you want to follow my paper in color, which shows the graphs much more clearly, just print it out from my webpage www.firstgravitymachine.com)

MEASURING THE TEMPERATURE DISTRIBUTION IN GAS COLUMNS

Roderich W. Graeff

International Licensing
Prof.-Domagk-Weg 7, D-78126 Königsfeld, Germany

102 Savage Farm Drive, Ithaca, NY 14850-6500, USA
E-mail rwgraeff@yahoo.com

Abstract. Late in the 19th century J. Loschmidt believed that a vertical column of gas in an isolated system would show a temperature gradient under the influence of gravity, cold at the top and warm at the bottom. L.Boltzmann and J.C. Maxwell disagreed. Their theories tried to prove an equal temperature over height.
Experiments with various test setups are being presented which seem to strengthen the position of Loschmidt. Long-term measurements at room temperature show average temperature gradients of up to - 0,07 °K/m in the walls of the enclosure, cold at the top and warm at the bottom.
The measured values can be explained by the conversion of the potential energy of the molecules into an increase of their average speed through gravity.

INTRODUCTION

In the late 19th century an animated discussion went on between L. Boltzmann [1], J.C.Maxwell [2], and J. Loschmidt [3]: Would a vertical column of gas under equilibrium conditions in a closed system show equal temperatures over height under the influence of gravity? The 2nd law of thermodynamics seemed to demand this. Only J. Loschmidt presented a different position. He argued for a temperature gradient over height, cold at the top and warm at the bottom. A. Trupp [4] gives a good overview over this debate. In a meeting of the Austrian Imperial Academy of Science in February 1876 Loschmidt declared [3]:

... With this the terroristic nimbus of the Second Law which makes it look like the destructive principle of any life in the universe would be destroyed. On the other side it opens up the comforting perspective that humanity is not solely dependent upon using coal or the sun to produce work out of heat, but that for all times an inexhaustible reserve of changeable heat will be available.

The author reports on experiments measuring the temperature distribution in vertical columns of rarefied and dense gases, primarily air and Xenon. The height of the columns varies from 0,2 m to 2 m. Special efforts were taken to create an isolated system as close to a "closed system" as possible. Thermocouples were used to measure temperatures in order to avoid any introduction of energy into the systems.

This introduction, I thought, was short but to the point. I quoted Loschmidt who in 1865 had already had the idea that a column of gas, isolated from the surrounding, should, under the influence of gravity, show a temperature difference, cold at the top and warm at the bottom. Surprisingly he claimed that such a temperature gradient would not contradict the Second Law, but he never explained why he felt this way.

The following section of my paper described my experimental set-up. The figure 1 was a hand drawing which I had to produce quite rapidly as the editor of the conference insisted on a short time limit for getting my paper. But together with the 2 photographs I felt the reader should be able to visualize the details of my test set-up:

TYPICAL TEST ASSEMBLY

Figure 1 shows a typical experimental setup. A Dewar insert of a commercial Thermos bottle (1) is mounted within a wide mouth Dewar insert of 1 litre size (2) which is covered by a similar Dewar insert of ½ litre (3). The space (4) between the innermost Dewar and the two outside Dewar inserts is filled with fine PET fibres.

The innermost Dewar (1) of ½ litre is filled with a fine powder in order to eliminate convection currents and radiation between the inner wall surfaces. A thermocouple (5) is arranged in the middle axle with a height distance between junctions of 170 mm. A second thermocouple (6) is mounted on the outside of the Dewar insert (1) with a vertical distance of 180 mm. A third thermocouple (7) is mounted on the outside of Dewar insert 2 with a vertical distance of 230 mm.

The Dewar inserts are held in place by fine PET fibres within a box (8) fabricated out of 40 mm thick polystyrene foam panels. This box is surrounded by 50 mm thick panels (9) consisting of pressed copper wires. The whole setup is insulated against the room air by 100 mm thick polystyrene foam panels (10).

The diameter of the thermocouple wire is kept as small as possible in order to reduce any heat conduction through these wires. The measuring instrument measures the voltage of the thermocouples with a resolution of 1×10^{-7} V corresponding to 0,0003 °K.

FIGURE 1
Typical test setup of Dewar glasses arranged in an isolated box

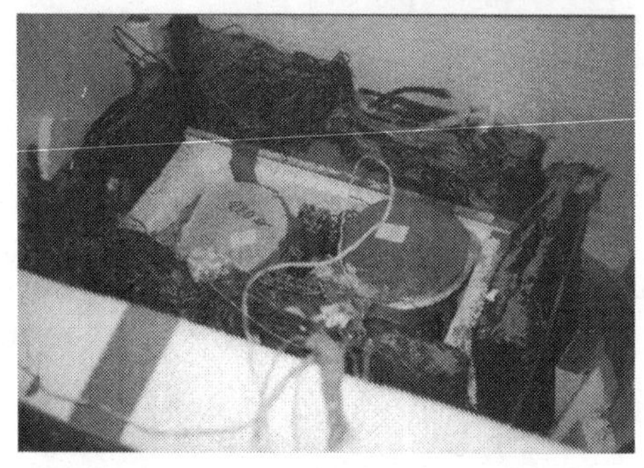

FIGURE 2
Assembly of Dewar inserts and highly insulated box with 3 test columns

Once I had felt in making my experiments that gravity was the force creating a temperature gradient the question very quickly arose: What would be the results, if I turned the test setup on its head, turning it by 180 degrees?

I had started to make such experiments for which I had designed and built a special rotating drum arrangement. This is shown in the figure 3 below. It allowed turning an experiment mounted in the inner part of this drum by 180 degrees while the outer drum covered by a very thick insulation kept rotating. As a result the turning happened without disturbing the insulation between the experiment and the environment.

I had begun to make quite a few of these experiments. The results seemed to be quite clear: When turning one of my experiments after about 12 hours the temperature gradient in the very centre of the experimental setup had switched so that again the top was cold and the bottom was warm. The same switch happened of course to the surrounding Dewar glasses. The bottom parts which had been colder than the top parts before were now, after the turning, again colder. The gradient on the outside of the inner Dewar was again warm at the top and cold at the bottom.

I wanted to claim a priority for this set-up which I felt was quite nice. Therefore I put the photographs of my turning drum into the paper even though I didn't mention or described the turning experiments.

TEST ENVIRONMENT

An isolated system demands that there is no material exchange between the system and the outside and no energy exchange between the gas and the surrounding area. While it is practically possible to fulfil the first condition it is impossible to exclude any exchange of energy. Even the best insulation will always transmit some energy based on the temperature difference between the inside and the outside.

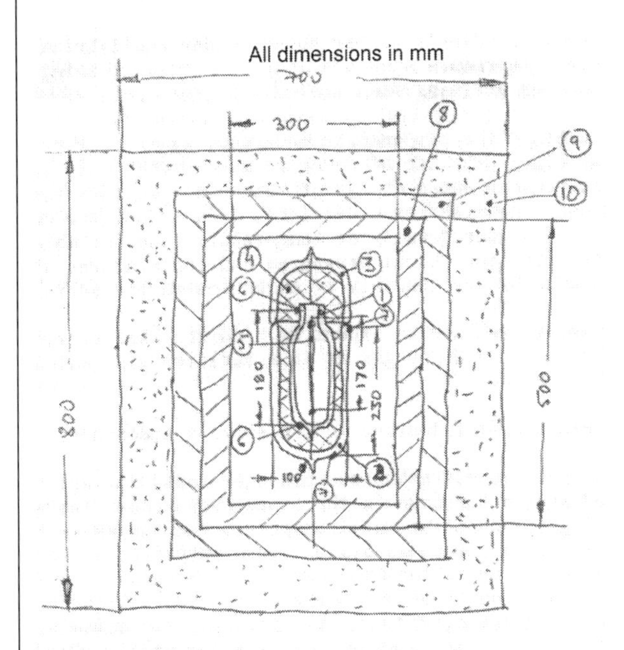

FIGURE 2

A typical test assembly of a number of Dewar glasses is shown in figure 2. Half of one assembly consisting of three Dewar glasses is shown. They would be mounted inside aluminum containers two of which are shown on the right side of figure 2. They are insulated from the outside by 40 mm thick polystyrene foam panels, 50 mm thick copper blocks made out of copper wires pressed together, and finally by 100 mm thick polystyrene foam panels.

Figure 3 shows a double drum apparatus [5]. While the inner drum remains in its vertical position the outer drum made of 25 mm thick aluminum plates rotates continuously around it. This creates a very small temperature difference, warm at the top and cold at the bottom, of only about 0,002 °K per the 50 cm of height of the inner drum. Another special feature of this apparatus is that while the outer drum keeps rotating, the inner drum with the test assemblies can be turned by rotating it on its head by 180° without interrupting the temperature measurements or the quality of the insulation.

FIGURE 3
Left: Apparatus with outer rotating drum and inner non-rotating drum
Right: 5 test assemblies arranged in the inner non-rotating drum

Of course, at least from my point of view, the test results were a sensation. The data which I could show as graphs were automatically recorded. Figure 4 showed the test results for a period over 9 months. In these graphs any point with a value above zero means that the gradient measured is a positive gradient meaning that the top thermocouple is warmer than the one at the bottom. You can see in figure 4 quite clearly that the outside of the outer Dewar – the highest curve in the graph - showed always a positive gradient, meaning that the top of the Dewar was always warmer than the bottom by about 0.05 degrees.

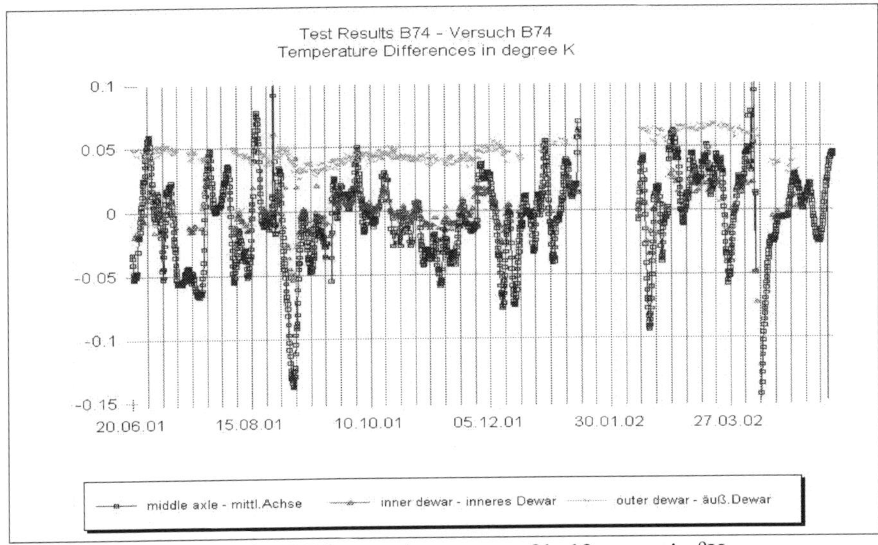

FIGURE 4 Test results B74 over time span of half a year in °K

In contrast to this the temperature gradient in the middle axle of the inner Dewar is quite jumpy showing sometimes values above zero but most of the time values below zero, meaning, that at least on average the top of the middle axle is colder than the bottom.

This result can be seen much clearer when you look at figure 5 showing average values normalized for a height of one metre in degrees K. Each point in any curve shows the average value from the very beginning of the experiment up to this point. Here my "sensation" can be seen much clearer. While the outside of the outer Dewar always shows a positive gradient, top warm and bottom cold, the average of the middle axle shows the opposite, top cold and bottom warm. This was not explicable by anything I had learned as an engineer. Within a surrounding which is warm at the top and cold at the bottom there cannot be any other inner part of the experiment showing an opposite gradient, cold at the top and warm at the bottom. I explained this in the paper as follows:

TEST RESULTS

It is very difficult to give meaningful values for the accuracy of the temperature measurements. I believe that values given are accurate to about ± 0,002 °K. Such a low value is possible because we do not measure absolute temperatures but temperature differences. For measuring small values we use five leg thermopiles instead of single thermocouples which increase the accuracy. For those measurements where the absolute values are not as important as the direction of the temperature gradient, each temperature as measured with switched polarity.

FIGURE 5
Test results B74 over time span of half a year: Average values in °K/m

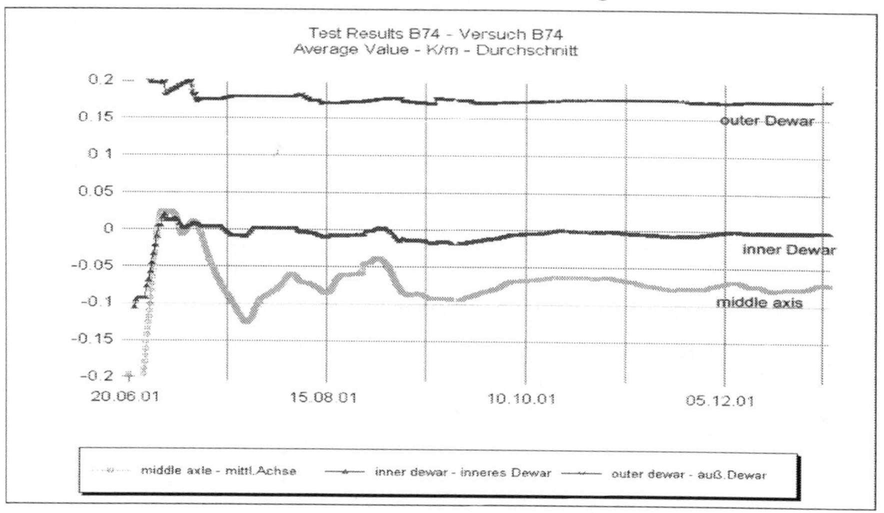

The temperature differences shown by the three thermocouples were recorded every 4 minutes over a time of half a year. Figure 4 shows a printout where each point represents the average of a 4-hour time span equaling the average of 60 temperature measurements.

Any point above the zero line means that the upper junction of the thermocouple is warmer than the lower junction.

It can easily be seen that the outside of Dewar insert 2 is always warmer at the upper junction than at the lower one by about 0,040 °K corresponding to 0,174 °K/m. This is the result from the temperature gradient in the surrounding room air, which is warmer at the top and colder at the bottom, typically by 1 °K per meter of room height. The insulation around the Dewar inserts and especially the thick copper plates are the reasons for reducing this gradient for the Dewar inserts.

Contrary to this, at the inner axle most of the time the temperature of the upper junction is lower than at the lower one.

This result is more clearly demonstrated in figure 5 showing average values from the beginning of the test until the date in question. Table 1 shows the average values for the half year test span for the actual distance between junctions of the thermocouples and normalized for a height of 1 m.

TABLE 1. Average Temperature Differences over 6 Months:

	from figure 5:	between junctions:	distance for 1 m:
Dewar 2	+0,040 °K	0,23 m	+0,174 °K
Dewar 1	-0,001 °K	0,18 m	-0,0055 °K
Middle axle:	-0,012 °K	0,17 m	-0,0705 °K

The average value over the test span of half a year for Dewar 2 shows clearly a positive value, warm at the top and cold at the bottom. The temperature difference for Dewar 1 was practically zero. The average for the middle axle shows a negative value meaning that on average the top was colder than the bottom.

In the following I tried to explain this surprising result by the effect of gravity. Gravity accelerated the molecules on their downward path. The increasing speed resulted in an increase of their impact on the lower wall creating an increase in the temperature of the walls. I tried to explain that this negative gradient and its value could be calculated by equating the potential energy of a molecule with the increase of its speed on its downwards travel.

INTERPRETATION OF TEST RESULTS

That the middle axle shows, on the average, a temperature gradient, cold at the top and warm at the bottom, cannot be explained by applying the normal formulas for the conduction of heat. Under these laws, heat can travel only from an area with a higher temperature to an area with a lower temperature. But my test results can be explained by assuming that the molecules travelling vertically are accelerated by gravity in their downward path and decelerated in their way upwards. On the average the molecules will hit the upper wall of the enclosure with a lower speed than the lower wall. Elastically bouncing off the walls with the temperature of the wall, under equilibrium conditions, the upper wall would show a lower temperature than the lower one. This temperature difference due to the influence of gravity I call T(Gr).

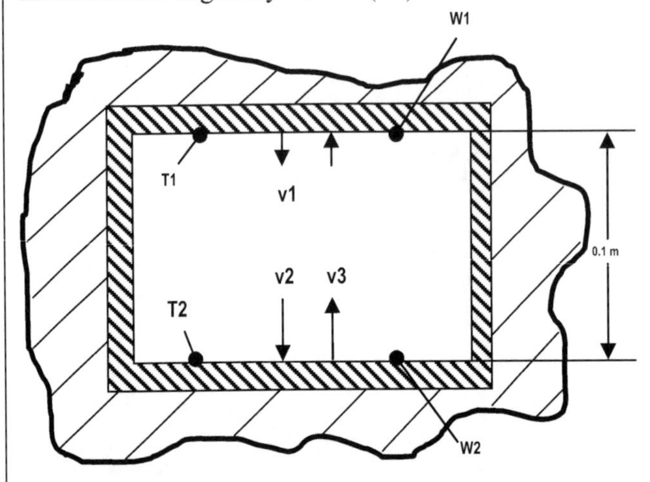

The value of T(Gr) can be calculated by equating the potential energy of the molecules with the increase of their speed which represents their temperature. At least for a rarefied gas no compression work has to be performed on the downward path. Therefore calculating with the specific heat c_p, for air we get

T(Gr) = - 0,01 °K/m

This value is typically found in the atmosphere [6]. With higher gradients a column of gas would be instable, as the colder gas in the top would be heavier the warmer gas at the bottom and replace it through convection with in the gas column. Therefore the innermost measuring area was filled with fine powder in order to avoid any such convection currents. Furthermore the higher values found in the experiments can be explained only by assuming that acceleration by gravity affects only the vertical speed component of a molecule negating the equipartition of energy.

It is believed that this argument for the conditions of a rarefied gas is equally valid for dense gases. Details are discussed in [7].

The value of -0,01 K/m is the value found in the atmosphere around the earth. The good news or equally a problem of my experiment had been that I measured higher values, namely -0.07 °K/meter. When I started to get these values I was very happy. If I had got only smaller values than – 0.01 °K/m you could have assumed that these values were created by small convection currents within my test setup which resulted in the same gradient as one finds in the atmosphere around the earth. I had realized right then that higher values than this adiabatic value would not be possible in my experiments as such a temperature gradient would not be stable. The molecules at the top of the column would become too heavy and start to move downwards destroying any value higher then -0.01 K/m. After this realisation I started to add convection-suppressing powders, hopefully eliminating any convection current. Only then was I able to measure values up to -0.07 as documented in my curves in figure 4. So at the end of this paragraph I explained this only in words and only in very short sentences:

T(Gr) = - 0.01 °K/m

This value is typically found in the atmosphere [6]. With higher gradients a column of gas would be instable. Therefore the innermost measuring area was filled with fine powder in order to avoid any convection currents. Furthermore the higher values found in the experiments can be explained only by assuming that acceleration by gravity affects only the vertical speed component of a molecule negating the equipartition of energy .

It is believed that this argument for the conditions of a rarefied gas is equally valid for dense gases. Details are discussed in [7].

This really summarized my theoretical understanding and interpretation of my measured results. But a reader of my paper would really have to be very attentive and focused to catch and realize the importance of these 2 short paragraphs. In addition, instead of

...affects only the lateral speed...

I should have written

...affects only the vertical speed component...

In the summary I still use very carefully words like "the result seems…" and "if correct…"
I realize that if these results were repeated and proven by a second and third party it would be a sensation. So I kept wondering: What would be the reaction of the people at the conference?

SUMMARY

Measuring the temperature distribution in isolated spaces filled with a gas and a powder a vertical temperature gradient was found, cold at the top and warm at the bottom as argued by J. Loschmidt. [3].

This result seems to be a contradiction to the 2nd law of thermodynamics. If correct, it would make possible the creation of work out of a heat bath.

So this was my paper. I realized flying over the Plaines of Middle America with our plane approaching San Diego that if these results were repeated and proven by a second and third party it would be a sensation. But this had not happened yet. So I kept wondering: What would be the reaction of the people at the conference?

The conference took place in the very lovely campus of the University of San Diego built in an historic Mexican style. For 3 days we listened to about 150 papers, all of them were theoretical discussions of the Second Law, most of them concerned with Quantum effects. There was only one exception discussing actual experimental results: My paper!

Maybe the atmosphere of the conference is best described by reprinting the preface which Prof. Daniel Sheehan as Chairman of the Conference wrote for the book "Quantum Limits to the Second Law", first and initial conference on Quantum limits on the Second Law, San Diego, California, 2002 [24], contained in the AIP Conference proceedings.

Over the last ten years an unprecedented number of challenges have been raised against the absolute status of the second law of thermodynamics, more than during any other decade of its 175 year history. In addition, several related foundational issues dating back to the late 19th and early 20th centuries have now persisted unresolved into the 21st, with unsettling consequences. Many of these have been exasperated by quantum mechanics.

This conference was born from the desire to explore these many challenges and unresolved issues in light of recent theoretical and experimental results. Many of the questions raised at this conference could, for more than a century, be only whispered in serious scientific circles, so powerful has been the second law's mystique. As evidence, this appears to be the first conference of its kind in at least 100 years, and perhaps ever.

Over 120 researchers, representing 25 countries, participated in QLSL2002. The program was eclectic. Talks spanned theoretical and experimental classical and quantum statistical thermodynamics, time and gravitation, as well as the history and philosophy of science. In fact, despite the conference title, less than half the talks spoke directly to quantum mechanical issues.

Given the contention that has historically surrounded second law discussions, this conference was fortunate to maintain a collegial and constructive atmosphere. The balance and diversity of opinions was heartening. This balance is perhaps best demonstrated by the outcome of an informal vote conducted by S. Braunstein at the end of the last conference session. For belief in the following proposition – "The second law is inviolable." – the vote Yes: No: Maybe: and Abstain/Absent was roughly 25: 25: 25: 45.

Although pains were taken to balance the conference slate, the evenness in this vote was surprising. Surely this was a select group, one not representative of the general scientific community. Were it so, we should properly consider ourselves in the midst of a fundamental paradigm shift involving the second law – which is clearly not the case.

But, of course, all storms begin with a breeze.

Daniel P. Sheehan
Chairman, Organizing Committee

As one of the 120 participants I enjoyed 3 days of listening to talks, getting to know people from various parts from the world, all interested in some aspects of the Second Law. As I had been late to sign up for the conference I could not give a talk about my experiments but was allowed to participate in the poster session. Here about 20 of the participants got a table each about 2 m long where we could demonstrate the core of our work and discuss it with the participants passing by and showing some interest. Within these 3 days maybe 10 people enquired about the curves I showed. But there were only 2 who stayed more than two or three minutes, got more involved and asked more meaningful questions. "Yes, your work sounds interesting, good luck!" More and more it became clear to me: all of these participants had an interest in the Second Law but they really were interested in a very narrow aspect of it, they were interested just in the specific work they were doing. There really was no serious response to my work!

One participant scheduled for giving a talk didn't show up and Prof. Sheehan asked me if I didn't want to fill the 15 minutes spot and talk about my experiments. Happily I agreed. About 20 people came to listen and during the second half even Prof. Sheehan showed up. There was time for 3 questions and that was it. Polite applause!

In summary, what was the response to my coming to San Diego, having participated in the poster session and even giving a 15 minute talk telling about my sensation? Basically: zero, silence, nothing!

All these participants were concerned with theoretical aspects of the Second Law. They felt that maybe on the microscopic and quantum level there was the chance for some discrepancy with the Second Law but they knew and were convinced deep inside, that on a macroscopic level it was inviolable. So if somebody suddenly showed up, like me, and claimed to have experimental proof of an exception to the Second Law, they knew consciously and unconsciously that this couldn't be true. There really was not much point in spending time with thinking about or discussing such hopeless experiments in order to find their basic flaws.

Still, one positive thing came out of my participation in the conference. I had gotten to know Prof. Daniel Sheehan, a real professor of physics, very much interested in possible exceptions to the Second Law, and a very pleasant and amiable person. He, as the very big exception, showed real interest in my work. He planned a vacation together with his wife early the following year visiting friends in Europe and maybe could interrupt it for 2 days to visit me in the Black Forest. He would be interested to look at my experiments in more detail.

And he came in June 2003. We met for a day in Königsfeld together with Alexander Haltmeier, Mariana Strümpel and Andreas Trupp. In the picture Daniel Sheehan is sitting on one of my aluminum tubes. A number of these I use in a vertical position surrounding each experiment for equalizing and reducing the vertical temperature gradient induced from the temperature gradient in the room.

We started with a lively discussion about the theoretical aspects of my experiments, the Boltzmann distribution, partitioning of energy, gravitation. After that we visited my laboratory located in 2 rooms in the cellar of my home.

From left to right: Dr. Alexander Haltmeier, Prof. Daniel S. Sheehan, Dr. Andreas Trupp, The author

Daniel Sheehan was very much interested in the details of my experiments. He took quite a few notes, as he planned to write a book about the Second Law.

After the visitors had left I found the following entries in our guest book:

July 13, 2003
The moon shone through the upper window on my feet as I fell asleep, strong and clear. Soon it shall be full and I will be far away.

Daniel Sheehan

July 11-13, 2003
For 2 full days we discussed the Second Law up to exhaustion. At least the thoughts of Roderich and my own will find its way into the book of Daniel Sheehan. Since my last visit in December 2001 the "Creative Mess" in this house has not changed. I hope I can say the same during my next visit here.

Andreas Trupp

July 11-14, 2003
With me it has been 2.5 days and as always I keep learning; but new questions always arise. Now I move partly away from the unlimited validity of the Second Law of Thermodynamics. Considering the "Creative Mess" I believe one can stand this only up to a certain age. After that one has to find orderliness, or one sinks into a "Total Mess."

Alexander Haltmeier

Daniel Sheehan published his book through Springer in 2005 under the title: "Challenges to the Second Law of Thermodynamics, Theory and Experiments" [10]. I feel he succeeded splendidly writing a wonderful book. But again if you look there for descriptions of experiments, not just theories, you can hardly find any. He covered my work as follows:

6.3.1 Gräff Experiments
R.W. Gräff has conducted roughly 50 individual experiments with the goal of putting quantitative limits on the magnitude of the Loschmidt effect Laboratories in Ithaca (US) and Königsfeld (Germany) simultaneously conduct multiple experiments.

He continuous his description in great details over 3 pages. He writes:

.

Gräffs experimental results support the existence of Loschmidts gravito-thermal effect in gases and liquids, but are inconclusive in solids, the phase for which Loschmidt made his original proposal. The most sophisticated long-term experiments reported average vertical temperature gradients of - 0.07 K/m. for air and -0.04 K/m for water (with glass micro spheres), cold ate the top and warm at the bottom of the sample columns. ....

In summary, Gräff's experiments are not yet conclusive and their theoretical underpinnings are disputed by other researchers. Nevertheless, his are the first to test and to support the Loschmidt effect. They deserve serious attention, and ideally, they should be replicated at independent laboratories.

This description I felt gave a fairly good and correct evaluation of my experimental work and I was very happy that he gave me so much space in his book. Now anybody interested in these challenges to the Second Law could read about it in his book or in the conference proceedings of the Conference in San Diego, judge the importance of my work, my results and contact me for further information. Or, as Daniel Sheehan wrote on page 205

"...Graeff's experiments...deserve serious attention and ideally they should be replicated in independent laboratories."

How many people have contacted me since these two books are available?

Do you want to guess? None, not one! And how many people have contacted me during the past 8 years with serious questions after reading my webpage www.firstgravitymachine.com, which has been available since the spring of 2002? Actually the interest in my webpage keeps astonishing me as about 80 people look at it each day. As I announced in it the availability of my book in the near future I also keep receiving quite a few orders. This shows me that there seems to be a serious interest in the subject but apparently without an interest in a follow up discussion. We all know that my results cannot be correct. Heat can never travel from cold to warm, every child knows that. So there is nothing to discuss or talk about.

This book tries to cover my experiments which I started eleven years ago. It took me then about half a year to change my original question into the new direction: Does gravity induce in vertical columns of solids, liquids or gases, a temperature gradient? Slowly, through hundreds of experiments I came to the conclusion: Yes, this statement is correct. Up to a year ago, I used to say: "The results of my experiments seem to indicate that the above statement is correct". But since spring 2009 I claim: "The statement is correct until somebody shows me a basic mistake in my experiments or their interpretation."

Chapter 17
Is there a Negative Temperature Gradient in Vertical Columns of Water?
The NET Journal Publishes my Paper

In the summer of 2002 it became clearer and clearer that columns of air filled with a convection-suppressing powder showed a temperature gradient, cold at the top and warm at the bottom. Other gases, especially xenon, should at least theoretically show an ever bigger value. I tried to demonstrate this in an experiment, but to use gases other than air was not so simple. A gas like xenon you purchase in a small bottle, like aluminum can, with a pressure of 16 bar. How do you get a powder inside the can in order to perform an experiment? And a thermocouple? Both of these tasks were too difficult for me to perform in my make shift "laboratory". So I just taped my sensors to the outside of the can, insulated the can towards the environment very well and started to measure. Well, I did record negative gradients, but I felt they were too small to be convincing. Of course this was casily understandable as the walls of the aluminum can, with its high heat conductivity coefficient, would tend to eliminate any temperature gradient generated by the gas.

Why not try water! I used tap water filled into a ½ liter glass bottle with a wide neck. It was not too difficult to mix the glass powder into the water and to arrange in the middle a thin plastic tube out of plastic film with my thermocouple inserted, well protected from the water. The water bottle I arranged in the middle axle of two aluminum tubes, both with a wall thickness of 10 mm. On the outside of the inner aluminum tube I taped my second thermocouple.

Based on my theoretical formula for T(Gr) I should expect to measure the following temperature gradients:

	air	Xenon	water
Degrees of freedom:	5	3	18
T(Gr) °K/m	-0,07	-0,311	-0,04

The next picture shows a graph of these two thermocouples for the test B890:

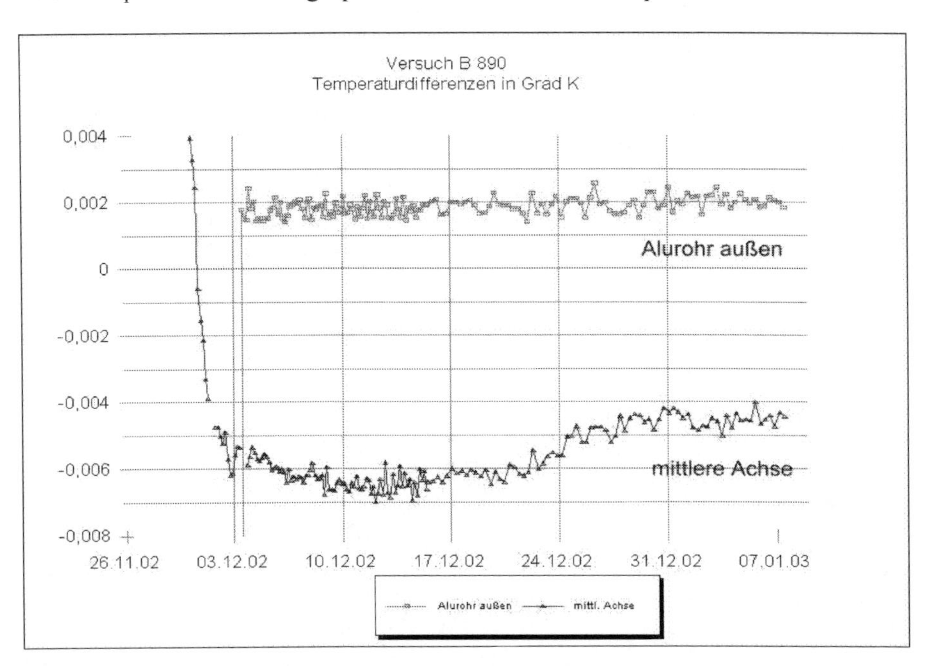

Each point of these curves is the average value of 60 temperature measurements which were taken every 4 minutes for 4 hours over a period of 5 weeks. Every point above the zero line means that the upper thermocouple point was warmer than the lower one, demonstrating a positive temperature gradient, and vice versa.

One can see easily that the temperature on the outside of the aluminum tube was always higher at the top than at the bottom by about 0.002 K. Contrary to this, the "mittlere Achse", the middle axis, is showing the opposite, a negative gradient, cold at the top and warm at the bottom.

The following table calculates the average values for the 5 weeks converting absolute temperatures into gradients per meter:

From Graph:		Distance between Thermocouple points:	
(Alu Rohr aussen) Aluminum tube:	+0,002 °K	0,21 m	+ 0,01 °K/m
(Mittl. Achse) Middle axis:	-0,0046 °K	0,18 m	-0,026 °K/m

Although the experiment B890 shows values only for 5 weeks it is obvious that this water test results in much more stable gradients than the values shown in the previous chapter 16 for air in test B 74. You can find a more detailed description of B890 in Appendix 2 in German. Of all the papers I sent to magazines asking them for a publication, this paper was the only one which was accepted. It was published by the NET-Journal, ISSN 1420-9292, Jupiter-Verlag, Zürich, in February 2003 under the title „Produktion von nutzbarer Energie aus einem Wärmebad"[9].

The NET Journal is not a peer-reviewed scientific magazine but concentrates on the idea of "Free Energy" and similar topics which are not part of the main stream of today's physics. Still, I appreciated that they printed my paper.

This first test with water and additional ones with some different sized bottles looked very promising. I therefore decided to try an experiment with water on a much larger scale. The following description is taken out of my paper covering B372 and is reprinted in Appendix 6. If you want to study this experiment in detail, you might want to find it on my webpage www.firstgravitymachne.com and print out the graphs in color for a much better viewing.

For this experiment I would fill 2 glass tubes with a height of 850 mm and a diameter of 40 mm, one of them with just tap water and the other one with tap water and glass powder. They would be arranged in the centre of a double-walled aluminum tube with a diameter of 500 mm and a height of 1500 mm. The space between the 2 walls having a width of 50 mm would be filled with water.

Hopefully the set up would provide a very stable environment with a pretty constant temperature reducing the impact of the temperature swings in the environment to a minimum. The following sketch shows the arrangement of the 2 glass tubes in the middle of the double-walled aluminum container.

EXPERIMENTAL SETUP

Fig. 1.

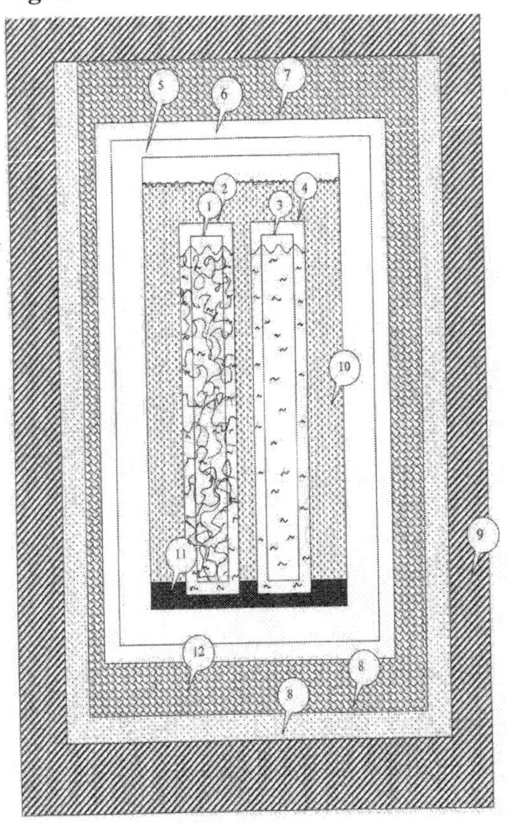

1: Glass tube 1, filled with water and glass powder
L=850mm, D=40mm

2: PVC tube 1,
L=910mm, D=50mm

3: Glass tube 2, filled only with water
L=850mm, D=40mm

4: PVC tube 2,
L=910mm, D=50mm

5: PVC tube 125mm,
L=1000mm, D=125mm

6: Aluminum tube 150mm,
L=1100mm, D=150mm

7: Aluminum tube 220mm,
L=1200mm, D=220mm

8: Double wall housing,
L=1500mm, D=500mm

9: Glass fibre insulation 100mm

10: Glass foam balls 1 mm

11: Brass Shavings

12: PET fibres

Test B 372, as shown in Fig 1, measures the vertical temperature gradient in two identical glass tubes of 40 mm diameter and 850 mm length. Each glass tube is individually surrounded by a PVC tube of diameter 50 mm and length 910 mm. Tube 1 **(1)** and its surrounding PVC tube **(2)** are filled with water and fine glass powder, while tube 2 **(3)** and its PVC tube **(4)** only with clear water.

These are arranged in a PVC tube of diameter 125 mm and 1000 mm length **(5).** The remaining space is filled with small balls of glass foam of 1mm diameter **(10)**. The bottom part is filled with small brass shavings **(11)** in order to try to equalize the bottom temperatures of the two 50 mm PVC tubes.

The assembly is inside a 150 mm diameter and 1100 mm long aluminum tube of wall thickness 5 mm **(6)**. This in turn is placed into another aluminum tube of 220 mm diameter, 1200 mm length and a wall thickness of 5 mm **(7)**. Each of these is closed at the top with round aluminum plates of the same thickness.

The aluminum tubes containing the test assembly are standing in the center of a double walled aluminum housing of height 1500 mm and an inner diameter of 500 mm with 50 mm between the two walls **(8).** This space is filled with water. The whole assembly is insulated on the outside with 100 mm of glass wool **(9).** The space between the larger aluminum tube and the inner aluminum housing, i.e. between **(7)** and **(8),** is filled with fine PET fibers **(12).**

Also new with this test was that the temperatures inside the test setup were measured not only by thermocouples but also by thermistors. Both type of sensors were mounted at the same locations at the tops and at the bottoms of the inner axes of the two glass tubes. Additional sensors are mounted on the outside of these glass tubes and on the outside of the two PVC tubes. The temperatures of the double wall aluminum housing are measured 3 cm below the top and above the bottom.

The test setup B372 was installed in May 2006. All sensors were connected to DMM Multimeter Keithley model 2700 and the data fed into a computer. Measurement results are reported from December 2006 through March 2007, a time period long after the setup, so that it can be expected that equilibrium conditions had been reached.

Fig. 2 shows 4 temperature gradients of the 8 values measured by thermocouples as temperature differences from December 14 through March 15. Each point of the curve represents a 10 value average (of a ten times repeated reading of the same sensor) measured every hour, using the scale on the left side of the graph. The smooth lines represent the 2 temperatures measured by thermistors of the 6 values measured hourly in centigrade, using the scale on the right side.

Fig. 2: Temperature gradients curve 1 – 4 and Thermistor temperatures 5 and 6.

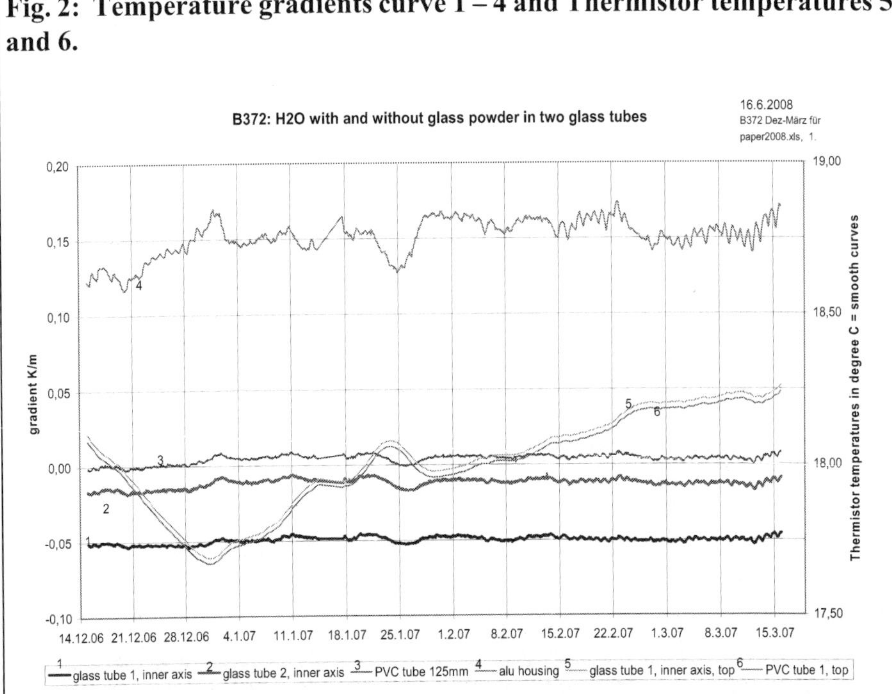

So I finally dared to use thermistors for measuring absolute local temperatures rather than only temperature differences as I was doing with thermocouples. I had avoided the use of thermistors as they needed a small amount of electric current sent by the multimeter through the thermistor head at the same time as the instrument measured the resistance of the thermistor to be able to calculate the local temperature. But tests convinced me that this very small amount of energy input did not affect my measurements in any negative way. With the knowledge of these local temperatures measured by these thermistors it became much easier to understand the fluctuations shown by the gradients measured by the thermocouples.

Environmental influences

Ideally, the measurements would take place in an isolated system not allowing the exchange of matter or energy across the boundaries. While the exchange of any matter can be eliminated, the exchange of energy cannot be avoided even with an optimal insulation. Because the temperature on the outside will always fluctuate to some degree, some energy will always pass in and out through the boundaries and influence the measurements.

The temperatures, measured by the thermistors at various locations within the test setup, give an indication of the amount of energy entering or leaving the system. From initial values around 18.10 C the temperatures all declined to about 17.50 C during the first 17 days (winter) and rose to a peak around 18.25 C during the following 50 days (spring). This gives a maximum change of only 0.75 C in a 13 week period that is caused by the temperature fluctuations of the environment.

Even an air-conditioned room can have such fluctuations. These experiments were carried out in a basement without air-conditioning, but with a thermostat-controlled heating system during the winter. The smooth parallel temperature curves (see curves 5 and 6 in Figure 2) indicate that the heat transfer took place uniformly in all parts of the test setup, not significantly disturbing the temperature differences that we tried to measure.

Originally I had only one method to trying to document a vertical temperature gradient. I just tried to measure it using a thermocouple connected to the top and the bottom to the inner axis and to draw up a graph showing this measurement over time. This gives the most direct result as shown in figure 2. Here it is: the lowest (blue) curve, the temperature gradient of the inner axis of glass tube 1, filled with water and glass powder. It is quite stable around a value of about -.05 K/meter; the minus sign indicating a lower temperature at the top than at the bottom.

The glass tube filled only with water, tube 2, curve 2, (lowest red curve) with its value of about -.01 K/m, has a less negative gradient than tube 1. This is plausible, because tube 2 contains only water containing no convection-hindering glass powder like tube 1.

Next further out is the PVC tube of diameter 125 mm, curve 3, enclosing both the PVC tubes 1 and 2 and the glass tubes 1 and 2. It shows a slightly positive gradient close to zero, which means that the top is warmer than the bottom.

Also very important is the gradient on the inner wall of the aluminum housing, curve 4 with a value of +.15 K/m. It is always positive, warm at the top and cold at the bottom. Only under these conditions with positive gradients at the outside locations the negative gradients at the inner axis of tube 1 or 2 -- cold at the top and warm at the bottom -- become meaningful.

In this test it is very helpful that we measure not only the temperature gradient of the inner axle but also the temperatures at various other points with thermistors represented by the smooth curves (5 and 6 in Figure 2. In comparing the measurements at different locations, one has to consider that the precision of a thermistor amounts to only +/- 1 C. But the measurements are very constant over time, as indicated by the smoothness of the curves, whereby the temperature change over time is measured to a much greater precision than the absolute values.

This fact becomes very important, when one looks at long time periods, during which the temperatures in a tube do not change. During these times one can find out, whether a temperature gradient $T_{(Gr)}$ exists under equilibrium conditions.

As a second method to determine the long-term value of the temperature gradient T(Gr), we calculate and graph all measured values as a long-term average. As shown in Figure 3.

Fig. 3

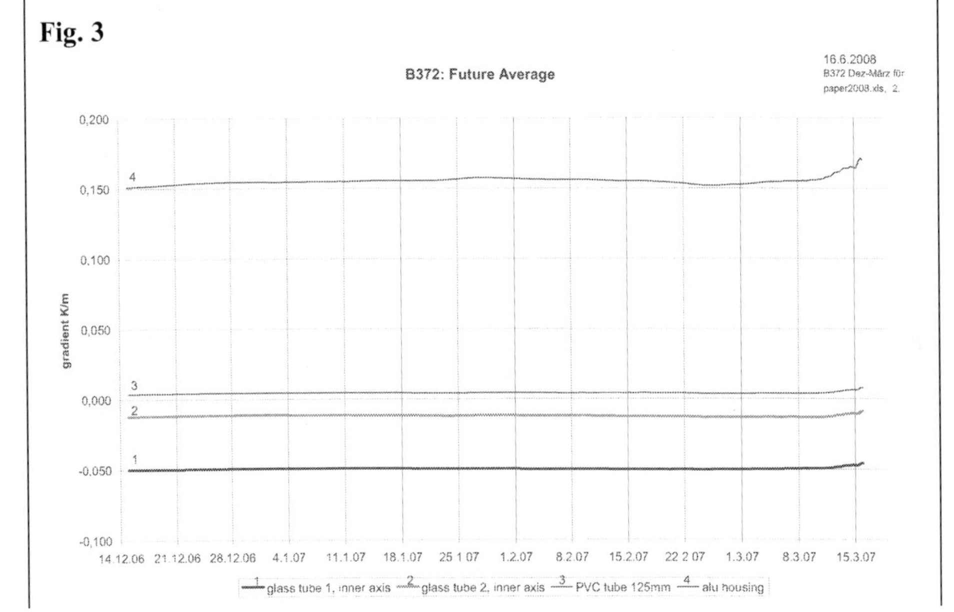

Fig. 3 "**B372: Future Average**" shows average values of all gradients over time. Each point on a curve represents the average value calculated for this gradient from that time through the last point of the curve to the right. For example, the values shown on December 14 are, therefore, the averages for the time from December 14 through March 15, while the last points on the right represent the values on March 15. Thus we can ignore the right end of the curves, where too few measurements are included in every point and the values are not meaningful. For the inner axis of tube 1 we get a steady average gradient of -.05 K/m.

Averages are typically shown as "past averages". Here each point on the curve represents the past average up to this point, from the time of the beginning of the test up to this point. This method has a disadvantage that it includes in this average the initial values measured during those times, when the temperatures still had to settle down to get close to an equilibrium status. Using "future averages" has the advantage that each point of the curve eliminates the initial values before equilibrium is reached.

As we have seen the measured values of $T_{(Gr)}$ fluctuate over time, because even the best insulation can not prevent small temperature changes in the test setup. In a regression analysis $T_{(Gr)}$ can be found very efficiently, when all measured temperature gradient values are plotted as a function of the rate of the temperature change (Figure 4). The correct value of $T_{(Gr)}$ can be obtained, whenever the rate of temperature change is zero. At these times no heat is flowing into or out of the system and we have equilibrium conditions with no temperature change over time. We measured these rates both for the top and at the bottom of the tubes, and found very similar results. The parallel nature of the actual temperature changes over time at different locations in the system was already observed in Figure 2. In Figure 4 the x-axis stands for the rate of temperature changes measured only at the top of the tubes in question.

Trend lines are calculated as least squares regression lines for the scattered values. The trend line for the lower triangles – blue in the coloured printout - for inner tube 1 (water with glass powder) show an intercept with the vertical zero line, where the rate of temperature change is 0, at -0.05 K/m. The upper markers give a $T_{(Gr)}$ value of -0.12 K/m for inner tube 2 (only water). Both of these values agree well with the long term average values, seen on curves 1 and 2 in Figure 3. The error bars correspond to +/-SDs for all measured values.

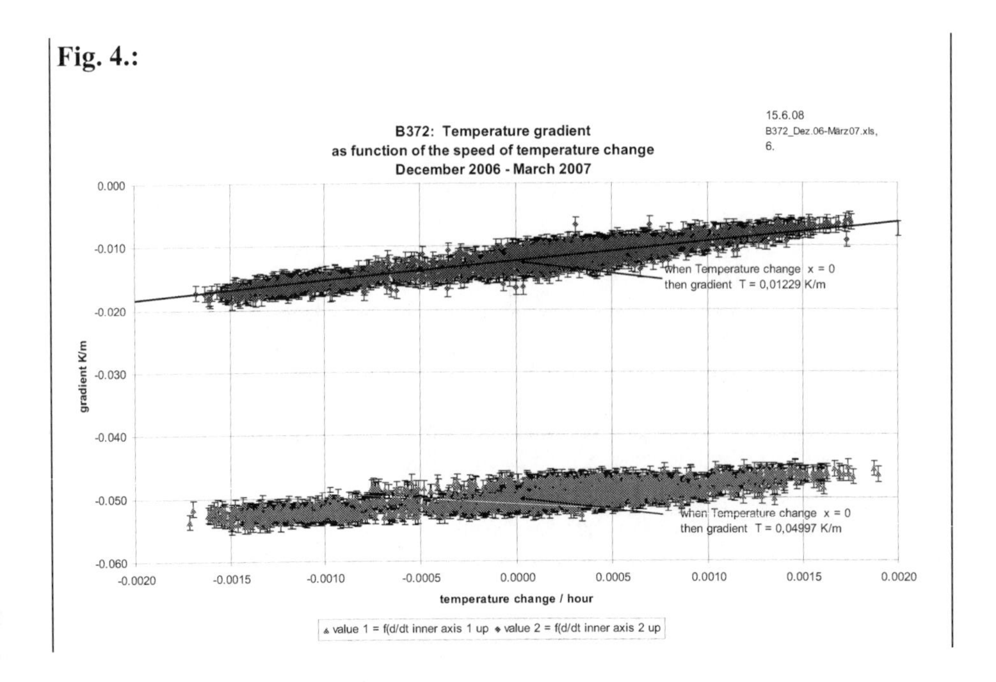

It seems to me that this is a very important result. It shows practically the same T(Gr) values developed through 2 totally different calculation methods. But the question comes up: how precise are my measurements for these 2 methods? And secondly: are there maybe other causes for these results rather than the influence of gravity? If you would like the answer for these questions check the complete paper in Appendix 5:

VIEWING THE CONTROVERSY LOSCHMIDT – BOLTZMANN/MAXWELL THROUGH MACROSCOPIC MEASUREMENTS OF THE TEMPERATURE GRADIENTS IN VERTICAL COLUMNS OF WATER

This paper I sent to a number of physics periodicals but they all refused to publish it. Typically I received the final refusal when I answered their question for my professional affiliation. No, I was not a professor at a well known University; I was just a private scholar.

Chapter 18
Patents

Patents played and still are playing an important part in my life. Patents are a method of protecting your rights to an invention you might have made. For me it was very exciting and pleasurable to think about new products and methods, trying to create something not only on paper but possibly in real life, something which had not existed before. This typically is called making an invention. Once you found something which is new and which has a commercial value then you can ask to protect this invention through a patent. Once you receive a patent you have the exclusive right to use this invention or to grant the right for the use of this patent to another party. Typically a patent can remain valid for 15 to 20 years depending on the country which grants the patent. If other people want to use this patent you can sell the patent to them or you can grant them a license or agree on payments of a royalty.

The procedure for granting patents, the requirements placed on the patentee, and the extent of the exclusive rights vary widely between countries according to national laws and international agreements. Typically, however, a patent application must include one or more claims defining the invention. It must be new, non-obvious, and useful.

Interestingly enough, during my studies to become a mechanical engineer, we had to take one small course in law, but I didn't hear once the word "patent". So when I started to work in my first job as an engineer, which was with a company producing window glass using the old fashioned Foucault method, I quickly had ideas how to improve the method of production. By adding secondary coolers at the point where the glass plate was formed, I could increase the production by about 20 % with only a small decrease in glass quality. I talked to the management proposing to apply for a patent, and they agreed. Actually, nothing came out of this invention. Even so my method had a great commercial value, my employer did not receive a patent, because there were too many previous patents too close to my invention.

When I got the job to start in the plastic department producing glass reinforced plastic panels, my inventive mind started working again. Very quickly I produced new products and improved production methods. Within one year I was able to apply for more than ten different patents. Soon I learned that Germany had a special law which gave the inventor, if he was an employee, some special rights. In practically all countries in the world the rights to any patent belong to the employer as long as the invention is close to the interest of the company. So the fruit of such an invention goes wholly to the employer. But in Germany the employee has a right to share in the benefits which the employer gains from such an invention. The law describes how to calculate this share which typically is a certain percentage of the royalty an outside company would have to pay in order to use such an invention. The employee continues to receive this share even after leaving such an employment. In my case, after I left this employer two years later, I continued to receive sizeable sums for about twelve different inventions for many years thereafter.

After four years of working as an employee in the glass industry, I felt it was time to start something different. One major reason was my marriage to Nancy who was an American. All along we had decided that we wanted to live on both sides of the ocean, some times in Germany and at other times in the United States. By now we had two children, Michael and Susanne, and we felt it would be positive for them to grow up in two different cultures. I decided to start work as an independent consultant. When I told my employer that I wanted to leave, he tried to keep me with a number of different and tempting offers. When he realized that I remained firm, he asked me, if I couldn't keep working for the company as a consultant, especially for the plastic department, which I had started and headed. I agreed happily as this made my start as an independent engineer and consultant much easier, providing me with a base income.

By that time I had made about ten different inventions which had been patented in the name of my employer in Germany and some of them in other European and overseas countries. I felt I would be able to find companies outside of Germany which might be interested to use these patents under a license agreement and, of course, against paying certain fees and royalties. I would perform this work at my own risk and expense, but if successful would get a good share of the fees and royalties, these licensees would have to pay. My former employer agreed to these somewhat high proportions as he didn't believe I could be successful.

Well, my judgement was right! It didn't take me too long to find companies in Europe, from Norway and Sweden to Spain, Portugal and Israel, and others in the United States, Canada and Japan. The financial return of this work made it possible for Nancy and me to establish residencies in Germany and the United States allowing us to live for longer periods on both sides of the ocean and for our children to grow up on both continents.

My inventive mind did not rest. I applied from now on for new patents in my own name, some in the field of building products and others in new fields I became concerned with, like solar energy and the drying of plastics. Of all these inventions and patents I will describe only one product, a very special plastic building panel. I like it because of the basic inventive idea necessary to produce this plastic panel, the beauty of the product, and because it has close connection to my Path to Peaceful Energy.

The main products of the plastic department which I had organized for my employer were flat and corrugated glass reinforced polyester panels. In these productions, glass fibre mats where impregnated with a liquid Polyester resin and cured through the application of heat in a flat or corrugated form. We designed machinery which could produce these products continuously, the flat or corrugated sheets left the machines in a rolled-up form and could be shipped to the customer in big roles. So when I designed a passive solar heating system for my plant in Weiterstadt, near Frankfurt in Germany, I was able to get one of these big rolls and to install one huge panel in a size of 4x15 meters on the south side of the main building. During the day-time hours it let in the sunshine from the south, heating our production building. During shady hours or during the night a huge insulating curtain would automatically come down and insulate this plastic window.

Huge light transmitting panel as part of a passive solar heating system

In 1958 our plastic department was quite able to produce thin flat sheets, one or two millimetres thick with a good light transmittance and two smooth surfaces. But they were quite flexible. The market asked for a stiff, insulating panel, maybe 20 mm thick, with an insulating air space in the middle. Could we produce it?

Here came my inventive idea: I would use thin plastic film tubes around which I would make the plastic lay-up out of liquid plastic resin and glass fibres, covered on both sides again with glass mats and resin and finally two cellophane sheets in order to be able to squeeze out all air bubbles. This lay-up would be put between two flat steel surfaces arranged at a distance of the final panel thickness. Then warm water would be introduced into the inner tubing, forming the inner channels and pressing the lay-up against the two flat steel surfaces.

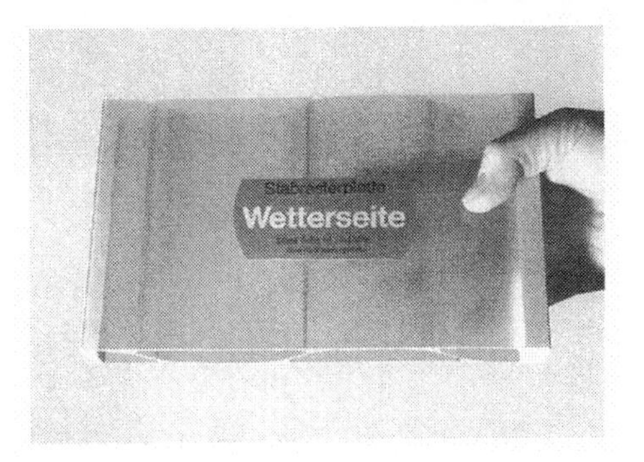

Cross section of our flat panels "Stabrasterplatten"

After about one hour, the Polyester resins would be cured and rigid and could be taken out of the form We had to search for a while to find the right plastic tubing to produce our lay-ups. They had to be resistant against water up to temperatures of 80 Centigrade and also against the liquid Polyester resin. We found them finally in the form of thin plastic film tubes which were used to make sausages.

My idea worked! We were able to produce panels up to 1.60 meter wide and 15 meters long in one piece

One big sized and rigid insulating panel

In order to use more solar energy, on the roof of the same building I had installed a solar warm air collector system. Our huge Stabraster panels were very well suited to be used as the light transmitting and insulating cover panels for our collectors. We designed them in four rows each in a size of 1.50 x 20 meters.

Employees of SOMOS in 1984 standing before the solar heated plant building

Considering that Weiterstadt is located pretty far north on our globe and therefore receives a limited amount of sunshine throughout the year it was pleasing to see, that our two solar heating systems covered on average 22% of our heating needs for the year.

Air view of SOMOS plant with 4 rows of warm air collectors and one huge vertical solar passive window

And now comes the connection to my experiments: If we wanted to measure the temperature gradient in liquids like water, we had to protect the thermocouples and the thermistors from the influence of water. I remembered these tubing for making sausages and Stabrasterplatten and realized that I might be able to use them once more. I could insert thermocouples and thermistors into these tubes and put them into the centre of my test assemblies, this way protecting them from the influence of the liquid water. I tried and it worked fine!

Late 2002 I started to wonder: should I apply for a patent trying to protect the results of my experiments in physics? One typical reason to proceed with this would be the same as it had been with all my previous patents. The patent should give me the rights for the exclusive use of the invention and hopefully provide some future income. For my experiments this reason would be only valid if a vertical temperature gradient really existed and if it could be used to produce work, for instance electricity. At this time, early 2003, I thought yes, there really existed a temperature gradient, which established itself under the influence of gravity. But this gradient was so small that it would be quite unlikely that it ever could produce usable amounts of energy.

To apply for a patent in one country is already expensive enough. But this gives you the rights only for one country, the country in which you have asked for a patent. It becomes very expensive as soon as you apply in additional countries. You have to do this within one year of the application date established with the first application. If you don't apply for additional patents in other countries within one year then your chancehas gone, everybody else would be free to use this invention in those other countries.

As in early 2002 it was totally unclear if this invention could ever be used in an economical way, it would make sense to wait with a first application for a while until a clearer answer could be established. But then, on the other side, somebody else might have similar ideas, somewhere in the world, and might find similar results. Right then, he or she could ask for a patent. I would be too late.

But there was one additional reason, the question of priority. I found out more and more that the magazines specializing in physics, refused to accept my papers for publication. I had succeeded only with the conference in San Diego which in 2002 had published the proceedings of the conference including my paper on the temperature gradient in air. To apply for a patent would have the big advantage that I could describe in it whatever I knew and had found out at that time. Under the rules of the patent law this text typically would be published within a few months after the filing date. Priority would be established by this method. Therefore I decided to go ahead with an application with the United States being the first country.

You can find the complete patent application in Appendix 3. I will describe here only the major points.

I started with describing the fields of the invention:

FIELD OF THE INVENTION

[0003] The present invention pertains to the field of energy production. More particularly, the invention pertains to a method and apparatus for creating temperature differences in columns of gases, liquids or solids in a closed system under the influence of gravity, and using the temperature differential to extract useful energy.

A little bit later I state:

SUMMARY OF THE INVENTION

[0012] In accordance with the invention, it was found that it is possible within the right apparatus to create temperature differences in a closed system without introducing work and thereby to decrease the entropy. The temperature differences so produced can be used to perform work outside the closed system.

This is a very critical statement because anybody familiar with physics will realize that this invention would contravene the Second Law of Thermodynamics. I had to assume that the examiner in the U.S. Patent Office would also come to the same conclusion and for many years the U.S. Patent Office had decided not to accept any more patent applications which would contravene the Second Law. They would make only an exception if the inventor could provide a working sample. At this time in early 2003 I felt that I could not really do that or if at all, only with a very high expense. At that time it was not clear to me that my "invention" really did not contravene the Second Law, it contravened only some statements of the Second Law.

Knowing all this, I decided not to mention the Second Law and to delay this coming discussion with the examiner as long as possible. I proceeded with my Summary of the Invention as follows:

[0013] The present invention is a method and apparatus for creating temperature differences in columns of gases, liquids or solids in a closed system under the influence of gravity. The temperature differences can be used to provide energy in the form of electricity or heat. The invention also pertains to apparatus for reducing temperature influences from outside the closed system.

[0014] In an embodiment of the invention, a temperature differential element, optionally a solid, liquid or gas, is suspended vertically in a chamber inside an enclosure. The chamber optionally is evacuated, filled with fibres, powder or small spheres, or otherwise arranged to minimize the effects of convection currents. Under the effect of gravity, the upper end of the temperature differential element becomes cooler than the lower end. A thermocouple can be used to generate electrical energy from the temperature difference between a vertical segment, for example the upper and lower ends, of the temperature differential element.

In this short summary my major ideas have been described, the use of solids, liquids or gas, and the idea to fill the chamber with convection reducing fibres or powders.

Next come eight figures from which I show figure 1:

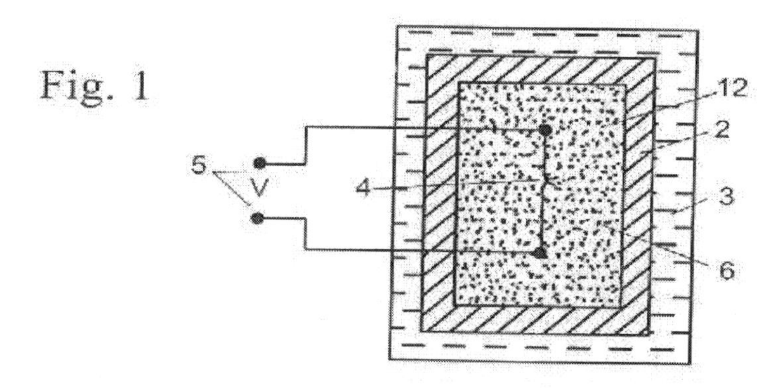

This figure one shows the very basic invention, a closed container with a vertical thermocouple mounted with in a convection reducing powder and surrounded by a great amount of insulation.

Interesting is also Figure 7:

Fig. 7

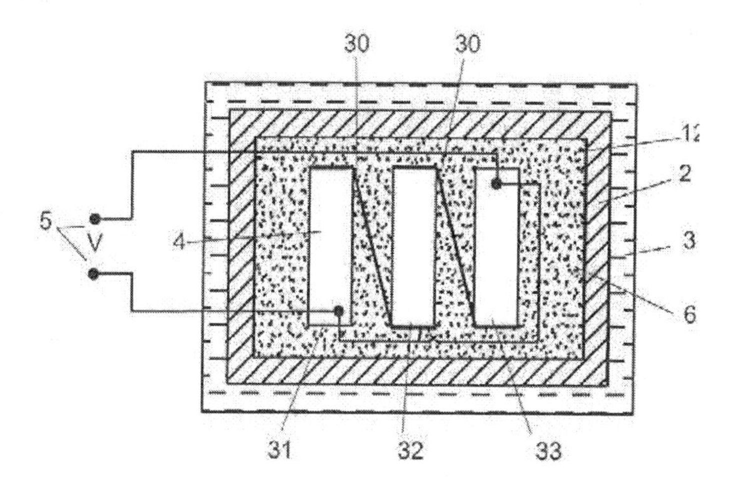

This picture shows the arrangement of the basic invention in sequence with each other trying to reach higher temperature differences. At the time when I wrote this patent application this was only an idea. I tried to prove it with my experiments in the following years but I did not succeed yet convincingly, even up to this date. It shows that you can ask for patents even if at the time it is only an idea, a possibility, but not proven yet.

In the text follows now a more detailed description of the invention which you can read through in the Appendix 3.

Here I only wanted to point out that I also used mathematical formulae when calculating the size of the temperature gradient for various materials. The formula

$$\Delta T = -g \times H / c(Gr)$$

with $c(Gr) = c/n$

where c is the specific heat divided by n the number of freedom of degrees.

This formula has really nothing to do with the invention. But I put it in to establish priority for it. To calculate the temperature gradient by multiplying by the number of degrees of freedom was a new idea. Only by describing it in this patent application I was able to publish it at that time.

Finally and very importantly one has to formulate a number of patent claims. This describes in short sentences what finally is patented. Typically one tries to formulate the first patent claim as broadly as possible in order to get a protection as broadly as possible. This was my first claim:

1. A method to create a temperature difference within a mass of solid, liquid or gas, comprising the steps of: a) providing a closed system in the form of an elongated insulated container having an interior; and b) enclosing the mass, comprising a temperature difference element, within the interior of the container in a vertical arrangement.

Of course, the broader you formulate your first claim the more trouble you will have with the examiner as he or she surely will find plenty of previous work which will force you to change the wording of your first claim into a narrower and narrower form.

With the help of my patent lawyer in the United States, this description was sent to the U.S. Patent Office on February 1, 2002. From that filing date on, I had now one year to decide if I wanted to apply for patents in other countries. I decided to do this for Japan and for Germany. The German application was extendable to other European countries.

What is the situation today, in 2010? In the United States, the Patent Office started the examination process in 2004. They rejected some claims because they contravened the Second Law. In April 2005, after some written exchanges between them and my U.S. patent lawyer, they sent a Final Office Action. Basically they rejected all claims because:

"The claimed invention is not supported by either a credible asserted utility or well established utility."

As a next step, I could request an interview with the examiner or file an appeal to the Board of Patent Appeals. Up to that point, I had spent a few thoudend US Dolllars on this patent application. Getting into personal discussions with the examiner, which would have to take place in Washington, or going through with an appeal, would become quite expensive. In 2005, my experiments had resulted in further evidence of the existence of a temperature gradient, but it was still totally open in my mind if this process would ever be an economical way to produce energy. So I decided not to spend more money on the US Application but to drop it.

In 2002, I had applied for a Japanese patent. Of course, everything for this application had to be translated into Japanese and any answer from the Japanese Patent Office had to be translated into English. Quite often these translations are hard to understand and it is difficult to really find out what the examiner is thinking. Already the original patent application had cost already more than the one in the USA, mainly to cover the costs of the U.S. and especially the Japanese patent lawyer. So in 2004 I also decided to give up this patent application.

This left my German-European patent application. Here the big advantage is that the examination of the application can be delayed for seven years, which I did. The process is starting just now. So at this point I am willing to continue and not to give up this application. It will be interesting to see if I will be able to finally get a patent here.

Judging from today, summer 2010, was it worthwhile to spend thousands of Dollars applying for a patent in 2002? I would say "yes": It established what I knew in 2002 about the results of my experiments. And thinking about the high expenses I have had up to now: today we start to offer small Gravity Machines for sale, gravity machines which could be used by the purchaser not to create a great amount of electricity, but to perform experiments increasing our knowledge about the existence of a temperature gradient in vertical columns of gas, liquids and solids. And with each machine sold we can claim:

COVERED BY EUROPEAN PATENT
APPLICATION Number 03 002 000.2-2315.

Maybe such a statement might help to convince an interested party to go ahead with a purchase!

Chapter 19
Tests in the Rotating Drum Confirm: Gravity Lets Heat Travel from Cold to Warm

In 1947, 2 years into my studying engineering in Darmstadt, Germany, I got my first textbook of thermodynamics. Here I learned: radiation gets reflected quite differently depending on the type of surface the radiation hits. A black surface absorbs practically everything, a blank surface reflects in a very peculiar way. I described this in more detail in Chapter 7.

This difference in the behaviour of black and blank surfaces let to a contradiction to the statements of the Second Law, at least in my mind. A discussion with my professor for thermodynamics, Prof. Krischer, could not clear up this problem. It took me until my 70th birthday in 1997 to decide that I wanted to measure this radiation effect myself and try to get clarity about this question.

In order to measure these minute differences in the effect of radiation potentially creating very small temperature differences, I needed a very well insulated small space in which to perform these experiments. This small space had to be very well insulated so that all inside surfaces would show equal temperatures. And trying to develop and design such a space I ran very quickly into a basic problem. I was not able to succeed. Whenever I measured the inside surface temperatures at different locations, I found temperature differences between the top and the bottom wall. Practically always the top was colder than the bottom. How could this be, especially as the surrounding space always showed a temperature gradient warm at the top and cold at the bottom? Such a temperature gradient is typical for almost all rooms which are either heated or cooled. In winter we heat rooms by various heating systems, during the summer we try to keep it cool. Always the result within any room is a positive temperature gradient, around + 1K / meter of height.

The container I tried to design with an equal temperature all over was going to have a volume of maybe 2 to 4 litres, or one half to one gallon.. I insulated it on the outside with Styrofoam panels in thicknesses of 10cm to 20cm. As an engineer I knew that with a positive gradient of about 1K/meter on the outside, the inside walls of this insulated space could also show only a positive gradient. Depending on the quality of the insulation, it only would be much smaller. The heat flux in the materials surrounded by the insulation would reduce the outside positive gradient in the wall material, but it always had to remain a positive gradient.

What could be the cause of this reversal of the temperature gradient? The only reason I could think of was an effect of gravity. It seemed to be quite logical, if I thought of a container filled with a diluted gas under a very low pressure. If the pressure was reduced low enough then eventually the molecules moving back and forth in a Brownian movement would get such a great free length, that they move only from wall to wall without ever hitting any other gas molecule. This kind of a gas is called a Knudsen gas. Here it seemed to me, molecules hitting the top wall and then moving downwards until they hit the bottom wall would get accelerated by the influence of gravity. They would hit, on average, the lower wall with a greater speed than the upper wall. Gravitation would slow them down whenever they moved up, while accelerating them on their downward movement.

And speed of molecules meant temperature. This, it seemed to me, was a clear explanation for a lower temperature at the upper wall and a higher temperature at the lower wall.

How could I prove this? In order to insulate my inner space very well, I had started to use glass inserts or glass flasks of thermo bottles. I had contacted the company ISOtherm in Karlsruhe, Germany, specialized in producing commercial Dewar glasses. Yes, they would be able to produce for me a glass container with a diameter of 10cm and a height of 30cm , and evacuate this unit to a pressure of

10^-4 millibar. Under these conditions the average free length would be about 30cm. They would also cover the inside walls with silver in order to eliminate radiation between the walls as much as possible. This I felt should create a temperature difference between the top and the bottom. I would insulate this container as well as possible and try to measure the temperature gradient through thermocouples glued to the outside of the top and the bottom of this glass container.

My closed glass container with an inner pressure of 10^-4 mb

The following graph shows the temperature gradient T(Gr) in K/m, The average value is negative, meaning that the top was colder than the bottom. But this value with -0,002 K/m was so small that it was not totally convincing.

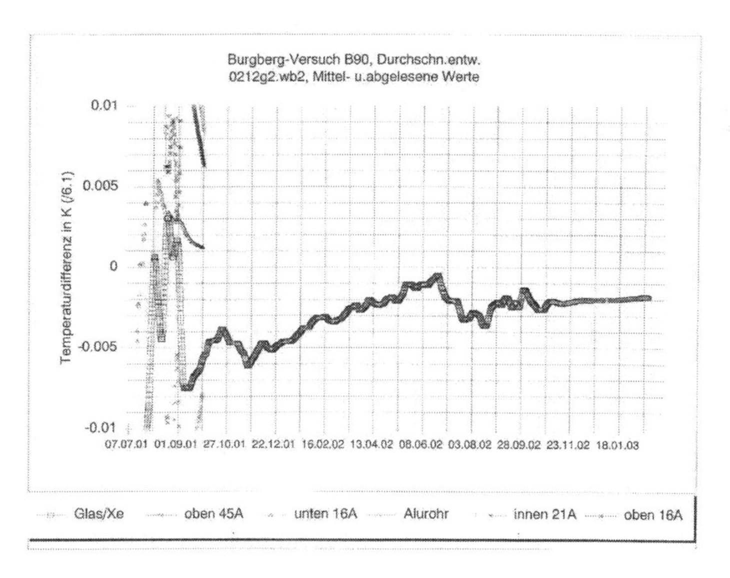

The problem was that I could attach my thermocouples not in the inside but only on the outside of the glass, attaching them there with tape. In addition, the low pressure inside of the glass meant that only a comparatively small number of molecules could transport energy from the top to the bottom while on the outside with the dense atmosphere a multitude of molecules would try to impose the positive gradient, warm at the top and cold at the bottom, as it existed in the room.

So I tried to calculate the amount of work a gas could perform under the influence of gravity. Here I state only the final result:

Q(Gr) max = 1/6 vm x g x rho x g (W/m3)

with vm = average speed of the molecules
Rho = density of the gas
g = gravity factor

This makes it very clear: The experiments with a dense gas with normal atmospheric pressures of 1 bar compared with my experiment B 90 within the glass container ball and an inner pressure of only 0.0001 mb would have an effect 1 million times as big. I really should concentrate my experiments on gases with a normal pressure of 1 bar. But would these experiments show any effect considering that in these higher pressures the free length would be only about 0.0001 mm? The molecules travelling from the top wall to the bottom wall would collide millions of times with other gas molecules before getting to the bottom wall. Would they still transport energy from the top to the bottom?

My experiments with dense gases or with water indicated in the years 1999-2002 more and more that a negative gradient did exist. But how could I show that the underlying cause for this was the effect of gravity? Maybe if I was able to turn my experiment on its head and keep measuring the gradient before and after? If gravity was the cause for the gradient, then after a turn by 180° the top should become cold and the bottom warm. Here a new problem arose. As the temperature difference I measured was very small I would have to turn my experiment in such a way that the insulation would also be turned at the same time without getting disturbed in any way. This was a difficult feat to perform. In order to achieve this I designed the following apparatus:

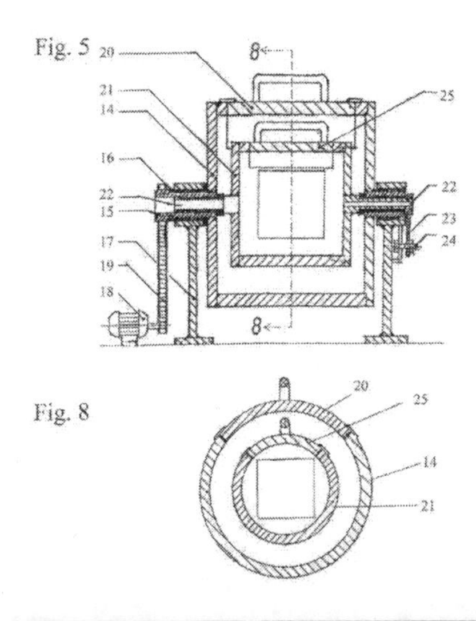

Basically two nesting drum-like containers were mounted on a horizontal axle. Both were built out of 20 mm thick aluminum. The outside drum would be connected through a chain to an electric motor and would continuously and slowly be turned about twice every minute. On the outside of this container 10 cm thick sheets of glass wool – not shown in the picture - were fixed to insulate it quite well against the influences of the environment.

Finally in summer 2001 the contraption arrived. To get it into our laboratory located in the cellar of our home was another problem. Certainly it was pretty heavy and too large to be able to negotiate our stairway. But our cellar had a direct exit to the garden with a big double door which we could use.

Whenever something heavy arrived and had to be transported into the laboratory neighbors were called to help like with moving this heavy aluminum tubing Gerhard Rapp, our upstairs tenant, tried to steer in the front while I in the middle and neighbour Norbert Speelmann in the back pushed. And even his 5 year old son Johannes got into the picture, can you find him?

And here is our turning installation, sitting at the entrance to the cellar. Norbert, son Johannes and even 3 year old Clara plan to help while Gisela Hoffman looks skeptical, thinking about the next move.

Turning drum on the way into the laboratory

Finally, we got it moved to its correct location and the motor was connected to the power supply. It was switched on and started to turn the outside drum about twice every minute while the inside drum mounted on the same axle remained stationary. This turning day and night lasted for 5 years until one day the gear in the drive motor gave way and stopped. Catastrophe! It would be very difficult for us to find a fitting replacement.

How would the stoppage influence the temperature in the inner drum which had been kept so nicely to a value of about +0.001 K/m? What a nice surprise it was: the gradient stayed practically at the same low value! The 20mm thick outer drum even while remaining stationary still equalized the higher temperature gradient of the room to the former small value in the inner drum.

The next picture shows both drums opened at the top exposing 5 experiments mounted in the inner drum:

Within this inside drum one could place a number of experiments. They were fixed in place by blocks of polystyrene foam, insulated against each other and against the inside wall by mats of PET fibers.

As I described in chapter 16 the rotating drum used as housing for my experiments had a number of advantages. Most of all, it provided a very small temperature difference over height of a few thousandths of a degree per meter, warm at the top and cold at the bottom, ideal for my tests. And this temperature difference was kept very constant over time. And compared to all other types of test housings I tried, I could turn any experiment on its head without interrupting the experiment or influencing its insulation from the environment.

Finally, in October 2002, we were ready. I loaded the inner drum with a number of experiments including the test B76, which at that time was still named V 76 for V= Versuche or Experiment, but was later renamed into B 76 with B standing for Burgberg, the location of my little laboratory in Germany. This enabled us to differentiate our testing in Germany with the parallel experiments I was performing in Ithaca, N. Y., which got numbered with a prefix "I" for Ithaca.

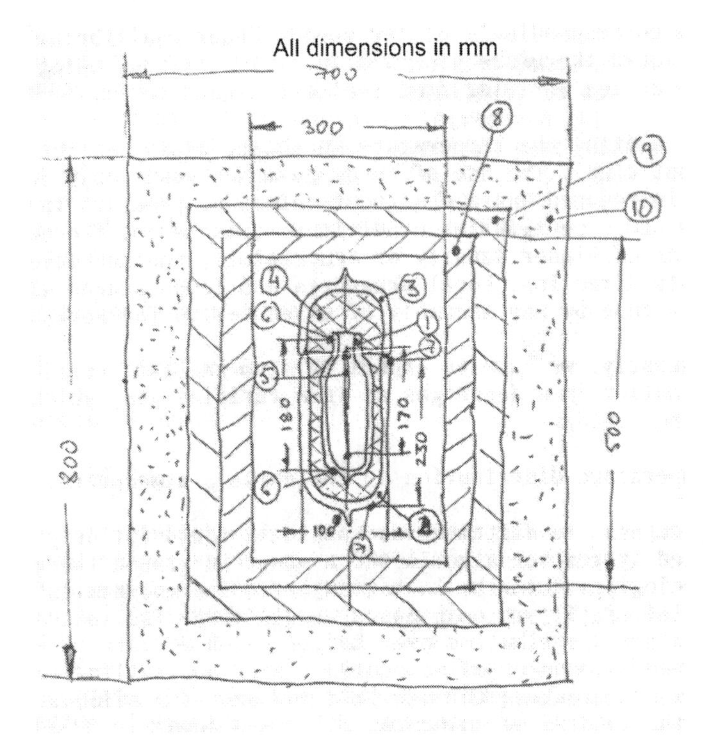

Figure 2

Our drum was ready for the first test in October 2002, which was actually a 5 year anniversary of the beginning of this endeavour. When I had started out at that time, I felt it would take 2 – 3 years to come to a conclusion!

My excitement increased. What could be the results of this turning experiment? If there existed a gradient before the turning, would warm areas stay warm and cold ones cold, unaffected by the turning? This would indicate no effect of gravity! In this case, may be an exothermic or .....process would possibly be the cause. But if the turning affected the local temperatures, this would indicate a strong influence of gravity!

Well, turning the inner drum about every 2 days, quickly a tendency became obvious:

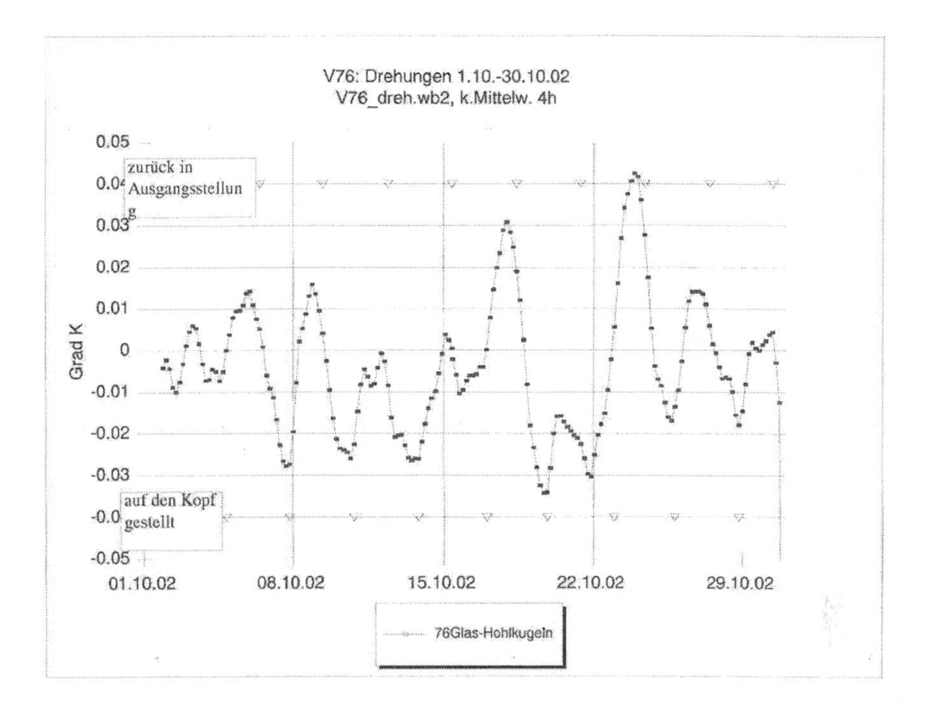

Each point on the curve represents the average of 60 measurements taken each minute over one hour. The triangles on the top and the bottom represent the times when the drum was turned by 180°. The triangles at the top of the graph show the times when the drum was turned back into its original position, the top being again on the top. One can see easily that starting at these times the temperature gradient measured in absolute values went to negative values. This meant that the temperature at the top becomes colder than the temperature at the bottom. When the drum was turned in the opposite direction, meaning the top becomes now the bottom, indicated by the triangles at the bottom of the graph, then the gradient went to positive values meaning the top, being now at the bottom, becomes warm and the bottom, being now at the top, cold.

Exciting and wonderful! The measurement clearly showed that the turning had a great effect on a vertical temperature gradient. But why were the absolute changes of the temperatures so different from turn to turn, sometimes more pronounced and sometimes showing only a small influence? The answer was close at hand. Obviously, my measurements would be influenced by the changing temperatures in our room. Even though in my laboratory in the cellar of our home, temperatures were pretty stable and did not change very fast, they still did fluctuate parallel to the temperatures on the outside. And at this date we could measure only gradients with our thermocouples, we had no possibility to measure the absolute temperature within our experimental setup, for instance by installing thermistors at the same places where we had attached the thermocouples. Very intentionally I had decided not to use thermistors because whenever a measurement would took place a small current would pass through the thermistor creating a small heating effect, which could potentially influence the local temperature. Maybe I should relent and install some thermistors after all in order to understand the variations of our curves?

The next question was a natural one: How did the gradient change on average over time in its absolute value when we turned the inner drum by 180°? Let's find this out for our test B 76 which we published in San Diego. Here we had measured a long term average of – 0.07 K/m. Would the turning experiments lead to the same results if we averaged the values of a sufficient number of turns?

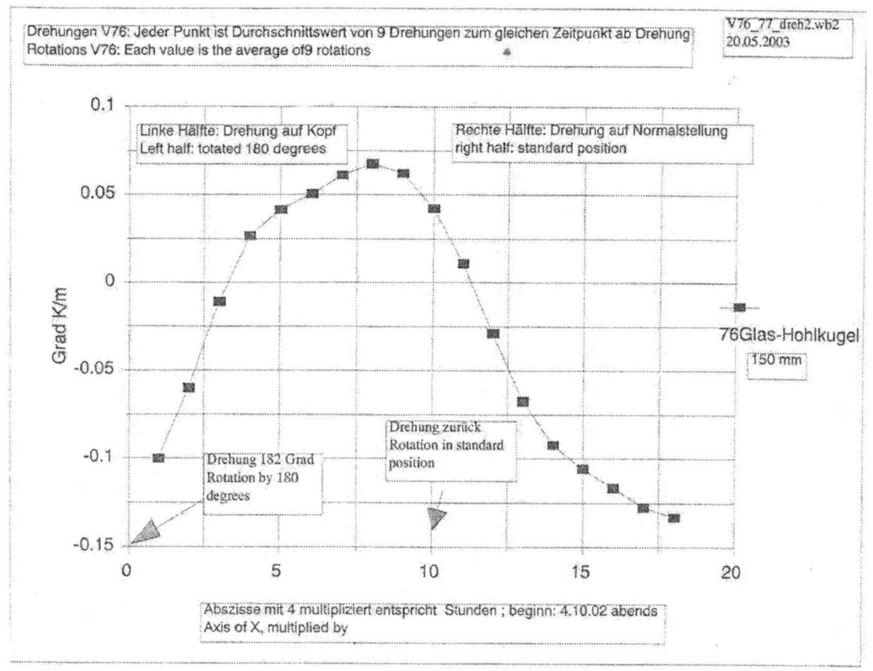

The curve shows the average of nine rotations, the temperatures averaged for equal times before and after the turn. If you look at the first point with the value of -0.1 K/m and the maximum point of +0,074 K/m then amazingly enough you find as a half way point 0.082 K/m, very close to our long-term average value of -0.07 K/m. This seemed to be a clear indication that gravity was the cause for the negative temperature gradient of the inner axle in the experiment B 76.

With the next turning experiments for B428 I would like to show you the curves of 13 turning tests performed over a period of 6 months. The test set-up B428 contains air plus glass powder in a Dewar glass quite similar to our test B 76 as shown in figure 2 three pages earlier. The individual curves taken over such a long time vary quite a bit from each other. This is understandable as during a period of half a year, the temperatures in the environment, which is the room of our laboratory, change considerably over this time span of half a year, even so we try through our thermostatically controlled heating system to keep this temperature as constant as possible.

The following graph shows these 13 curves of the temperature gradient of the inner axle each one in the normal position for 48 hours before and for 48 hours after the turning took place. The thick line is the calculated average of these values:

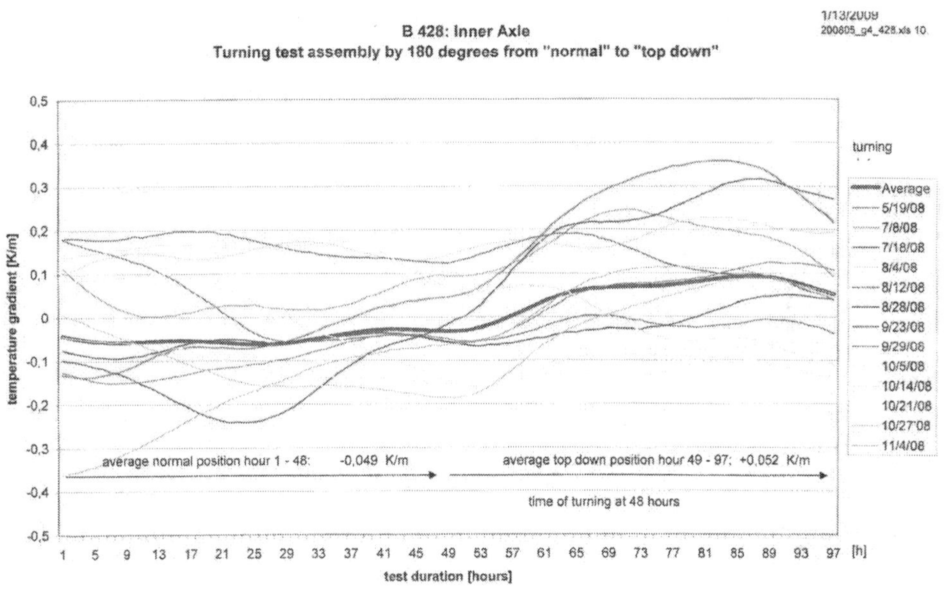

Clearly, while each curve of a single event does not tell us much, the average value of all curves does. The calculated average value for the 48 hours before the turning is – 0.049 K/m while the average value after the turn is + 0.052 K/m.

These values are in the same "Ball Park" as the value of -0.071 K/m for our test B 76.

Question: What happens to the temperature gradient measured on the outside of the inner glass as shown in picture 2, four pages earlier? This is demonstrated in the following graph showing the temperature gradient for the inner axle as discussed above when being turned from a "normal" position in a "top down" position and the values measured outside of the inner Dewar glass for the same number of turns. While the change of the inner axle amounts to a total value of 0.11 K/m, the gradient on the outside of the glass is much smaller amounting only to a value of 0.024 K/m.

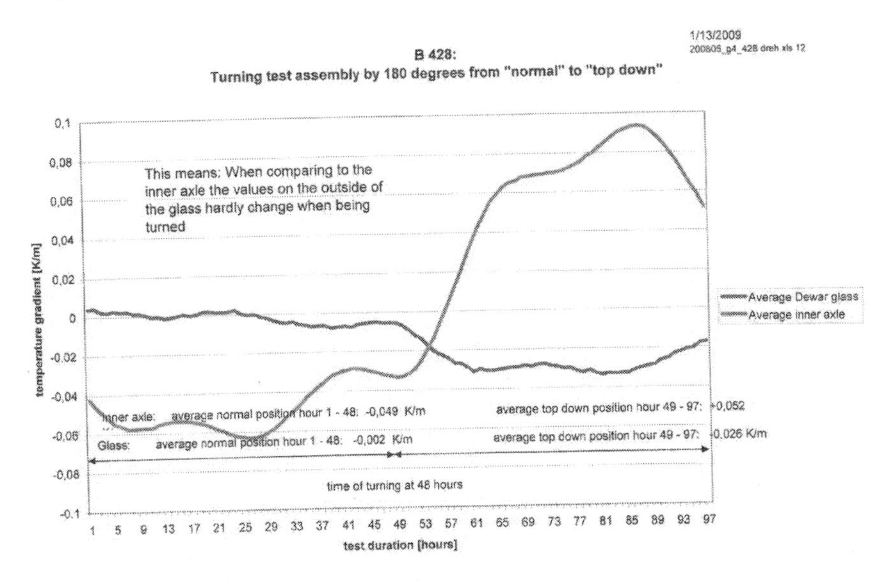

In summary: because of the temperature change in the environment many identical turning tests have to be done over time in order to get meaningful results by calculating average values for these tests. Then they seem to show clearly that the average values calculated from these turning tests come very close to the average values found through measurements over longer time spans without turning. Also, the changing values before and after the turn on the outside of the Dewar glasses are much less pronounced then the change of the T(Gr) values of the inner axle. This seems to be a very good confirmation that a negative vertical temperature gradient exists in the inner axle in these experiments and it is caused by the effect of gravity.

In 2005 I started to install thermistors at the same locations where the thermo elements were located. As discussed already in Chapter 17 the use of thermistors to measure local temperatures does not seem to affect the basic value of that temperature. Temperature differences measured by thermistors between the top and the bottom are very close to those measured by thermocouples.

This is confirmed by the following table comparing the results for 3 turning tests for B 428. One can see looking at the temperature gradients measured on the outside of the inner Dewar glass measured by thermistors in row 1 with the same values measured by thermocouples in row 2, how well they compare. As shown in row 4, already 3 turns performed in May of 2008 show a change in the value of the inner axle in the normal position from negative value to a positive one, namely from -0.1648 K/m to a value of + 0.0657 K/m, when turned on its head.

B 428: Ergebnis Drehversuche — 30.05.2008 — 2008_g4_425.xls

	aus Thermist. 13-14 geeicht	aus Thermoel. glass aussen	aus Thermoel. glass innen	aus Thermoel. Mittl achse
richtig stehend:	K/m	K/m	K/m	K/m
10.-13.5.	0,0993	0,0724	-0,0909	-0,4419
16.-19.5.	0,0002	0,0026	-0,0284	-0,0604
22.-25.5.	-0,0128	-0,0094	-0,0156	0,0062
Durchschnitt	**0,0290**	**0,0218**	**-0,0449**	**-0,1648**
auf dem Kopf stehend	K/m	K/m	K/m	K/m
13.-16.5.	-0,0127	-0,0089	-0,0180	-0,0394
19.-22.5.	-0,0674	0,2185	0,0246	0,2185
25.-26.5.	-0,0041	-0,0014	-0,0205	-0,0462
Durchschnitt	**-0,0338**	**-0,0275**	**-0,0009**	**0,0657**

Dieses Ergebnis zeigt sehr schön die Parallelität von Thermistormessung (Spalte 1) mit Thermoelement-Messung "Glas außen" (Spalte 2) und die Gegenläufigkeit mit den Thermoelement-Messungen "Glas innen" (Spalte 3) und "Mittlere Achse (Spalte 4).

Außerdem sieht man, wie man schon nach drei Drehungen für die mittlere Achse einen TGr-Wert von -0,16 K/m in der richtigen Stellung und von +0,066 K/m für die auf Kopf-Stellung erhält.

I discussed already in Chapters 16 and 17 that the measured values of T(Gr) for gases and for liquids are quite pronounced and with our ability to measure small temperature differences correctly, they are big enough to be very believable. With solids like steel, lead, aluminium or copper the story is different. Yes, I measure negative gradients, but their values are too small to be convincing. A test for copper, steel or aluminium might show a long-term average of – 0.003 K/m. This value is considerably smaller than the zero point correction of the voltage meter. Here, turning experiments seemed to give a clearer picture.

The following shows test B 429 with thick copper wires arranged within a copper tube:

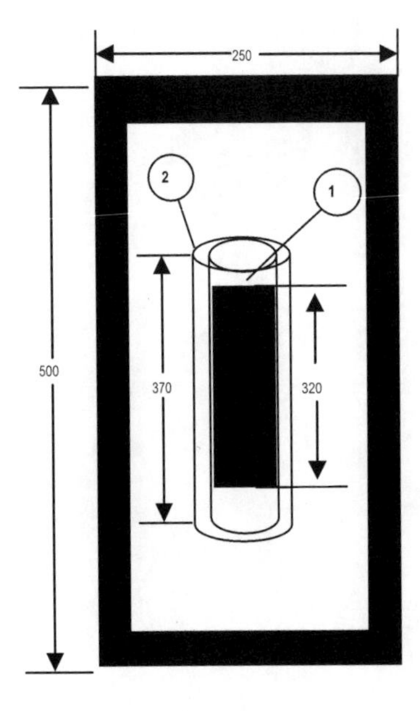

With this set-up of B 429 we performed 13 turning tests during the second half of 2008. The following graph shows the result when the copper tube was turned from the "normal" position to a "top down" position. Here the average value before the turning was – 0.00047 K/m and in the "top down" position + 0.00222 K/m. This indicates an average T (Gr) value of - 0.0023 K/m.

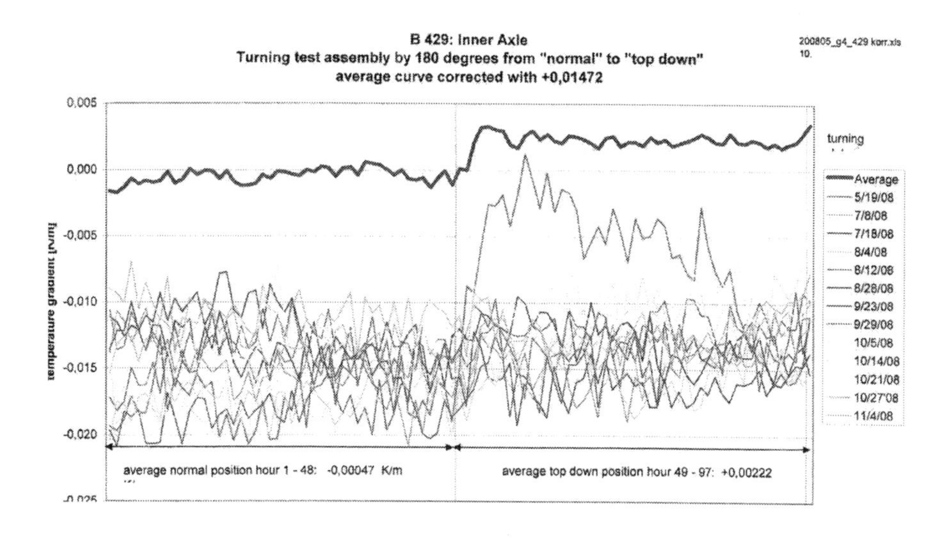

When turning it back, the opposite happens:

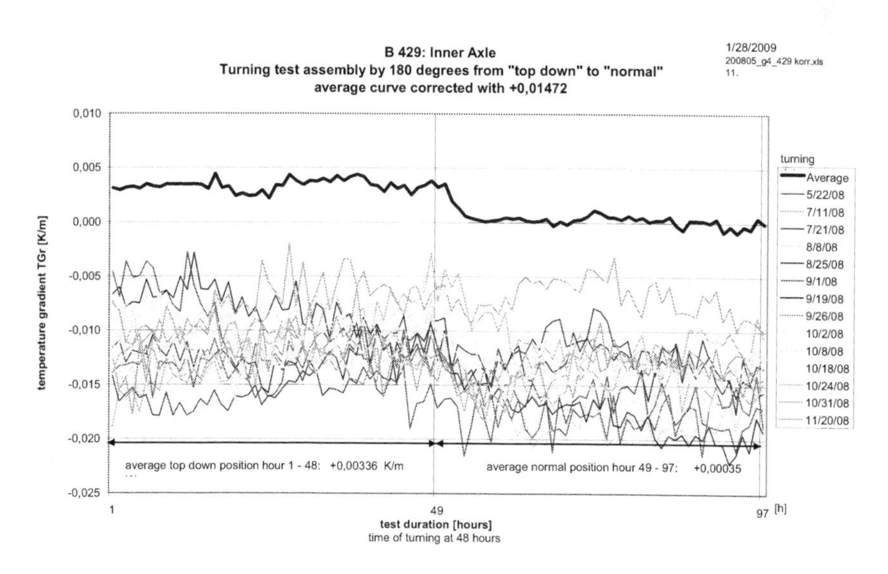

In these graphs, the thin lines, showing the results of individual tests, don't indicate a clear picture. But the thick lines, the averages of all the individual turning tests, demonstrate a definite result. Within 3 to 10 hours after a turn the temperature gradient rises, or when we turn in the opposite direction, it falls. This is a very clear indication of the influence of gravity.

In summary: many, many turning tests show the basic value of this test procedure. They confirm to my satisfaction:

> When a test setup is turned on its head, turning it by 180 degrees, within about 12 hours, – in metals in around 3 hours -, the gradient switches around: 12 hours after the turning the tops are cold and the bottoms warm, just as they were before the turning..

The change in temperature when being turned by 180 degrees amounts to twice the value of the long-term average gradient, equalling a change from +T(Gr) to –T(Gr).

> These changes show up in gases, liquids and solids. For solids these turning tests might be the only type of test which can confirm a vertical temperature gradient. The normal tests without turning show in solids values of the gradients which are too small to be convincing.

After trying to measure temperature gradients in a wide variety of test set-ups I came to the conclusion: Yes, there exists a negative temperature gradient in very well insulated vertical columns of gases, liquids and solids, located within a surrounding environment with a positive gradient. And yes, when turning these tests by 180° these values show up again with temperatures cold at the top and warm at the bottom. And finally, these measured results are very close to calculated values.

These turning tests confirm the influence of gravity as the basic cause for getting negative temperature gradients, eliminating other potential explanations.

Chapter 20
Nancy Leaves

Ones more Nancy had to be moved to the hospital, fighting of three different kinds of cancer over eighteen years had left its mark. The possibility to communicate went down to zero. She was kept alive with the help of various machinery, connected to her by a number of tubes. She clearly did not like her oxygen mask. So I took it of. Not being able to communicate any more and her breathing becoming more laborious, did it make sense to continue with the intravenous feeding?

If we stopped it she could go back to Kendal which was our home. This seemed to be more appropriate for her last days and much more natural for me and the arriving children to be with her in her normal surroundings. And the greatest benefit, Zuko could visit with her.

When I proposed this move the doctors and hospice nurses were quite against this. The drive in an ambulance would be quite strenuous; she might die on the way. Strange reasoning, I thought.I insisted and the move took place with out incident. Now everybody felt very positive, we could be close to Nancy, uninterrupted by trips between home and hospital. A cot was provided so I could sleep in her room. Zuko came often and licked her hands and arms. Did she still feel this?

Morphine kept her comfortable. Without the intravenous feeding her fluid levels slowly decreased. Her breathing slowed down. So finally on Thanksgiving morning with only I with her in the room, her breathing slowed down with increasing pauses, it finally stopped. My wife of 54 years had died.

I stayed with her quietly for another hour. The door opened, and Nancy's very special friend through many, many years, Benedictine Monk Brother David came for his daily visit. He said a prayer and joined me in silence.

54 years of togetherness floated through my mind, warmth, togetherness and closness in good and in difficult times.

Silence.

Quietness.

Memories.

The shared past, experiences, plans, discussions, common actions…

3 years before we married I had met Nancy in Madison, Wisconsin, dancing in a square dance group. We became friends and she introduced me to the beauties and also the darker sides of her country. Exploring the local churches, a different one each Sunday, we were mostly touched by the Quakers or Friends. We participated in a weekend workshop, trying to help some families in the slums on the South Side of Chicago with repainting some rooms in their apartments. Have you ever seen an apartment in the slums of an inner city? The lady of this apartment was confused about our working there. "Are you trying to learn how to paint and needed a place to experiment?"

Our children arrived, first Michael, 2 years later came Susanne. They were a major focal point of our marriage. Thirty years ago we travelled in our Volkswagen Camper all the way across this huge continent of North America. We stopped at the Niagara Falls where Michael demonstrated that he hardly ever could sit still.

Passing one of the peaks in Colorado Susanne asked me:

How many people live in Ann Arbor, our home town?

"50000," I answered.

"And in the United States?"

"About 200 Million," I guessed.

"And in the whole World "she wondered?

"Well, maybe 3 billions"!

And 5 year old Susanne looked at me: "Including me?"

Yes, Susanne, it includes you, and Nancy, and Michael and me, our whole family.

Memories. One late evening Nancy stepped with Michael through our back door into our yard and admiring together the star filled sky, Michael suddenly cupped his hands before his mouth and called out:

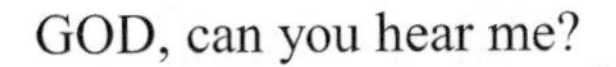

GOD, can you hear me?

A very close connection for 55 years, is this just history now, gone for ever? With whom would I discuss in the future my thoughts, work and plans? Was there any point in continuing with my work in the Foundation for Violence-free

Living, my plans to convert the German Armed Forces into a Peace Corps, my physics experiments, my writing a book about them?

Quietness.

Emptiness.

I finally informed the staff:

Nancy had died.

Chapter 21
A Six Year old Girl Sends a Message: Disgrace and Hope, the Armed Forces and Circability

An amazing thing happened 3 months ago. A 6 year old girl, Helena, living in Germany decided to create a newspaper as a present for her mother's birthday. Apparently she looked through the paper "Süddeutsche Zeitung" (South German Newspaper), the daily paper read by her parents. She found a picture with some soldiers standing in front of a helicopter, cut it out and pasted it on an empty page. Then she put down her ideas about this picture, demonstrating her limited writing and spelling ability as a normal six year old, maybe close to the end of her first grade:

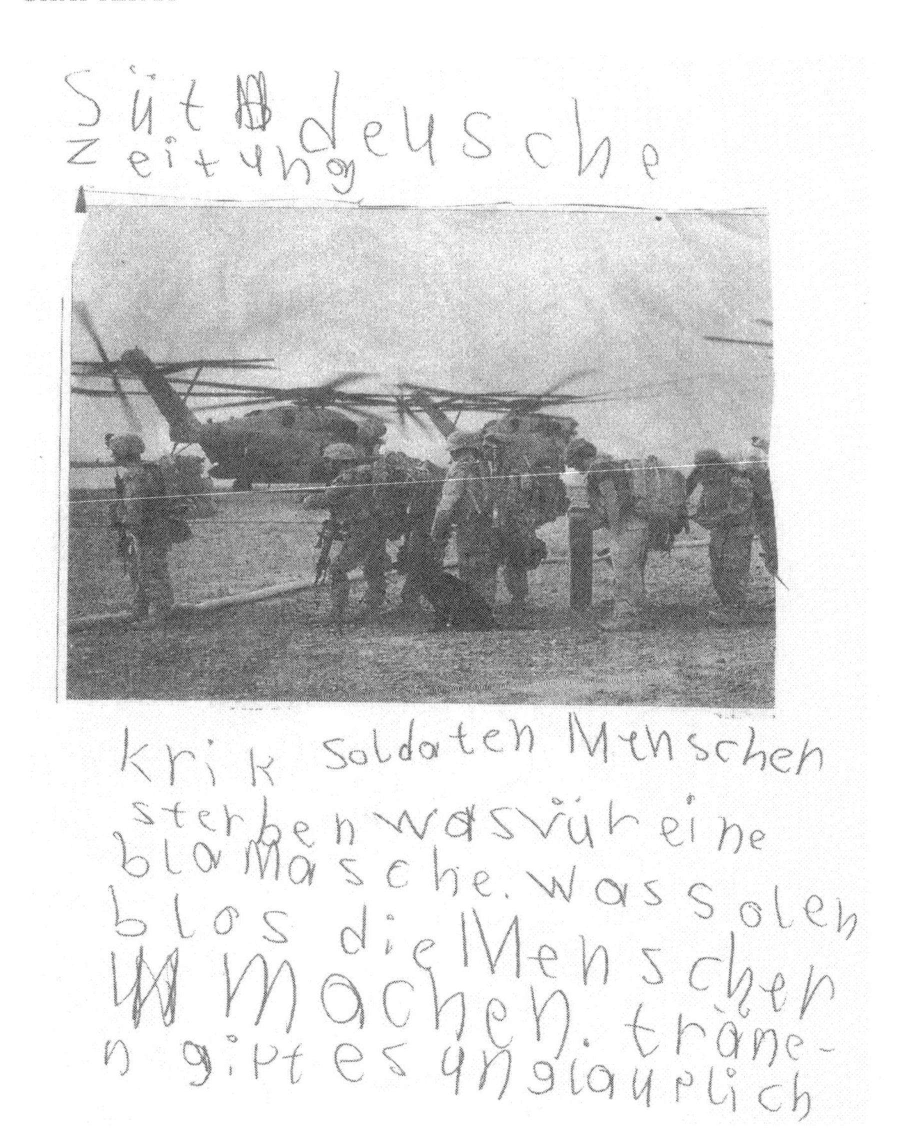

Spelled correctly in German:

Krieg. Soldaten, Menschen sterben.
Was für eine Blamage.
Was sollen bloß die Menschen machen.
Tränen gibt es unglaublich

And translated into English:

Sueddeutsche Newspaper

War. Soldiers, human beings die.
What a disgrace.
What on earth should the people do.
Tears there are unbelievable.

How is it possible that a 6 years old girl creates such a text? That living in peaceful Germany in 2009 she has these thoughts and is able to express them in such a meaningful, even poetic language? Her parents tell me that she created the newspaper totally on her own, that the ideas expressed here are not typical subjects discussed between them or with their daughter. They themselves can not understand how the daughter got these ideas and was able to assemble and write this text.

I am very moved and affected by it:

War. Soldiers, human beings die.
What a disgrace.
What on earth should the people do.
Tears there are unbelievable.

I copied this first page and sent it to friends, relatives, former classmates, and to a group of people whom I invited to discuss it. Specifically I asked all these people what would be your answer to this question of 6 year old Helena

What on earth should the people do?

12 of these people met and discussed primarily two questions:

1. Should the German parliament decide to change the armed forces into a Peace Corps?

2. What kind of answer can we give 6 year old Helena?

Most participants thought that the basic idea was a very good one: Germany, without any commitment by any other country to do the same, would totally disarm. And equally important: instead of Armed Forces there would be a Peace Corps of equal size formed, which would assist any in need in this world.

The Peace Corps created by using the money formerly used by the Armed Forces would offer to help anybody, anywhere in the world, who needed help and asked for it. This could mean that we send hospital ships to the poor areas in the world to treat patients and train healthcare personnel. Areas without enough clean water could be visited by specialists bringing along the knowledge and the needed funds to install a clean water system. The list of calamities inflicted on so many people unable to help themselves is endless. The basic idea is that the protection of the German people against invading enemies would be much more effective through such a Peace Corps than through the existence of an armed military force. Yes, we would keep a police force giving citizens protection against people who might want to hurt their neighbour.

The opinion of the majority in this group about this plan was such: the idea is very nice, very positive, but it is utopia. To try to reach this goal is totally hopeless. Also, would not most people feel that we lost all of our protection, a form of protection society is familiar with and has been accustomed to for the past thousand years?

Concerning the second point, which is to answer Helena's question:

What on earth should the people do?

There was a broad range of answers to this question. A protestant minister declared that humans are born in sin; there always would be bad deeds by others for which we would need protection by an Armed Force. Others felt pretty much the opposite. One should approach Helena with warmth and encourage her to look forward to a peaceful life, which would be possible if we all tried harder to achieve it.

I had thought about this question 20, 30 years ago when my children, then in their teenage years, told me that the future of their generation looked very dark. They feared there would suddenly be an atomic war, an atomic holocaust, or we humans would affect our environment in such a negative way that life would not be worthwhile anymore. Was my generation leaving such a world behind, one which appeared to be so negative to my own children? And did I myself internally not agree with their pessimistic outlook towards the future?

DISGRACE

What on earth should the people do.

What on earth could and should I do?

I felt that in this age, with a threat of total annihilation of the human race, we humans needed a new ethical view. I called it

KREISFAEHIGKEIT - CIRCABILITY -

and I started a foundation in Germany under the name of GEWALTFREIES LEBEN - "Violence Free Living" - . This foundation would study this new ethic of CIRCABILITY and if, eventually, we found this to be a meaningful way of life we would try to encourage other people to consider these norms for their own actions. If only enough people would follow the ideas of CIRCABILITY I felt the major problems of the world could be solved, those problems, which gave my children such a pessimistic outlook towards their future, namely:

- a continuing threat of global annihilation through atomic war
- the threat of a drastic change of our environment through the burning of fossil fuels or the storage of waste materials generated through the production of atomic energy
- overloading the earth with humans
- destroying the variety of nature by killing off animal and plant species
- by not caring about the future of our children and our children's children.

.

Let me try to explain the meaning of CIRCABILITY.

CIRCABLE ACTION means in view of every single individual and every animal and plant species, to perform only those actions which could within one generation return to their original condition on its own or could be returned to it.

CIRCABLE ACTION allows all activities as long as their consequences make it not impossible for future generations to recreate their original conditions.

In other words: We human beings should not perform any actions which binds the next generation and which principally cannot be undone by them. Let's look at some examples:

To cut down a tree is a circable action because we can plant a new one when we or our children decide that we should not have cut it down.

To burn coal or oil is not a circable action because we cannot recreate these materials.

The creation and use of atomic energy is not a circable action because we leave waste material behind which will emit dangerous radiation for thousands and thousands of years.

And there is one action, one direct consequence, which has directly to do with Helena's statement:

In war soldiers, human beings, die:

To physically hurt or kill any human being is certainly not circable as there is no way to recreate this human being.

The idea of Circability leads to 5 directives:

Each of these directives eliminates actions whose results could not be rectified and returned to their original conditions by future generations.

Physical hurting a human being or killing one is not retractable, therefore:

I. The killing or physical hurting of a human being can not take place anymore:

Killing is out, killing is passé.

In order to fulfill this directive we have the following consequences:

- Peace preservation instead of war preparation
- Dialog instead of condemnation
- Development of violence-free methods for the negating of violence by un-peaceful people
- Changing the military to peace preserving police and trained assistance-giving specialists
- Defending against threatening strangers not by building walls or preparing for the use of force, but by building bridges and offering local help, education and assistance.
- Making possible local CIRCABLE LIVING instead of population movements
- Building of schools and hospitals instead of prisons
- No more sending of one's troops to secure one's own energy supply.

Not retractable are the changes in the atmosphere due to the burning of fossil fuels or the creation of radiating atomic waste materials through the generation of atomic energy, therefore

II. The burning of fossils fuels or the use of atomic energy is off limits.

Consequences:

- Speedy switch from the use of atomic energy or the burning of fossil fuels to an energy production based only on renewable energy sources.
- Learning, acting, creating and helping instead of consuming.
- Striving for education, creativity and sharing in place of increased consumption.

The increase of the number of people on this earth is not retractable, therefore:

III. Families are to have on the average not more than two children.

Consequences:

- Striving for couples not to exceed 2 children instead of advertising for larger families.
- Instead of asking for more children actively pursuing minimizing child mortality
- Equal rights and equal education for men and women world wide.

The elimination of an animal or plant species is not retractable, therefore:

IV. Actions which endanger the survival of any animal or plant species are off Limits.

Consequences:

- Encouraging all actions which promote the survival of all animal and plant species.

Without responsible actions of all people within the limits of CIRCABLE ACTIONS non-retractable actions would take place, therefore:

V. Responsible personal action.

Consequences:

- Based on a feeling of responsibility for other human beings, animals, plants and especially for the next generation, everyone should plan and execute all their actions within the idea of CIRCABILITY.

A more detailed description of these five directives can be found in Appendix 6.

Maybe this idea of Circability is my answer to Helena's question:

"…What on earth should the people do…."

In view of her age, I believe I could explain each of these 5 directives in words appropriate for a 6 year old. I would tell Helena that each of us, depending on our age, education and interest, could try to concentrate our energies on any one or more of these consequences in order to create peaceful and non-violent ways to solve our present problems and to avoid future problems. I would tell her that I personally would plan to concentrate my energies on the goal of changing the German Armed Forces into a Peace Corps. If we could leave the **blame** behind, the **disgrace** would be changed to a positive feeling of being helpful. If she agreed that this was a good idea, she could assist me by allowing me to use her statements.

HELENA in 2009

Working for changing the Military Force of Germany into a Peace Corps and developing a Gravity Machine for creating peaceful energy, what a meaningful and wonderful combination on my Path for Peaceful Energy!

Chapter 22
Nobody Is Interested, Nobody Wants to Listen, Nobody Believes me;   A Last Try??!!

On October 13, 1997, on my 70th birthday I decided to try to find an answer to a question which had arisen 50 years before when at the University of Darmstadt, Germany, I tried to learn more about the field of thermodynamics. I was going to try to find out if, in an isolated system, a temperature difference would develop by the radiative exchange between the surfaces of a black sphere and a blank surrounding enclosure, as discussed in chapter 11.I would perform actual measurements of temperature differences on test specimens in my homes in Germany and Ithaca, NY. It would take two or three years at the most, I estimated, to get an answer, confirming or disproving my ideas.

Very quickly after starting with my first experiments I ran into a problem: I was not able to get an equal temperature over the height on the outside of my experimental set-up which was a necessary precondition to get meaningful measurements of my original inquiry. So the aim of my experiments changed to a totally different question: In insulated systems, does the influence of gravity create a temperature gradient, cold at the top and warm at the bottom in vertical columns of gases, liquids or solids? It was such a gradient which I kept finding. But such a gradient is contrary to what today's physics teaches. Such a gradient would be contrary to today's understanding of the Second Law of

Thermodynamics. So whenever I talk about these findings to anybody knowing some physics very quickly they stop listening. They know that my results cannot be correct because, in their opinion, they would be contravening the Second Law.

Well, there are a few exceptions. Gisela Hoffmann, who works hard taking care of the data collection and creating the printouts of the experiments, not only listens but discusses the implications of each new test result. Alexander Haltmeier, an engineer, but at heart a very knowledgeable physicist, not only listens but tries to understand each detail of my experiments and is extremely helpful through our discussions in my developing new ideas and designing new test set ups. Marianne Struempel, a crystallographer, knows physics and mathematics and is not only a good discussion partner, but extremely helpful in sprucing up my papers written in my broken English into a readable scientific paper. And there is one real Professor of physics, Daniel Sheehan. He teaches at the University of San Diego, CA, and is very much involved in all questions of the Second Law. He was the major organizer of the conference QUANTUM LIMITS TO THE SECOND LAW in San Diego in 2002 [24]. He even interrupted his vacation in France in 2003 to visit me in Koenigsfeld in Germany to view my experiments and discuss them in his book CHALLENGES TO THE SECOND LAW OF THERMODYNAMICS, 2005 [10]. And when I ask them from time to time: What do you think about the impact of my experimental results onto the Second Law, they Hmm and Haw and indicate: In spite of your measurements and their surprising and often amazing results, the Second Law cannot be challenged, it has to stand.

But there is one exception: Andreas Trupp. He is a lawyer with an immense interest and knowledge on mathematics and physics. He is a theoretician and he claims: You, Roderich proved by your experiments a temperature gradient contrary to the writings of Boltzmann and Maxwell. And actually there are two more persons: A follower of Buddhist teachings and a Benedictine monk: Both declare that they don't see any contradiction with their belief systems that heat can travel from cold to warm!

Outside of these five people, hardly anybody listens, nobody believes me! Why? For more than 10 years I have been trying to tell the world: Heat can travel from cold to warm, it is possible to produce work out of a heat bath, you can build a Perpetuum Mobile of the Second Kind. These, I agree, are all very excitable, sensational statements. They are not based on some difficult theory, difficult to understand and difficult to disprove. They are the results of very straight, down

From left to right: Dr. Marianna Strümpel, Prof. Daniel Sheehan, Dr. Andreas Trupp, Dr. Alexander Haltmeier, Gisela Hoffmann.

to earth experiments, not secret experiments, but experiments described in all details, basically very simple experiments, easily repeated by anyone who wanted to do it! And if correct, this result means that potentially humanity has at its hand an unlimited amount of energy, accessible without the use of fossils or atomic fuels.

So, why is it, that today not even one person in the whole world jumps at this possibility, repeats my experiments and either disproves them or starts to work on an economical realization of this newly found unlimited amount of energy? Why are the specialists in this field, the professors of Physics or Engineering refusing to talk to me? Why do the scientific magazines decline to publish my papers?

After being rebuked for 10 years I believe now to have the answer: All these people who look at my work know very quickly: The results claimed by me cannot be correct, they contradict all what they have learned about the movement of heat, of the Second Law, everything what they know by heart of being correct. Therefore it makes no sense to concern oneself with these experiments; it would be a waste of time. And why do they come so quickly to this conviction?

Since 1998, one year after I started with my experiments I have tried to discuss my findings with real physicists. In Ithaca, where I live for half of the year and in Darmstadt, where I studied and received my degrees, I wrote to and phoned the professors of Thermodynamics, Engineering and Physics. I sent them my papers describing my experiments, I asked for a meeting. The results were always the same: This is not our field of expertise, we are very busy with our own work, we don't have students who could duplicate your experiments. I must have written to more than 70 Professors of Physics in Germany, the United States and some other countries sending them my papers to be published, offering to give a seminar on the results and to provide a sample gravity machine for them to repeat my measurements, all with no result. One professor answered very openly that he did not want his name to be connected with such experiments. But not one of them informed me of any reason why my measurements might be wrong or what could be the mistakes in the underlying theory I had developed.

There was only one exemption: In 2007, I was invited to give a seminar at the Department of Theoretical Physics at the University of Würzburg, Germany. 25 people participated, we had a good, when short discussion, but sadly, only theoretical physicists took part and there was no follow-up.

Still now, today on March20, 2009, the friends introduced above still listen but none of them really believes in the deeper meaning of my results which in my opinion prove that gravity creates a temperature gradient, cold at the top and warm at the bottom.

So finally I decided to try once more at my own Alma Mater, the Technische Universität Darmstadt, by writing to its President on December 2, 2008:

Dear Dr. Proemel: (Translated from German)

In January 1946, I believe I was the youngest student in Darmstadt at the age of 18 years, who was admitted, with a stamp "finished the 8th grade of high school", to a preliminary course for 8 weeks and afterwards to the first term of Mechanical Engineering. Twice a week we shoveled rubble in Darmstadt and on the grounds of the TH to clean streets, pavements and rooms and make them usable again. We tried to find some left over sugar beets on the harvested fields at least on one day of the week for syrup production and collected beechnuts in the woods for squeezing out some oil. Occasionally we studied. On the first day, we were dressed in our military coats, gloves and hats, standing in room 201, which was the only room still usable, even though without chairs, tables or any heating, when Prof. Walter arrived and started his first lecture.

,

In spite of this: I was just extremely happy to have survived and to be able to study. After 5 years I received the degree of Dipl-Ing. (Master degree of engineering) and a few years later a PhD degree of engineering. All of this created a feeling of connection and thankfulness to the University of Darmstadt, especially towards the professors at that time. I especially think of the professors Walther, Krischer, Waeschle, Vieweg and Scheubel.

Today I am writing to you as on my travel from the USA to the Black Forest I will pass through Darmstadt during the morning of December 16 and I would be very grateful if I could drop in for a short visit... For a number of years I have been trying to get into contact with some professors of engineering and physics. The reason for this you will find most easily if you look at my website

www.firstgravitymachine.com.

Sadly up to now I have had no success in getting such a contact. My invitations to look at my experiments in the Black Forest or to let me come for a visit or to give a seminar were all declined.

For 10 years I have been performing experiments in Koenigsfeld, in the Black Forest and in Ithaca, New York, trying to measure the temperature distribution in vertical columns of gases and liquids. The results are very surprising and provided they would be confirmed, would necessitate a number of changes in the statements of the Second Law. Now I am searching for a university which would be interested in repeating my experiments. I would be most happy if this would be the University of Darmstadt.

I would like to discuss this with you. I will arrive on Tuesday, December 16, coming from the USA at 6 a.m. in Frankfurt, so a visit with you after 10 a.m. would be possible. Would you have time for such a meeting?

I hope it will work out.

Greetings
Roderich Graeff

Very quickly, only 4 days later, I received a reply:

Saturday, December 6, 2008
From: Prof: Dr.-Ing. R. Anderl
Subject: Your visit at the Technical University in Darmstadt

Dear Dr. Graeff,

Many heartfelt thanks for your email addressed to the president of the Technical University Darmstadt, Prof. Proemel. My name is Reiner Anderl and I am the vice president of the TU Darmstadt, responsible for the subjects Research Exchange and International.

The president, Prof. Proemel, asked me to write to you.

Sadly neither Prof. Proemel nor I can meet you on December 16 as we are travelling on official business. We have a great interest to get to know you. I, myself, am a professor in the Department Mechanical Engineering and responsible for computer assisted design and have read your publication with interest. Even though I am not close to that special subject, I am fascinated by your ideas. I would like to talk to the head of the department of engineering and other colleagues close to this subject. If you agree, I would like to organize a meeting with you and these special colleagues. By the way, I was responsible between 1995 and 2001 for the student exchange between the TU Darmstadt and Cornell University. Sadly this exchange was ended due to financial problems, in spite of the success as the students from Darmstadt received the Outstanding Award for 5 years running.

In case you are in Germany for a little while I would be glad if we could organize such a meeting at another time.

For your information, I am enclosing our last research report and would be very glad if you find in it something of interest to you.

Kind Regards
Yours
Prof. Dr.-Ing. R. Anderl

Great, this sounded like a very positive reply. I answered right away:

Saturday, December 6, 2008

Dear Prof. Rainer Anderl,

I would enjoy coming for a special visit to Darmstadt which is easily possible as I will stay for some weeks in Koenigsfeld. Maybe we can organize around such a visit a seminar in which I describe my experiments and results, followed by a discussion. This way every interested person could form very quickly and economically a picture in his mind to what degree my experimental results could be of importance for your work in Darmstadt. Once my results are confirmed in Darmstadt, which I would expect, Darmstadt could move to the head of their further development.

Your research report is impressive. I found the following headings

.....TU Darmstadt Energy center...., and
...how the microscopic structures of surfaces influence the transport of heat liquids and gasses...., and
...Nano is the name of the key...
All of these have a close connection with my work.

I would be pleased to hear from you soon about possible dates as this would be helpful for my own planning.

With greetings, still coming from Ithaca, which you probably know quite well
,
Yours
Roderich Graeff

I will be in Koenigsfeld after 17th December, phone no. 07725 91191

Christmas passed, the New Year arrived. No answer came from Professor Anderl. Finally I called and reached him around January 15:

"Yes I have not forgotten. I found out that you contacted already Prof. Stephan, who heads the department of Thermodynamics. But today I will meet the Professor of Structural Engineering who is also a specialist in the use of energy in building construction. Maybe he will be interested. You will hear from me in a day or two."

This was bad news. I had tried for the last 6 years unsuccessfully to meet with Prof. Stephan who really would be the right person to work with. He always gave me lots of different reasons why such a meeting could not take place. And I could see no reason why a Professor of Structural Engineering should be interested in my work.

Silence. No news from Prof. Anderl. So I started another, a final try. If this would be unsuccessful I wondered if I should send an open letter to the students of Darmstadt telling them about the reaction of their Professors.

1st March 2009 10:06
Subject: Visit at the TU Darmstadt

Dear Rainer Anderl,

On Tuesday, March 17, I will travel by train to Darmstadt. Would a visit with you in the afternoon be convenient? I have an especially good connection with an arrival time at the station at 2 p.m.

Of course, a PowerPoint introduction into my work or a discussion with students about my studying in the first years after the war and the following life of an engineer would be possible at any time during that day.

I will try to reach you by phone tomorrow.

Greetings
Roderich Gräff

Already 2 days later I received a replay:

March 2, 2009
Subject: Visit at the TU Darmstadt

Dear Mr. Gräff,

Many thanks for your email. I am looking forward to welcoming you on March 17. I propose that we meet at 2.30 p.m. in my office on the campus Lichtwiese, building L1/01, room 10 Building of Engineering). I will also try to invite interested parties and it would be nice if you could give a presentation.

With best Greetings,
Prof. Dr.-Ing. R. Anderl
Vice president

I wondered: Who would come from the „interested parties"? I answered right away:

March 3, 2009
Subject: Visit at the TU Darmstadt

Dear Reiner Anderl,
As proposed by you, I will show up on 17th March. It will surely become an exciting afternoon. In my PowerPoint demonstration, I will report about my 10 year long experiments and results concerning the dispute Loschmidt-Boltzmann. I will also talk a little about my student life in Darmstadt 1945 – 1950 and about the solar energy installations in my factory in Weiterstadt, which led to 8 master degree papers.

In order to demonstrate my work, I will bring along one of my heat bath/ gravity machines which produces electricity without any additional energy supply.

It will be important for everybody present, that we have enough time for discussions. Is it possible to also invite younger assistants, PhD candidates and students of older semesters?

I would be helpful if you could give me an idea about the expected participants and your time plan. I assume that in addition to my memory stick, I won't have to bring my computer. I would like to bring along Dr. Haltmeier who studied with me after the war and after receiving his PhD, was an assistant to Prof. Schreiber for 4 years. He has participated in my experiments during the last years.

See you 17th March, 2.30 p.m.

Greetings
Roderick Gräff

Sadly, no reply arrived. Should I call? I did not dare. I felt it would be too easy for Prof.Anderl to cancel. Then the day before my visit an email arrived. Was it the half expected Cancellation?

Dear Mr. Graeff,

I am pleased to return to your visit tomorrow, Tuesday, 17th March, at 2.30 p.m. Sadly, I cannot join you right away at 2.30 p.m. The reason is that out new minister, Mrs. Kühne-Hörmann,is coming for her fist visit to the TU Darmstadt.

Of course, the presidium has to welcome Madame Minister. This means that I can join you only after the end of the visit of Madame Minister. I assume that this will be around 4:00. or 4.30 p.m. In spite of this, I propose that we proceed with the program as planned.

As meeting point I propose the engineering building, Campus Lichtwiese, building L1/01, room 10. (My office)

In the following I propose this program:

- Welcome through Mrs. Krickow and Mrs. Kuntsch
- Presentation of the TU Darmstadt (Mrs. Krickow and Mrs. Kuntsch)
- Around 3.30 p.m., lecture (Mr. Roderich Graeff)
- Discussion

I would be very happy, if you agree to this plan.

Very sincerely,
Prof. Dr.-Ing. R. Anderl

Great, the visit was still on. I quickly replied - I wonder if he caught my slight sarcasm:

Hello Reiner Anderl,

Many thanks for your news, we will arrive tomorrow at your office at 2.30 p.m. as proposed by you.

Maybe you can bring Madame Minister to our discussion in order to demonstrate how open the TU is concerning new ideas!

Greetings until tomorrow

Roderich Gräff

The next day after a 3 hour train ride from the Black Forrest to Darmstadt, Alexander and I drove towards the Lichtwiese, the location where the extension of the TU Darmstadt following the last war took place. When I was a student in the years 1945 to 1950 in Darmstadt the Lichtwiese was just an empty field covered with grass. On its edge the University sports stadium was located, together with a modern outdoor swimming pool and tennis courts. But these facilities were confiscated by the US army and we students were not allowed to use them. Change came finally 2 years after the end of the war in 1947: Students, the army declared, would be allowed to use the tennis courts. This was great progress, but there was one hitch: We would be allowed on the courts only from 5 to 7 o'clock, not in the afternoon, no, only in the early morning hours. It was hardly light enough for playing at 5 in the morning, but we came anyway, feeling that we demonstrated with this action our protest.

Two ladies in charge of alumni relations welcomed us at the office of Prof Anderl. It was still unclear who would come to my lecture. Shortly we entered a small seminar room and found 15 young men and women, all from the department of Prof. Anderl, the Department for Computer Assisted Design. All were introduced as "wissenschaftliche Mitarbeiter", scientific assistants or lecturers.

There was not one Professor present or anybody from the Departments of Thermodynamics or Physics. I adjusted my talk accordingly. It ended after one hour with a lively discussion covering just the surface of the thermodynamic questions involved. Finally Prof. Anderl arrived. Right away he presented us with copies of a book covering the history of the Department of Mechanical Engineering and thanked us for our coming. But then it happened: When I repeated my wish that somebody in Darmstadt would repeat my experiments he declared that he would find a student for this purpose. Seeing my somewhat skeptical face he repeated his statement, adding, probably the measurements would come to a different result but this would not really matter. Right now it was between semesters, but in 2 weeks he would contact Prof. Hampe in charge of Thermal Process Technology, where these tests really belonged.

In thanking Prof. Anderl I presented him with a graph mounted in a glass frame and showing the energy output of one of my gravity machines. "Please hang it up in your office reminding you of your promise to organize the repetition of my measurements. We have all heard it and will not forget."

So our visit ended on a high note. Prof. Anderl is not a specialist in the field of thermodynamics, but he makes a forceful impression of someone who will work hard to fulfill his promise. I will certainly do everything in my power to make him succeed.

Is this the breakthrough I am hoping and working for? Only the future will tell.

But there was one positive result of our visit right away:
Prof. Anderl introduced me to Prof. Hampe. He heads the department Thermal Process Technology which of course is very close to my subject. He invited Dr. Haltmeier and me to his office. When we arrived for our first meeting, he was very well prepared. He had studied my webpage and the detailed information about one of my tests intensively. He felt that the reported test results were very convincing and for the time being he assumed them to be correct. But the results contradicted the Second Law of Thermodynamics. Therefore there had to be some kind of explanation which would make these results possible within the present laws of Physics. "I keep trying to find an explanation for your test results, but I have not been successful yet." he explained.

Over a period of half a year, we met 3 times. Finally I asked: "Couldn't you give a student the task to repeat one of my experiments? I would be happy to provide you with one of my gravity machines." Prof. Hampe thought for a second and answered "No, I cannot give a task for an experiment to a student as long as I don't know or understand the results!"

So I reached this very positive response by Prof. Hampe who, in spite of his work load which includes frequent trips to foreign countries, involved himself in the question of a temperature gradient in vertical columns. And I have not lost my hope that my former Alma Mater will eventually become really interested and will repeat some of my experiments in the foreseeable future.

But then a really amazing news arrived: In October 2009 Prof. Chuanping Liao, informed me about his paper which one can find under [12]: Chuanping Liao, *Temperature Gradient Caused by Gravitation,* International Journal of Modern Physics B, Vol. 23, No. 22 (2009) 4685-4696,
or at <http://www.worldscinet.com/ijmpb/23/2322/S0217979209052893.html>

Apparently, there existed somebody in this world after all who also tried to measure vertical gradients similar to what I was trying to do. He actually not only quoted my paper given in San Diego [8], but he also described results with a negative temperature gradient.

Prof. Chuanping Liao measured the temperature gradient not only in three meter high, standing columns, but performed also tests on a centrifuge trying to increase the gravitational force. He always finds a negative gradient with a value of -.02 K/meter. This, of course is interesting news, somebody in this world is also measuring and finding a negative gradient in vertical columns of gases and solids. But these results are only meaningful as long as the environment shows a positive gradient, opposite to the negative gradient measured for the inner axle. This question Prof. Chuanping Liao does not discuss in his paper. Therefore more data are needed to fully understand the work of Prof. Chuanping Liao.

So, what do we need? More testing by well established physics laboratories!

Chapter 23
What Do I know in summer 2010 about a Temperature Gradient in Vertical Columns? Who Was Right: Boltzmann or Loschmidt?

Experiments performed during the last 13 years showed me that gravity induces a negative vertical temperature gradient, cold a the top and warm at the bottom in isolated vertical columns of liquids and gases, and maybe also in solids. This demonstrates the transport of heat from cold to warm, allowing the production of work out of a heat bath. It represents a Perpetuum Mobile of the Second Kind. It does not contradict the Second Law of Thermodynamics.

These are daring statements, as they contradict in the mind of physicists today's understanding of a very basic law, the Second Law of Thermodynamics. I allow myself to make these statements because they are not based on one single, specific experiment, one kind or just one calculation, but on a great number and variety of them. I will try to list my results and insights here , known to me after 13 years of experiments and calculations, so you, the reader, can draw your own conclusion about the existence of this strange gradient T(Gr), not based on one or two of these but of all of them. I will list them in the sequence I have described them in the prior chapters and give a short summary for each.

1. Measured values of the temperature gradient T(Gr) for air: Experiment B76

2. Measured values of the temperature gradient T(Gr) for water Experiment B372

3. The gradient in Solids

4. Equilibrium

5. Correctness of measurements: Thermocouple und Thermistor

6. T(Gr) measured at times when Temperature change is zero

7. T(Gr) confirmed through turning experiments

8. Production of electricity out of a heat bath

9. An graphically explanation for T(Gr) (CHAPTER 15)

10. T(Gr) calculated (Chapter 24)

11. Arguments against the existence of a temperature gradient T(Gr)

12. Heat travels from cold to warm (Chapter 24 and 25)

13. Commentaries to T(Gr) (Chapter 22)

In detail:

1. Measured values of the temperature gradient T(Gr) for air: Experiment B76 . (Chapter 16 and Appendix 4)

The experiment B 76 published in [8] and in my web page www.firstgravitymachine.com shows for air filled with glass powder a negative gradient of T(Gr) = - 0.07 K/m, averaged over time, over a period of nine month within an environment with a positive gradient. The use of a convection depressant medium like glass powder is necessary as the measured value of T(Gr) is 5 times greater than the adiabatic lapse rate of -0.014 K/m, the limit for a stable vertical column. Without a convection depressant air currents would develop and negate such a great gradient.

Figure 1 demonstrates the basic setup of the experiment B 76. The temperature gradient T(Gr) is measured in position 5 as the middle axis within the inner Dewar 6.

Figure 1

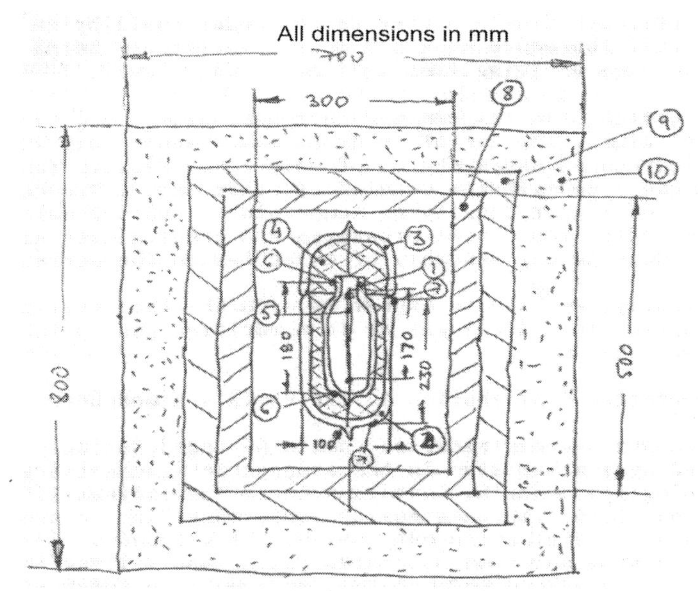

The middle axis within the inner Dewar shows a negative gradient of T(Gr) = -0.07 K/ m, averaged over time – lowest curve - , over a period of 7 month within an environment with a positive gradient of +0.17 K/m - highest curve-.

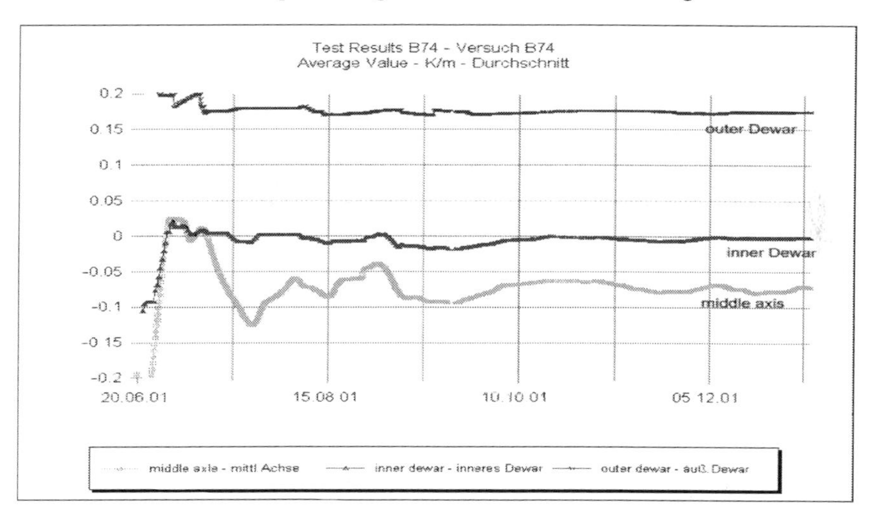

The complete paper can be downloaded from the web page www.firstgravitymachine.com or viewed at Appendix 4

Summary:
Experiment B 76 shows for air with glass powder a negative gradient of -0.07 K/m within a positive gradient in the environment

2. Measured values of the temperature gradient T(Gr) for water Experiment B372. (Chapter 17 and Appendix 5)

The experiment B 372 published and in my web page www.firstgravitymachine.com shows for a water column filled with glass powder a negative gradient of T(Gr) = - 0.05 K/m, as an average value over time, measured over a 3 month period within an environment with a positive gradient. The use of a convection depressant medium like glass powder is necessary as the measured value of T(Gr) is much greater than the adiabatic lapse rate for water of about -0.0001 K/m, the limit for a stable vertical column.

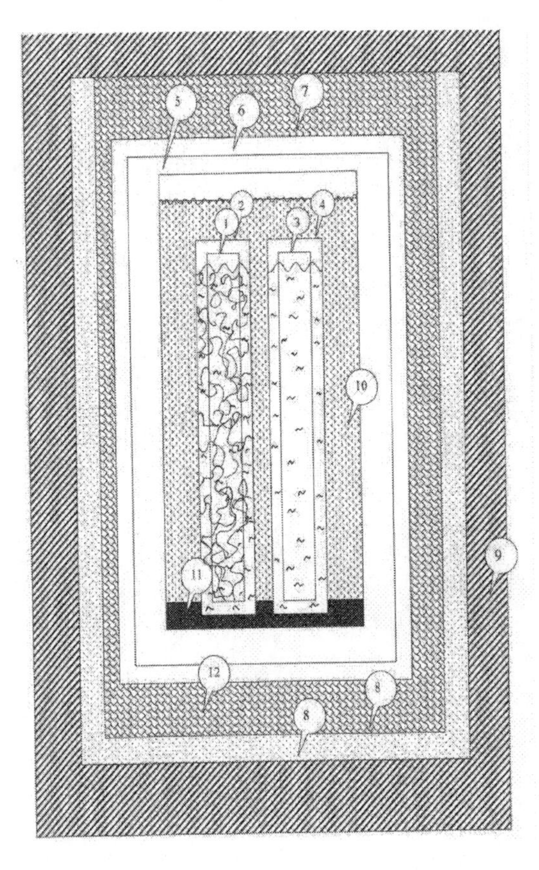

1: Glass tube 1, filled with water and glass powder
L= 850 mm, D= 40 mm

2: PVC tube 1,
L= 910 mm, D= 50 mm

3: Glass tube 2, filled only With water
L= 850 mm, D= 40 mm

4: PVC tube 2,
L= 910 mm, D= 50 mm.

5: PVC tube 125 mm,
L= 1000 mm, D= 125 mm

6: Aluminum tube 150 mm,
L= 1100 mm, D= 150 mm

7: Aluminum tube 220 mm,
L= 1200 mm, D= 220 mm

8: Double wall housing
L= 1500 mm, D= 500 mm

9: Glass fiber insulation 100 mm

10: Glass foam, balls 1 mm

11: Brass shavings

12: PET fibers

Test B 372 measures the vertical temperature gradient in two identical glass tubes of 40 mm diameter and 850 mm length. Each glass tube is individually surrounded by a PVC tube of diameter 50 mm and length 910 mm. Tube 1 **(1)** and its surrounding PVC tube **(2)** are filled with water and fine glass powder, while tube 2 **(3)** and its PVC tube **(4)** only with clear water.
These are arranged in a PVC tube of diameter 125 mm and 1000 mm length **(5).** The remaining space is filled with small balls of glass foam of 1mm diameter **(10)**. The bottom part is filled with small brass shavings **(11)** in order to try to equalize the bottom temperatures of the two 50 mm PVC tubes.

The assembly is inside a 150 mm diameter and 1100 mm long aluminum tube of wall thickness 5 mm **(6)**. This in turn is placed into another aluminum tube of 220 mm diameter, 1200 mm length and a wall thickness of 5 mm **(7)**. Each of these is closed at the top with round aluminum plates of the same thickness.
The aluminum tubes containing the test assembly are standing in the centre of a double walled aluminum housing of height 1500 mm and an inner diameter of 500 mm with 50 mm between the two walls **(8).** This space is filled with water. The whole assembly is insulated on the outside with 100 mm of glass wool **(9).** The space between the larger aluminum tube and the inner aluminum housing, i.e. between **(7)** and **(8),** is filled with fine PET fibres **(12).**

The temperatures inside the test setup are measured by thermocouples and by thermistors. These are mounted at the tops and at the bottoms of the inner axes of the two glass tubes. Additional sensors are mounted on the outside of these glass tubes and on the outside of the two PVC tubes. The temperatures of the double wall aluminum housing are measured 3 cm below the top and above the bottom.

Fig. 3: Temperature gradients curve 1 – 4 and thermistor temperatures 5 and 6

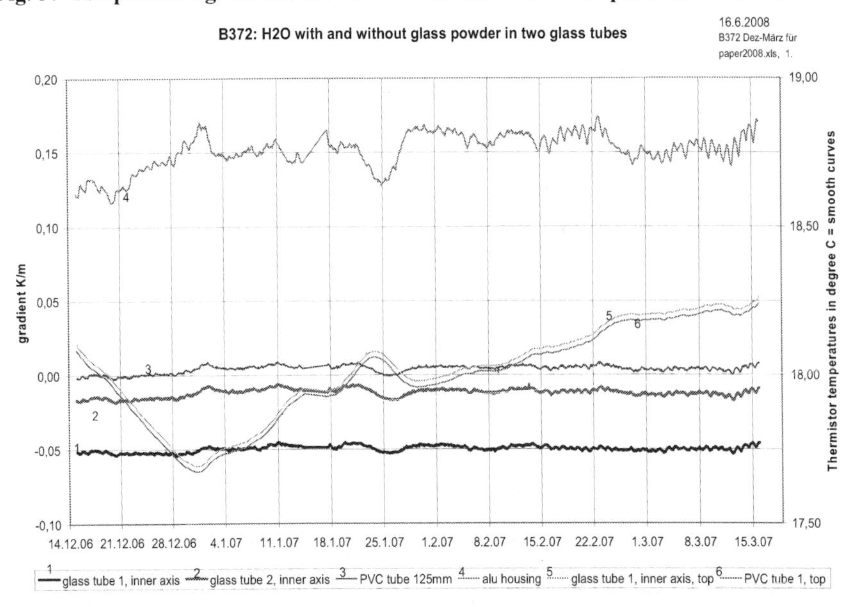

Fig. 3 shows 4 temperature gradients of the 8 values measured by thermocouples as temperature differences from December 14 through March 15. Each point of the curve represents a 10 value average (of a ten times repeated reading of the same sensor) measured every hour, using the scale on the left side of the graph. The smooth lines represent the 2 temperatures measured by thermistors of the 6 values measured hourly in centigrade, using the scale on the right side.

The complete paper can be downloaded from the web page www.firstgravitymachine.com or viewed at Appendix 4.

Summary:
Experiment B 372 shows for water with glass powder a negative gradient of -0.05 K/m within a positive gradient in the environment

3. Solids

Long-term experiments with copper, copper powder, aluminum and especially lead seem to indicate that a negative temperature gradient develops in an isolated vertical column but the values are too close to zero to make a valid argument.

The graph shows Experiment B367 with an Aluminum rod

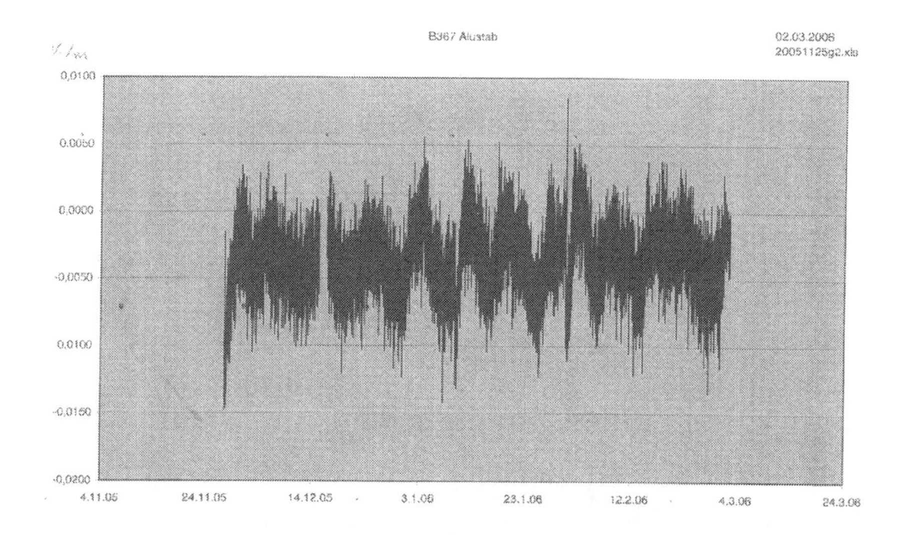

Summary: The T(Gr) average value of -0.003 K/m seem to indicate a negative gradient for solids but are to close to zero to be considered as prove.

4. Equilibrium

A total equilibrium, meaning constant temperatures over time, cannot be reached as the temperatures in the environment always fluctuate to some degree and the best insulation of the experiment cannot avoid some temperature fluctuations within the inner most part of the test setup. But the gradient can be measured at short term equilibrium situations et times when Thermistor temperatures show no heat influx or outflow, as explained under [6]. Secondly, long-term averages show a negative gradient during periods when the initial and final inner most temperatures are equal.

Summary: Under equilibrium conditions, when there is no heat influx or outflow, measurements show a negative gradient, confirming the long term measured averages.

5. Correctness of measurements: Thermocouple und Thermistor

We are measuring temperature differences as we are primarily interested in the gradient, not in absolute temperatures. And with a gradient in most cases the direction of the gradient is more important than its absolute value. The direction is easily discernable when using a thermocouple.

In addition, on many occasions we measure the temperature gradients by 2 independent methods, namely with thermocouples, measuring voltages, and with thermistors, measuring resistance values of a sensor. This insures to quite a degree that we are really measuring temperatures and not some other electrically induced value and, provided we get the same results measured by these two independent methods, that the measured results and the signs are correct.

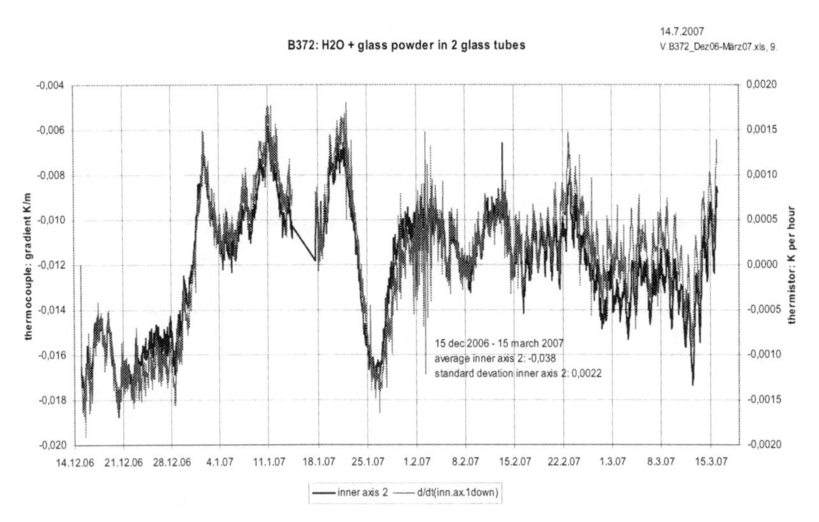

This shows the gradient of the inner axle experiment B372 measured by thermocouples with the scale on the left side comparing it with the gradient calculated as differences of two thermistor measurements with the scale on the right side. Over a period of 4 months the results are practical identical up to a few ten thousands of a degree K.

Summary: Using two independent measuring systems for measuring the gradient and still getting practically identical results prove that our curves are really showing temperatures and that the signs are determined correctly.

6. T(Gr) measured at times when temperature changes are zero (Chapter 17 and Appendix 5)

The measured values fluctuate around a long-term average as the effects of the temperature variations in the environment cannot be eliminated even with an optimal thermal insulation. But if there does exists a vertical gradient it should show with its correct value at those times whenever there is no influx or outflow of heat into or out of the innermost experimental space. This can be demonstrated by graphing all measured values of the gradient as a function of the temperature changes over time of the innermost vertical axis.

Trend lines are calculated as least squares regression lines for the scattered values. The trend line for the blue, lower triangles for inner tube 1 (water with glass powder) show an intercept with the vertical zero line, where the rate of temperature change is 0, at -0.05 K/m. The red, upper markers give a T(Gr) value of -0.12 K/m for inner tube 2 (only water). Both of these values agree well with the long term average values, seen on curves 1 and 2 in Figure 3. The error bars correspond to +/- 1 SDs for all measured values.

Figure 4:

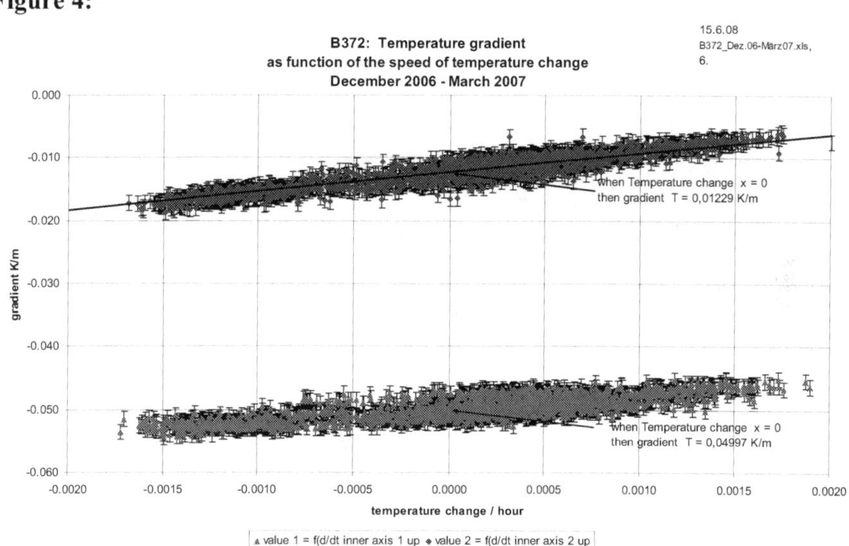

Trend lines are calculated as least squares regression lines for the scattered values. The trend line for the blue triangles (the colours seen only in the printout of the webpage.) for inner tube 1 (water with glass powder) show an intercept with the vertical zero line, where the rate of temperature change is zero, at -0.05 K/m. The red, upper markers give a T(Gr) value of -0.12 K/m for inner tube 2 (only water). Both of these values agree well with the long term average values, seen on curves 1 and 2 in Figure 3. The error bars correspond to +/- 1 SDs for all measured values.

The complete paper can be downloaded from the web page www.firstgravitymachine.com or viewed at Appendix 4.

Summary: The values of T(Gr) at times of no temperature change can be taken as an independent confirmation of the existence of a temperature gradient.

7. T(Gr) confirmed through turning experiments (Chapter 19)

This especially build turning drum allows to install an experiment in the inner drum which keeps standing still while the outer drum is rotating once every minute. This creates a very equal temperature gradient in the inner non-rotating aluminum drum. (The insulation around the outer drum is not shown). This setup allows turning the experiments installed in the inner drum by 180° on their head without disturbing the outer insulation.

The following graph of turning experiment V76 as described under [8] shows the average gradient after being turned on his head 9 times by 180°, first from the normal position onto his head and after 40 hours back into his normal position.

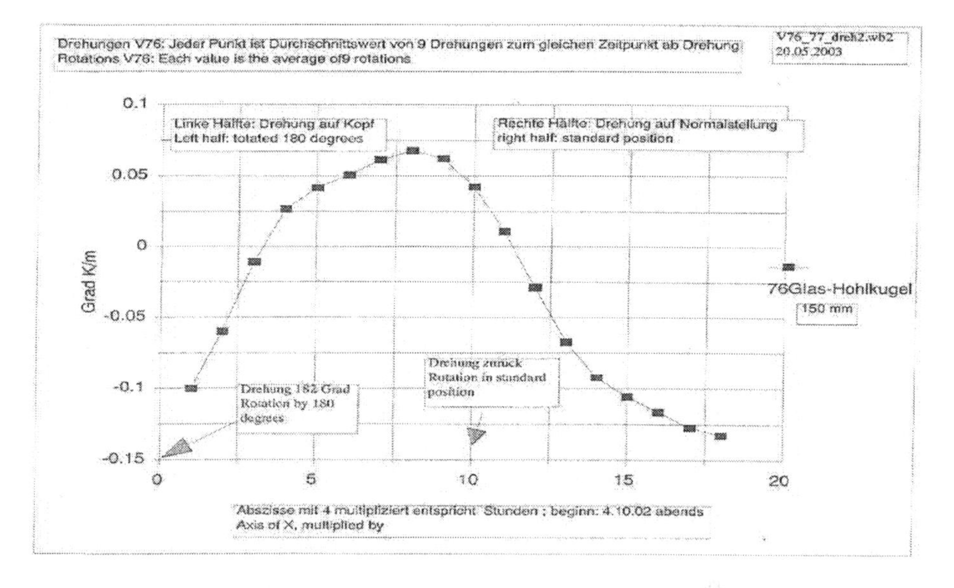

As the graph shows it takes about 30 hours after the turning for the gradient to re-establish itself, cold at the top and warm at the bottom.

Summary**:** Turning experiments confirm that a negative temperature gradient develops in vertical columns of air under the influence of gravity.

8. Production of electricity out of a heat bath (Chapter 26)

The following graph shows the power of a gravity machine B411. The setup is very similar to B76 described under [8]. The gradient of its inner axle is measured by a thermopile consisting of 5 pairs of thermocouples, which in turn is connected to a 100 Ohm resistor.

The two upper curves show the measured values of voltage and current (left scale in Volt and in Ampere/100). The lower curve is the calculated power output (V x A) in Watt shown on the right scale.

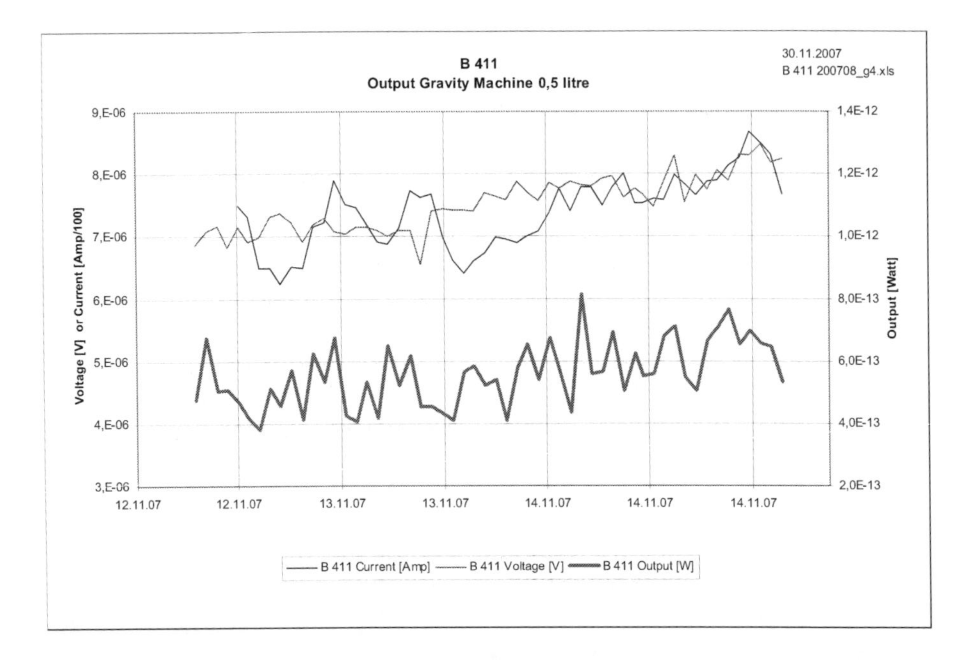

Summary: The graph represents the production of electricity out of a heat bath without the addition of outside energy

9. An graphically explanation for T(Gr) (Chapter 15)

A graphical interpretation explains how T(Gr) can have values larger than based on a calculation of the potential energy of a molecule over the same heigt.

Summary: The measured values of T(Gr) can be interpreted by a different definition of temperature for molecules in a gravity field.

10. The temperature gradient T(Gr) calculated (Chapter 24)

As discussed in detail in Chapter 24, the vertical temperature gradients T(Gr) can be calculated as

$$T_{Gr} = \frac{-g \cdot H}{c_{Gr}} = \frac{-g \cdot H \cdot n}{c}$$

with c = specific heat; n = number of degrees of freedom, H = height

With this formula for a height of 1 meter and taking the number of degrees of freedom for water as 18 and for air as 5, we obtain

for water T(Gr) = -0.04 K/m

for air T(Gr) = -0.07 K/m

These values correspond well with measured values published by the author in [5] and [8, 9] as discussed in Chapters 16 and 17. This closeness of measured with calculated values are another strong indication that temperature gradients are being induced by gravity in isolated columns of gases and liquids.

Summary: Calculations with the proposed algorithm results in values for T(Gr) whch are very close to the measured values.

11. Arguments against the existence of a temperature gradient T(Gr)

Historically, the major argument against a temperature gradient is a theoretical one. Boltzmann [1] published a number of papers in the 19teenth century demonstrating equal temperatures over height under the influence of gravity. But he calculated this only for ideal gases. As a theoretical physicist explained to me 3 month ago: "We cannot calculate real gases, therefore we, as theoretical thysicists, are not interested in this question."

It is of limited value to compare theoretical arguments valid for ideal gases with experimental results performed with real gases. Only experimental evidence can finally decide the question if a gradient exists in real gases.

A second argument, used by Maxwell [2] states: A temperature gradient would allow the production of work out of a heat bath. This would contradict the Second Law, therefore there cannot exist a gradient under the influence of gravity

Summary: The arguments against the existence of a gradient are not based on experiments but on theoretical arguments and these are not valid for real gases.

12. Heat travels from cold to warm (Chapter 24 and 25)

Chapter 24 discuses in detail why heat can travel from cold to warm under the influence of gravity, and in chapter 25 I argue, that this does not mean a contradiction of the Second Law.

Summary: The basic meaning of the Recond Law does not forbid the movement of heat from cold to warm.

13. Commentaries to T(Gr) (Chapter 22)

Commentaries from physicists about T(Gr) are very typically:

Your experimental results look very convincing, we cannot point out any mistakes; your theoretical interpretations are interesting, we cannot point out any obvious mistakes!

BUT: *Your results and interpretations cannot be correct because they contravene the Second Law of Thermodynamics, therefor it does not make sense for us to spend more time with this question.*

In 2007, I was invited to give a seminar at the Department of Theoretical Physics at the University of Würzburg, Germany. There was no follow-up.

In 2009, after a seminar at the Department of Engineering of the Technical University Darmstadt, Germany, an ongoing discussion started with considerations for a follow up experiment.

In October 2009 Prof. Chuanping Liao, informed me about his paper which one can find under [12]: Chuanping Liao, *Temperature Gradient Caused by Gravitation,* International Journal of Modern Physics B, Vol. 23, No. 22 (2009) 4685-4696,
or at <http://www.worldscinet.com/ijmpb/23/2322/S0217979209052893.html>

Prof. Chuanping Liao measures the temperature gradient not only in three meter high, standing columns, but performed also tests on a centrifuge trying to increase the gravitational force. He always finds a negative gradient with a value of -.o2 K/meter. This, of course is interesting news, somebody in this world is also measuring and finding a negative gradient in vertical columns of gases and solids. But these results are only meaningful as long as the environment shows a positive gradient, opposite to the negative gradient measured for the inner axle. This question Prof. Chuanping Liao does not discuss in his paper. Therefor more data are needed to fully understand the work of Prof. Chuanping Liao.

Summary: Repetition of my experiments in well established laboratories are urgently needed.

With this overview I tried to provide a list of experiments and theoretical arguments, each of which can be interpreted as pointing to the existence of a negative, vertical temperature gradient T(Gr). But for each one separately it might be possible to find an opposing argument. But in my view, this becomes extremely difficult or impossible if you consider all of them together. That seems to indicate: Loschschmidt was right in his prediction!

Would you like to settle the question, who was right, Boltzmann or Loschmidt? Which laboratory will repeat my experiments?

Chapter 24
How does Heat travel from Cold to Warm?

In chapters 16 and 17 I described my measurements from 2002, which for columns of air and water showed a vertical temperature gradient, cold at the top and warm at the bottom. In my "Pittsburgh experience" I, for the first time, calculated the magnitude of this gradient by converting the potential energy of the "dancing" molecules into a temperature increase based on their specific heat - recounted in Chapter 14. This seemed to me to be the maximum possible and explainable temperature increase. I expected my test results to show a somewhat smaller value, because, there always would be some heat loss to the environment. Certainly higher values could happen, but only for a short time, based on the temperature fluctuations of the environment, and not as a long term average.

The trouble was that I measured considerable bigger long term average values for the temperature gradient than predicted by calculation. The measured values were 5 times higher for air and 18 times larger for water Progressing in my experiments and in my thoughts step by step through the years, I only recently found a solution for this discrepancy: My calculation first written down in my "Pittsburgh experience" was basically correct, but incomplete. Correct was the first part:

"Any vertically falling molecule would fall, between collisions with other molecules under the influence of gravity, with the speed first given by Newton as the free fall speed."

But the insight was missing:

"Under free fall conditions gravity affects only one degree of freedom, namely only the vertical component of the speed."

Calculating with these two insights, I found a temperature increase by the square root of 5 times larger for air and the square root of 18 times larger for water, than previously calculated. These numbers are the respective values of the number of freedoms: 5 for air and 18 for water.

In the following I describe in detail my present understanding, which explains, how heat can travel from cold to warm, that is, from one location with a low temperature to another with a higher one. But, I run right away into a problem: What is the definition of temperature? One would think that there exists a clear cut answer, but as I found out there doesn't.

Looking at textbooks and related references and even websites for a definition of temperature can be very frustrating. Let's consider the following list collected by specialists and start to wonder:

What is Temperature?

-Courtesy of the references cited below but mostly based on the operational definition on the HyperPhysics pages at Georgia State University by R. Nave and the discussion in "Traceable Temperatures" Second Ed. by Nicholas and White- See the discussions and definitions below for our rationale.

HOW DOES HEAT TRAVEL FROM COLD TO WARM?

DEFINITION: (For this website usage):

Temperature: A measure proportional to the average translational kinetic energy associated with the disordered microscopic motion of atoms and molecules. The flow of heat is from a high temperature region toward a lower temperature region.
-Courtesy of the references cited below but mostly based on the operational definition on the HyperPhysics pages at Georgia State University by R. Nave. and the discussion in "Traceable Temperatures" Second Ed. by Nicholas and White-See the discussions and definitions below for our rationale.

Looking at textbooks and related references and even websites for a definition of temperature can be very frustrating. Many talk around the subject and never get to the point or never state a clear definition. Oh, yes, we can find information about the fact that temperature is not heat and what temperature scales are. We know it's related to heat and there is a difference between them. What is it? It's got to be on the Web.....

Bottom line. It is. However it is like finding the needle in a haystack; just like many topics on the web. Search engines do not help much; it takes digging by someone who understands the subject.

A serious group of experts at The American Society for Testing and Materials (ASTM) publishes standards about devices and information on how to calibrate, test and specify temperature sensors. But their standard on Teminology, ASTM E-344, is notably missing a definition for "Temperature"! Surprise!

Is it just us, or have you, too, noticed that there seems to be a real difficulty in defining or agreeing the meaning of the word and concept of "Temperature".

(Bottom line: It is not a simple concept, despite all the work done on it over the past few hundred years. Yes, it can be defined, but not easily. In many cases simplified explanations, not true definitions, are used. See the HyperPhysics Web site at Georgia State University for the clearest and best explained definition we have seen to date. Their definition appears at the bottom of this page.)

Temperature-Do we know what it is?

Here are some samples of definitions that we have found:

1. "Temperature is the degree of 'hotness' of a body: more precisely it is the potential for heat transfer. In our everyday lives, we are aware of different temperatures through the sensation of touch, but how hot or cold something feels is subjective. We can say that the kettle is hotter than the ice-cream, but not by how much. Measurement, on the other hand, must be objective and a thermometer is used."
   National Physical Laboratory, Teddington, Middlesex, UK, TW11 0LW © Crown Copyright 2001. Reproduced by permission of the Controller of HMSO.

2. **"What is Temperature?"**
   "In a qualitative manner, we can describe the temperature of an object as that which determines the sensation of warmth or coldness felt from contact with it. It is easy to demonstrate that when two objects of the same material are placed together (physicists say when they are put in thermal contact), the object with the higher temperature cools while the cooler object becomes warmer until a point is reached after which no more change occurs, and to our senses, they feel the same. When the thermal changes have stopped, we say that the two objects (physicists define them more rigorously as systems) are in thermal equilibrium . We can then define the temperature of the system by saying that the temperature is that quantity which is the same for both systems when they are in thermal equilibrium."
   From The Popular Website "About Temperature" (since 21 Nov 1995).
   "About Temperature" Disponible en espanol

3. Temperature, when measured in Kelvin degrees, is a number that is directly proportional to the average kinetic energy of the molecules in a substance. So, when the molecules of a substance have a small average kinetic energy, then the temperature of the substance is low.

   From The Physics and Mathematics Web Site "Zona Land"
   NOTE: This website also has an interesting page that graphically illustrates the relationship between molecular motion and the temperature of a gas.

4. "Temperature can be defined in macroscopic terms, using concepts of thermodynamics, as an intrinsic property of matter that quantifies the ability of one body to transfer thermal energy (heat) to another body..."Temperature can also be defined on a microscopic scale as proportional to the random kinetic energy of an assemblage of molecules or atoms."

   "Industrial Temperature Measurement" T.W Kerlin and R.L. Shepard, The Instrument

Society of America, *Research Triangle Park, NC, 1982 (ISBN 0-87664-622-4)*

5. **"Temperature: A convenient operational definition of temperature is that it is a measure of the average translational kinetic energy associated with the disordered microscopic motion of atoms and molecules. The flow of heat is from a high temperature region toward a lower temperature region.**

"The details of the relationship to molecular motion are described in kinetic theory. The temperature defined from kinetic theory is called the kinetic temperature. **Temperature is not directly proportional to internal energy since temperature measures only the kinetic energy part of the internal energy, so two objects with the same temperature do not in general have the same internal energy (see water-metal example).** Temperatures are measured in one of the three standard temperature scales (Celsius, Kelvin, and Fahrenheit).

"**A More General View of Temperature:** When a high temperature object is placed in contact with a low temperature object, then energy will flow from the high temperature object to the lower temperature object, and they will approach an equilibrium temperature.

"When the details of this common-sense scenario are examined, it becomes evident that the simple view of temperature embodied in the commonly used kinetic temperature approach has some significant problems.

"The above illustration summarizes the situation when the kinetic temperature gives a reasonable general description of the nature of temperature. For monoatomic gases acting like point masses, a higher temperature simply implies higher average kinetic energy. Faster molecules striking slower ones at the boundary in elastic collisions will increase the velocity of the slower ones and decrease the velocity of the faster ones, transferring energy from the higher temperature to the lower temperature region.

"With time, the molecules in the two regions approach the same average kinetic energy (same temperature) and in this condition of thermal equilibrium there is no longer any net transfer of energy from one object to the other. The concept of temperature is complicated by internal degrees of freedom like molecular rotation and vibration and by the existence of internal interactions in solid materials which can include collective modes.

" The internal motions of molecules affect the specific heats of gases, with diatomic hydrogen being the classic case. Collective modes affect the specific heats of solids, particularly at low temperatures. Complications such as these have led to the adoption of a different approach to the concept of temperature in the study of thermodynamics.

"Schroeder's proposal for a theoretical definition of temperature is:
"* Temperature is a measure of the tendency of an object to spontaneously give up energy to its surroundings. When two objects are in thermal contact, the one that tends to spontaneously lose energy is at the higher temperature."(Thermal Physics, Ch 1.)

"The kinetic temperature for monoatomic ideal gases described above is consistent with this definition of temperature for the simple systems to which it applies. In that case the equilibrium reached is one of maximum entropy, and the rate of approach to that state will be proportional to the difference in temperature between the two parts of the system. Noting that the equilibrium state of a collection of particles will be the state of greatest multiplicity, then **one can define the temperature in terms of that multiplicity (entropy) as follows:**

"Temperature is expressed as the inverse of the rate of change of entropy with internal energy, with volume V and number of particles N held constant. This is certainly not as intuitive as molecular kinetic energy, but in thermodynamic applications it is more reliable and more general."

From the website HyperPhysics at Georgia State Univerity, author R. Nave.

6. "For most materials, temperature can be considered to be a measure of the density of heat in a body."...

"Kelvin was also able to show that this definition (The ratio of heat taken in at a high temperature to the heat taken out at a lower temperature for an ideal, Carnot, heat engine depends on the ratio of a function of the temperatures) leads to an equation for ideal gases of the form

$$PV = constant\ x T$$

so that Kelvin's definition of temperature is equivalent to the the gas scale originally proposed by Amontons, and implemented by Chappuis in 1889.....

"When applied to an ideal box they obtained (by considering a material model based on the movement and collisions of individual atoms in a closed box, Kelvin and others were able to show that thermal equilibrium requires the mean kinetic energy of the atoms to be the same) the result

$$PV = constant \times (mv^2/2)$$

where $(mv^2/2)$ is the average kinetic energy of each atom in the gas. Comparison of this equation with Equation (1.3) (the previous equation) shows that the *temperature* is proportional to the average kinetic energy of each atom...."Note that the total kinetic energy of molecular gases is higher than that for monatomic gases because they can rotate and vibrate; in that case the temperature is proportional to the mean translational kinetic energy."

"Traceable Temperatures" Second Edition, by J. V. Nicholas and D. R. White, John Wiley & Sons, LTD, Chichester, 2001
ISBN 0-471-49291-4 .

SUMMARY:

Temperature can be defined in several ways, many of which are not completely accurate, but the **first definition of Number 5, above, seems to almost fit our understanding and work reasonably well as an "Operational Definition" for the term**. However, the detail provided in the last reference (Number 6) gives a more complete understanding and shows that the word "*proportional*" really belongs in the definition. At least that's the opinion of this website editor and publisher.

.

The summary suggests as the best definition:

5."Temperature: A convenient operational definition of temperature is that it is a measure of the average translational kinetic energy associated with the disordered microscopic motion of atoms and ***molecules. The flow of heat is from a high temperature region toward a lower temperature region.*** ....

.... ***A More General View of Temperature:*** *When a high temperature object is placed in contact with a low temperature object, then energy will flow from the high temperature object to the lower temperature object, and they will approach an equilibrium temperature.* ....

And what is now my trouble with this definition? It does not fit to my experimental results! My results show that because the Earth has a gravitational field, heat can flow here **from a low temperature region toward a higher temperature region. Furthermore, an "equilibrium" condition can exist simultaneously to a temperature gradient over height.**

As a consequence of this, I can define two kinds of temperatures:

1. The "conventional" temperature, measurable through a thermometer and showing this conventional temperature, when all internal energy is equipartitioned, which means, that the energy is distributed equally between all degrees of freedom and correct for an assembly of molecules which are in constant contact and interchange with each other.

2. The second one, which I call "gravitational" temperature, or temperature based

on the effect of gravity T_{gr} , is valid for molecules between collisions. This gravitational temperature is in effect, when the internal energy is not equally distributed between all degrees of freedom. It can be higher than the "conventional" temperature, when a molecule is falling in a gravitational field and the vertical speed component collects more energy, than the rest of the degrees of freedom, or it can be lower, when a molecule moves against a gravitational field and the opposite happens.

In a container filled with a gas we can find both types of temperatures existing at different locations. How does this understanding of temperature affect the temperature gradient existing between the top and the bottom of this container?

Let us assume that our vertical container is isolated very well against the environment and has three sections.

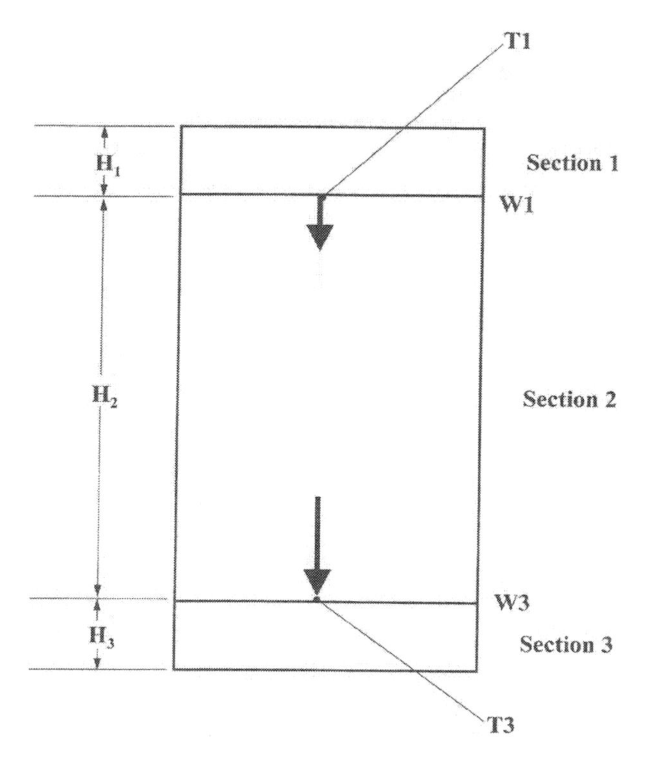

The middle section has a height of H2 and is filled with a Knudsen gas, meaning that the free length of the molecules in it is large in comparison to the height H2.

The molecules, hardly hitting each other, bounce between the walls of the container.

The top section 1 and the bottom section 3 of the container are filled with a dense gas, whose molecules have a Maxwellian speed distribution. The heights H1 and H3 are small compared to H2. There is a good heat transfer between the gas in section 1 and the dividing wall W1 and between the gas in section 3 and dividing wall W3. Therefore, these walls have the same temperatures as the gases in section 1 and section 3.

Presently in physics textbooks it is being taught that under equilibrium conditions, the temperatures in all sections, including all walls, would be the same, based on the theoretical arguments by Boltzmann and on the argument by Maxwell, who declared that these have to be identical based on the understanding of the Second Law. Contrary to this, my measurements show that, even starting out with a condition of equal temperatures, a temperature difference will develop between the top and the bottom. The temperature in the top will decrease and the temperature in the bottom will increase, getting to a new equilibrium condition after a while, with a temperature gradient, which I call T(Gr). This means that energy, in the form of heat, had to travel from cold to warm, from the top section to the bottom section. How can this happen?

Let us assume that the top and the bottom walls, W1 and W3, have rough surfaces. Let us further assume that a molecule hits the top wall and, because of the rough surface, gets thermalized, meaning that its energy becomes distributed equally between all degrees of freedom. We assume that our molecule gets reflected from the upper wall W1 and flies vertically down to the lower wall W3. On its way gravity will accelerate this molecule and its potential energy, mgH, will be changed into kinetic energy. Because the gravitational field creates a force only in the vertical direction, this increase of kinetic energy will show up only in the vertical speed component. Our molecule will impact the lower wall W3 not in an equipartitioned condition, that is, energy will not be distributed equally into all degrees of freedom, as it was, when it started out from the upper wall. There, the temperature was defined as above under 1 and called "conventional" temperature. The molecule arrives at the bottom wall W3 with a temperature defined under 2 as "gravitational" temperature.

For a better understanding, let us compare this "heating up" of a molecule through the acceleration in a gravity field to a heating up of the same molecule through an electrical heating element. In this second case the energy entering the gas would be quickly thermalized by the collisions with other molecules. Through this process the energy introduced would be distributed over all degrees of freedom. In a diatomic gas, like nitrogen or oxygen, having 5 degrees of freedom at room temperature, 2/5 would go to the rotational degrees of freedom and 3/5 to the translational ones. Only these 3/5 influence the "conventional" temperature of the gas, which can be calculated using the specific heat of the gas in question.

When our "vertically accelerated molecule" arrives at the lower wall, it will be caught up in the roughness of the surface and becomes thermalized. It will leave the wall with the temperature of the wall. This means that its vertical speed component will be smaller, when it leaves than what it was, when it had arrived at the wall. So it delivered energy to the wall, heating it slightly, thus energy was transferred from the upper to the lower wall. This means: Heat was transferred from a colder to a warmer space, i.e. from a wall with a lower to a wall with a higher temperature.

On the way up the opposite takes place. Our molecule, leaving with a speed corresponding to the temperature of the lower wall and being thermalized, will arrive at the upper wall decelerated by the effect of gravity. Gravity again will have affected the vertical speed component by reducing its magnitude. It arrives at the upper wall W1 with a "gravitational temperature" lower than the "conventional temperature" of the wall. After impinging and getting thermalized in the rough surface of the wall, it will bounce back from it having received some energy from the upper wall in this thermalizing process. Thus, the molecule has cooled the upper wall.

In the walls W1 and W2 the molecules follow a Maxwell-Boltzmann distribution which also means, that their energy is distributed equally over all degrees of freedom. An equilibrium condition will have been reached, when the "conventional" temperatures in the upper and lower walls will be equal to the "gravitational" temperatures of the arriving molecules. At this equilibrium, the temperature of W1 is lower than that of W3.

In Boltzmann's argument for equal temperatures over height, those molecules are of special importance, which leave the lower wall with such a low speed that they return to it, before reaching the upper wall, or another gas molecule. His arguments for equal temperatures are based on statistical considerations and on the calculation of local temperatures. My argument is based on the energy transport from the top to the bottom. Here these "returning" molecules have no influence on the energy transport, because they return to the bottom wall with the same speed, with which they have left. Therefore, no energy transport to or away from the lower wall takes place through them. I don't have to consider these, when calculating the temperature gradient based on energy transfer.

We can calculate the temperature gradient based on this energy transfer as follows:

The potential energy is

$$E_p = -M \cdot g \cdot H$$

with M = mass; g = constant of gravitation; H = height difference

(Negative, because g and H are measured in opposite directions)

We equate this potential energy E_p with the amount of energy available for a temperature increase of this mass

$$E_{avail} = M \cdot c_{Gr} \cdot T$$

with c_{Gr} = effective specific heat; T = Temperature increase

We now can equate E_p wit E_{avail} or

$$E_p = E_{avail} = M \cdot g \cdot H = M \cdot c_{Gr} \cdot T$$

or

$$T = \frac{g \cdot H}{c_{Gr}} = T_{Gr}$$

c_{Gr} is not the normal specific heat of the gas or liquid in question, because the acceleration through g affects only the vertical speed component of the molecule.

The potential energy of a molecule is converted only into an increase of its speed in its downward direction. No energy is added to the other degrees of freedom, like the remaining two lateral directions (left to right and front to back) or to the rotational energy in molecules that consist of several atoms. Therefore,

$$c_{Gr} = \frac{c}{n}$$

with c = specific heat; n = number of degrees of freedom

We therefore get

$$T_{Gr} = \frac{-g \cdot H}{c_{Gr}} = \frac{-g \cdot H \cdot n}{c}$$

With this formula for a height of 1 meter and taking the number of degrees of freedom for water as 18, we obtain

T_{Gr} = -0.04 K/m for water

and taking the number of degrees of freedom for air as 5, we obtain

T_{Gr} = -0.07 K/m for air.

Another way for this calculation starts with the thought that a molecule will accelerate on its downward path in accordance with the law for free fall, first formulated by Newton:

From the equation $mv^2/2$ = mgH we obtain for the velocity:

$$v = \sqrt{2 \cdot g \cdot H} \text{ or } v = \sqrt{2 \cdot 9.81 \cdot 1} = v = \sqrt{19.62} = 4.43 \text{ m/sec}$$

The gravitational temperature would then be for air with

Boltzmann constant k = 1.38×10^{-23} J/K

Molecular weight $M = 29$ kg/Mol

Avogadro's Number for (Mol) Kilomol $N= 6.02 \times 10^{26}$

$T(Gr) = mv^2 / 2/(k/2) = 29/(6.02 \times 10^{26}) * 4.43^2/(1.38 \times 10^{-23})=-0.07$ K/m

And for water with
Molecular weight $M = 18$ kg/Mol

$T(Gr) = mv^2 / 2/(k/2) =18/(6.02 \times 10^{26})* 4.43^2/(1.38 \times 10^{-23})=-0.04$ K/m

These are the same results as calculated above.

The significance for me lies in the fact that I found through my experiments throughout the years practically identical results as published in [8] and [9] in 2002 and 2003, for two such different molecules, for air and water.

Chapter 25
Perpetuum Mobiles of the Second Kind Are Possible, Consequences for the Second Law

The great physicist Sir Arthur Stanley Eddington declared in [30] *The Nature of the Physical World* , in 1927 , – it probably was on the day of my birth, on October 13, - :

The law that entropy always increases-the Second Law of thermodynamics-holds, I think, the supreme position among the laws of Nature. If someone points out to you that your pet theory of the universe is in disagreement with Maxwell's equations-then so much the worse for Maxwell's equations. If it is found to be contradicted by observation-well, these experimentalists do bungle things sometimes. But if your theory is found to be against the Second Law of thermodynamics I can give you no hope; there is nothing for it but to collapse in deepest humiliation. [30]

Well, Sir Eddigton, I don't worry that much about my day in humiliation, as even if my experimental results or their interpretation turn out to be correct or not, I never claimed in this book hat they contravene the Second Law of Thermodynamics. But I do claim, provided they are correct, that they contravene the presently taught and accepted statements of the Second Law. Let me explain:

Yes, the Second Law is sacrosanct, it has been always correct, it is always correct, it will be always correct! And yes, the Second Law of Thermodynamics forbids my experimental results.

So what is the basic meaning of the Second Law, which is so important? The trouble is that there is not one statement of the Second Law which all encompassing would explain it, not like it is with the First Law of Thermodynamics which can be expressed in one sentence like Claudius pronounced in 1850:

"In any process energy may be changed from one form to another – including heat and work -, but newer be produced or annihilated."

This statement is pretty much universally believed as correct. Already in 1775 the French Academy has forbidden consideration of any process or apparatus that purports to produce energy ex nihilo, a perpetuum mobile of the first kind.

But it is different with the Second Law, as there is not one statement describing all aspects of it, but a multitude. As the philosopher and physicist P.W. Bridgman describes the situation:

…………There are almost as many formulations of the Second Law as there have been discussions of it…………

Sheehan describes it very similar in discussing the difference between the First. and the Second Law in [10]:

The Second Law of thermodynamics was first enunciated by Clausius (1850) and Kelvin (1851), largely based on the work of Carnot 25 years earlier. Once established, it settled in and multiplied wantonly; the Second Law has more common formulations than any other physical law. Most make use of one or more of the following terms – entropy, heat, work, temperature, equilibrium, perpetuum mobile – but none employs all, and some employ none. Not all formulations are equivalent, such that to satisfy one is not necessarily to satisfy another. Some versions overlap, while others appear to be entirely distinct laws. Perhaps this is what inspired Truesdell to write, "Every physicist knows exactly what the first and Second Laws mean, but it is my experience that no two physicists agree on them."

Despite – or perhaps because of – its fundamental importance, no single formulation has risen to dominance. This is a reflection of its many facets and applications, its protean nature, its colorful and confused history, but also its many unresolved foundational issues........

Sheehan then continuous describing 21 formulations, the first 3 ones being:

The first explicit and most widely cited from is due to Kelvin [32][33]

***Kelvin-Planck** No device, operating in a cycle, can produce the sole effect of extraction a quantity of heat from a heat reservoir and the performance of an equal quantity of work.*

In this, its most primordial form, the Second Law is an injunction against perpetuum mobile of the second type (PM2). Such a device would transform heat from a heat bath into useful work, in principle, indefinitely. It formalizes the reasoning under-girding Carnot's theorem, proposed over 25 years earlier.

The second most cited version, and perhaps the most natural and experientially obvious, is due to Clausius (1854) [38]:

***Clausius-Heat** No process is possible for which the sole effect is that heat flows from a reservoir at a given temperature to a reservoir at higher temperature.*

In the vernacular: Heat flows from hot to cold. In contradistinction to some formulations that follow, these two statements make claims about strictly nonequilibrium systems; as such, they cannot be considered equivalent to later equilibrium formulations. Also, both versions turn on the key term, sole effect, which specifies that the heat flow must not be aided by external agents or processes. Thus, for example, heat pumps and refrigerators, which do transfer heat from a cold reservoir to a hot reservoir, do so without violation the Second Law since they require work input from an external source that inevitably satisfies the law.
Other common (and equivalent) statements to these two include:

***(3) Perpetual Motion** Perpetuum mobile of the second type are impossible.*

As I described in this book, my experimental results contradict all of these statements, do they therefore contradict he Second Law? My experiments have shown:

…a device is possible which extract heat from a heat reservoir and produces an equal amount of work….,

...heat can flow from cold to warm… and

...Perpetuum Mobiles of the Second Type are possible…..

So to all of those people who point out this contradiction, my answer is: My experimental results do not contradict The Second Law, they are in agreement with it! But I have to admit, they are not in agreement with practically all established statements of The Second Law. But we can resolve this contradiction:

Everybody agrees The Second Law is correct only for isolated systems. Therefore the above statements by Kelvin-Planck, Clausius and about the impossibility of a perpetuum mobile of the second kind are also correct only for isolated systems. “Isolated” means here, that no transfer of energy or mass across the boundary around the system takes place. But on earth we cannot exclude the effect of the gravitational force. This gravitational force inserts and exerts energy in and out of my experimental setups. So in discussing the influence of my results onto the Second Law we have to view them as part of an isolated system. We have to visualize our boundary to include the mass of the earth which is the source of the gravitational force affecting my experiments.

The combination of a heat bath with a vertical column of gas, liquid or solid material with the effect of gravity makes possible the movement of heat from cold to warm, the extraction of work out of a heat bath. This contradicts commonly used and taught statements of the meaning of the Second Law. So I believe, we need a new statement of the Second Law, but we have to make sure that any such new statement of the Second Law makes clear, that it is concerned only with isolated systems.

So I feel that this could be a new general statement of the Second Law:
In isolated systems – with no exchange of matter and energy across their boundaries AND WITH NO EXPOSURE TO FORCE FIELDS - initial

differences of temperature, densities, and concentrations in assemblies of molecules will disappear over time, resulting in an increase of entropy.

Conversely:
In isolated systems - with no exchange of matter and energy across its borders - FORCE FIELDS LIKE GRAVITY can generate in macroscopic assemblies of molecules temperature, density, and concentration gradients. The temperature differences may be used to generate work, resulting in a decrease of entropy.

Of course, these new statements live or die with the correctness of my experimental results and their interpretations. So where are the physicists and engineers who will study and repeat my experiments in order to contradict or confirm them?

Chapter 26
From the Maxwell Demon to a T(Gr) Fairy: The Production of Energy out of a Heat Bath, Gravity Machines

In the middle of the 19th century thermodynamics was an important topic for the physicists and chemists of the time. More and more they became convinced that the atmosphere was filled with billions of gas molecules, moving around in a Brownian (random) motion, their speed being an indication of their internal energy and of the temperature of the gas. So the total of their internal energy represented a huge amount of energy. Transferring this idea to the water in an ocean, the question arose, whether we could use this energy, for instance, to drive an ocean liner across the seas?

The answer was "No, we couldn't"! The Second Law of Thermodynamics, which was formulated during those years, forbade it. In order to produce work out of a heat bath, one always needed a temperature difference. And thinking about it, Maxwell and others invented the idea of a "Maxwell Demon". This "Demon" was somehow able to measure the speed of individual molecules in a heat bath and to separate the fast moving ones from the slow ones, shown here in a figure from Wikipedia:

Maxwell devised a thought experiment as a way of furthering understanding of the second law. He described the experiment as follows [39]:

... if we conceive of a being whose faculties are so sharpened that he can follow every molecule in its course, such a being, whose attributes are as essentially finite as our own, would be able to do what is impossible to us. For we have seen that molecules in a vessel full of air at uniform temperature are moving with velocities by no means uniform, though the mean velocity of any great number of them, arbitrarily selected, is almost exactly uniform. Now let us suppose that such a vessel is divided into two portions, A and B, by a division in which there is a small hole, and that a being, who can see the individual molecules, opens and closes this hole, so as to allow only the swifter molecules to pass from A to B, and only the slower molecules to pass from B to A. He will thus, without expenditure of work, raise the temperature of B and lower that of A, in contradiction to the second law of thermodynamics....

Xin Yong Fu describes the Maxwell Demon very nicely in his paper "Realization of Maxwell's Hypothesis, An Experiment against the Second Law of Thermodynamics" [37]. In the following two figures, the Demon performs two different tasks. In the left picture he is separating the slow moving molecules from the fast moving ones into compartments A and B, thereby creating a temperature difference. On the right picture he is collecting the majority of the molecules into compartment B, thereby creating a pressure difference between A and B.

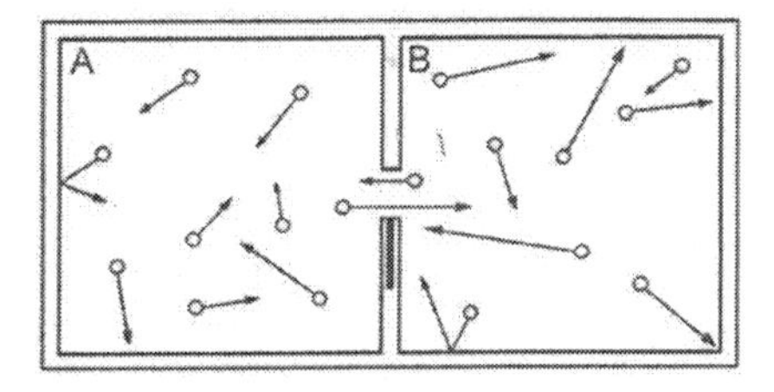

(a) By the first method, demon makes An inequality in temperature

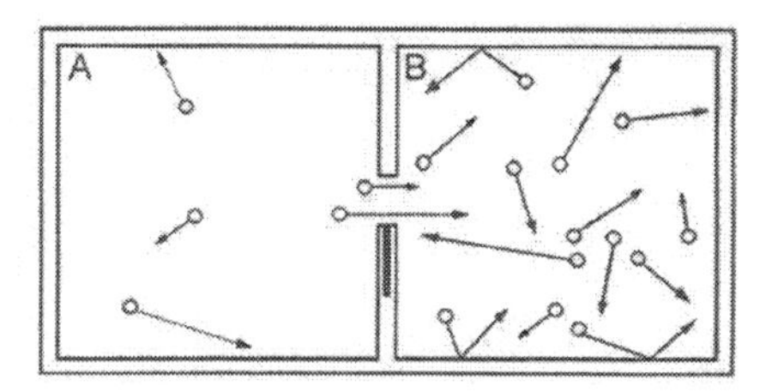

(b) By the second method, demon makes an inequality in pressure

The trouble is that up to now, nobody has found such a "Demon". This Demon would have to measure the speed of individual molecules arriving at his dividing flapper, make a decision as to what side of the dividing wall it should be placed, and then open his gate at the right moment to let this molecule pass from one side to the other. Slowly the fast molecules would be separated from the slow ones, producing compartments with differences in temperature or in pressure. These differences could then be used to produce work, like driving our steamer across the ocean. Again, such a demon hasn't been found. One reason is that even if one could think of such a contraption, it probably would need more energy than it could create through such a separation process.

The results of my experiments during the last 13 years could change this. I didn't find a demon, but I realized that gravity has the effect of a demon. In a vertical column of gas or liquid, gravity makes sure that on the top the average speed of the molecules is lower than at the bottom. This is demonstrated in the following two pictures. On the left, we have a large heat bath with a narrow height, where in all parts the average speed of the molecules is identical. On the right side, the container is a vertical column of gas or liquid, where the effect of gravity had created a vertical temperature gradient. This gradient can be used to produce electricity through a thermocouple, connecting the top and the bottom. When we generate electricity out of this container, then the temperature in it will decline. By making a heat-transmitting connection between the heat bath and this vertical column, we can raise the temperature of the gas or the liquid in the column again, up to the temperature of the heat bath. This arrangement represents the model for a continuing production of electricity out of a heat bath.

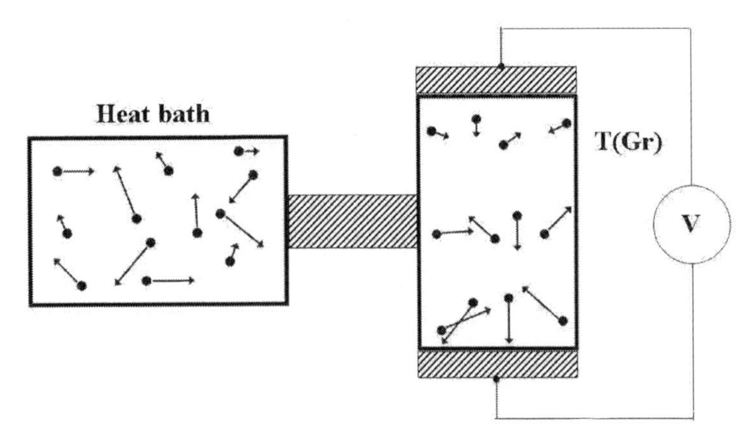

Production of electricity out of a heat bath

This is not just a theoretical option! This production of electricity really works!

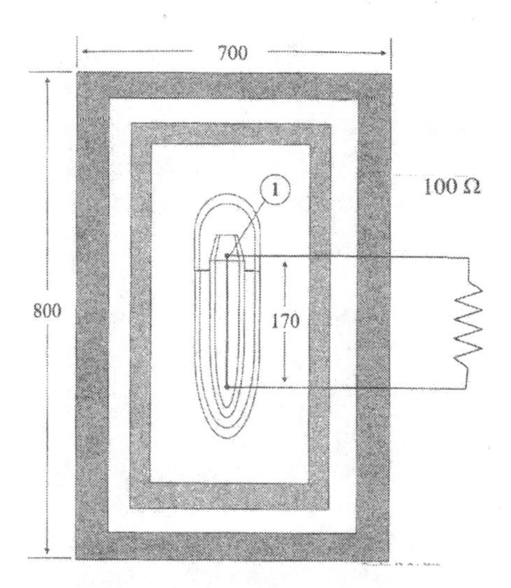

The above picture shows the sketch of a typical setup that I call "Gravity Machine". In the inner one of two nested Dewar flasks, filled with glass powder and air, a thermopile (1) is mounted along the vertical axis. This arrangement is heavily insulated on the outside, in order to reduce the influence of the temperature from the outside environment. Under the influence of gravity, a negative vertical temperature gradient of about -0.07 K/m develops in the air within the inner Dewar flask, very similar to experiment B76, described in Chapter 16 and in Appendix 4.

The thermopile is connected to a 100 Ohm resistor. As shown in the following graph we measured the current through the resistor and the voltage of the thermopile continuously for 2 days in November of 2007, the values being shown on the left side of the graph.

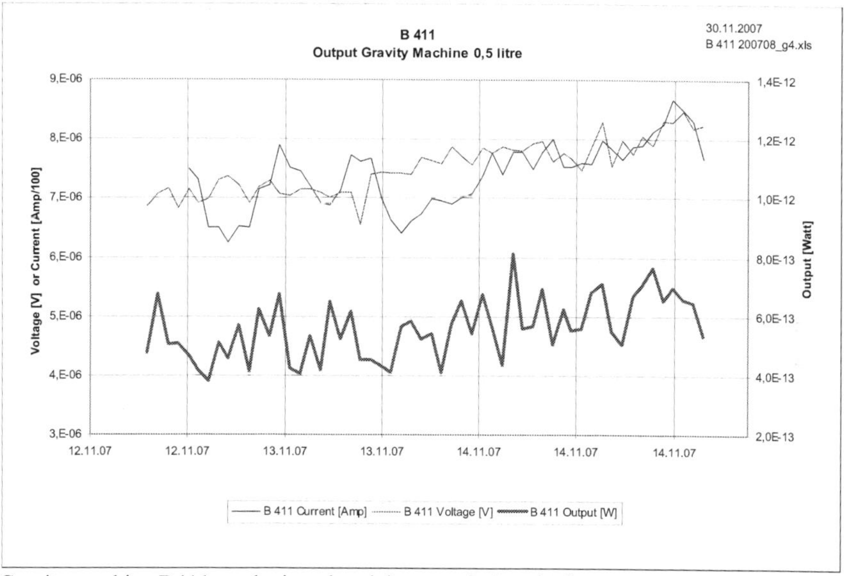

Gravity machine B411 producing electricity out of a heat bath.

The wattage, obtained as Ampere times Volt, is shown on the right hand scale. As can be seen, the output of this machine, around 6 times 10^{-13} W, is extremely small. Nevertheless, it is still very meaningful, because this power has been produced without the addition of outside energy, based on a negative temperature gradient generated by the influence of gravity. Furthermore, all of this took place within an outside environment having a positive temperature gradient.

This gravity machine has only a working volume of 0.5 liter and is, therefore, too small to generate enough power to drive a machine or an electric light. But it demonstrates the principle that heat can flow under the influence of gravity from a cold reservoir to one with a higher temperature and produce work.
The following setup might be even more impressive or useful in convincing the skeptics: I arranged two identical containers to stand next to each other on the

same thick metal plate, representing a heat bath. These containers are filled with different gases or liquids, which, due to their specific heats and numbers of degrees of freedoms, have each a different temperature gradient T(Gr). While the temperatures are equal at the bottom, they are different at the top. Now we can connect these tops through a thermo-element or, for instance, a Stirling engine, and so produce work.

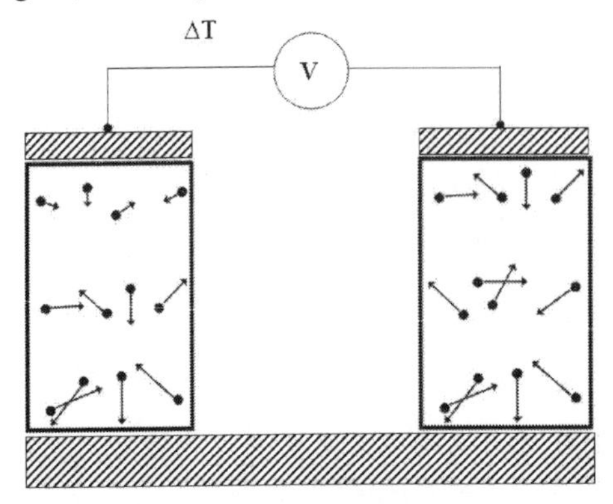

In order to motivate others to repeat my experiments, we now offer seminars, where we explain the building of these electricity-producing Gravity Machines in all details. The participants can purchase all parts necessary for themselves to put together such a machine, or take one along that is already fully assembled.

The dream of Loschmidt has become reality.

Gravity Machine type 1 - 2010

EPILOGUE
The Future of Peaceful Energy

Today, another birthday. It is a good day to finish this book, writing an epilogue, 13 years after I started with measurements to get clarity about the Second Law of Thermodynamics. Did I really cover the topics I wanted to write about, the Second Law, Non-Violence, Circability, as I indicated in my Introduction?

I would like to finish with a happy thought. How about this one:

Main article in Wikipedia: Heat death of the universe

According to the mathematical statement of the second law, the entropy of an isolated system never decreases. Since the whole universe is obviously isolated, its entropy should increase until it runs into a state of no thermodynamic free energy to sustain motion or life, that is, the heat death.

Even so this would happen only in a few billion years, there are people who worry about it. But you are one of the few, who, after reading my book, know: My Gravity Machines perform the opposite feat, they reduce Entropy. So you can relax the universe will continue to exist in a few billion years.

But I am not as sure about the very short tern outlook as it concerns the survivability of us humans on this earth: Maybe I should have chosen a different title for this book and called it

Is it too late?

Are we past a tipping point for the survival of humankind? I just changed the first picture in the book showing me as a three year old, playing safely and happily in our fenced-in yard. I added two more boys, because today, eighty years after that picture was taken, 3 times as many people are living on this earth, competing for clean water, food, and energy. Is it clear to us that we, the 20% living in the Western world, are already using the resources of 1.3 earths -- calculated from the point of view of sustainability -- and thereby leaving nothing for the other 80% of humankind? From what are the other 80% of humanity supposed to live from?

What is the Future of Peaceful Energy for our world ?

A week ago Israel announced the purchase of 20 stealth bombers for US $2 billion. The majority of its citizens believe, I suppose, that their country needs them to survive in a hostile world, to prevent a second holocaust. Three weeks ago the United States proudly announced the biggest single order ever for military goods. Saudi Arabia, not considered to be a democracy, is purchasing US $60 billion worth of military airplanes, drones, and all kinds of other military hardware. At best during the next 30 years these goods will only be used for training and exercises, and never for the killing of other humans, before they are finally thrown onto a scrap pile. They will be paid for by shipments of oil, oil which pollutes the soil when it is extracted, pollutes the oceans when it is transported and pollutes the air when it is finally used. And we allow all of this to happen, knowing that our children will be overburdened with tremendous problems created by our present day actions!

How can we, how can I, realizing and understanding these actions, allow them to happen, to continue to happen, to slowly overwhelm us? How can I accept that 6 year old Helena writes under the picture she clipped from a newspaper for her mother:

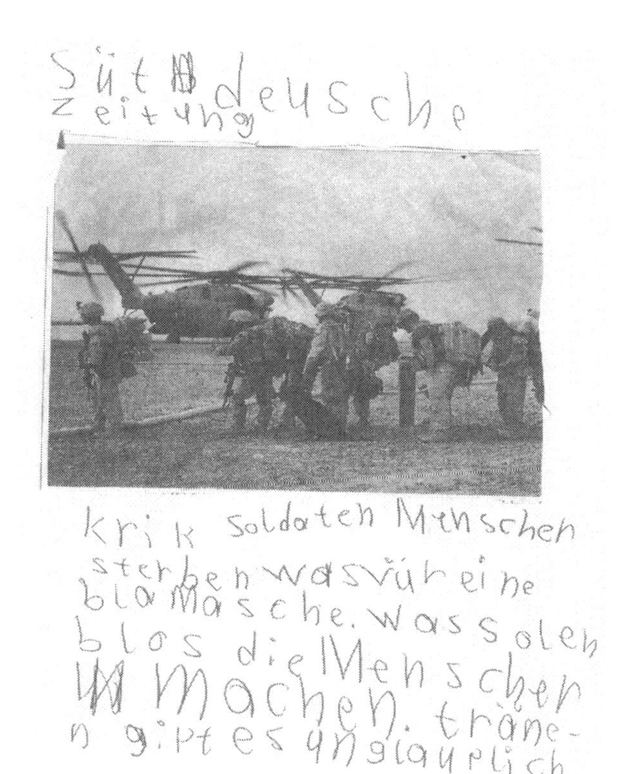

War. Soldiers, human beings die.
What a disgrace.
What on earth should the people do.
Tears there are unbelievable

My answer to Helena is a change of a main emphasis in my Path to Peaceful Energy: I will work for a conversion of the German Armed Forces into a Peace Corps.

I will need lots of encouragement and assistance!

Will you help me?

Let me close with a small story from the life of Hans Scholl [29]. I dedicated this book to him and to his group around the White Rose. It summarizes the message I learned in small steps throughout my life and which I have tried to address and describe in a number of ways.

When the war began Hans Scholl did not have to become a soldier because he was studying medicine. But the day came when he was drafted into the medical corps. On his way to Russia his train, filled with soldiers, stopped in the middle of Poland. Looking out of a window he saw a group of women working hard on maintaining the roadbed, all of them looking quite haggard and exhausted, wearing the yellow Star of David on their chest. He jumped out of the compartment window and approached the first one, a girl, maybe 18 years old. He wanted to give her a present and the only thing available to him at that moment was his emergency ration, a small package with some chocolate, nuts and raisins. Smiling he gave it to her. She took it, but with a proud look threw it down at his feet.

"I only wanted to give you a present," he explained.

He noticed a wild flower growing between the tracks. He picked it, attached it to the parcel, and placed the present at her feet. At that moment the train started to move, he ran alongside and jumped on. Leaning out of his window, he saw the girl. She was watching the departing train, having put the flower into her hair.

Isn't that a wonderful way to demonstrate peaceful energy? Is there still time for us to wake up and to learn to live this way, non-violently, allowing the life of humans to continue on this earth?

APPENDICES

My Path to Peaceful Energy
Appendix 1

I'm Not Ready Yet

Play in one act
by
Roderich W. Graeff

Madison, Wisconsin, USA
1952, during the Korean war

This play takes place between two countries -- the WE'S and the OTHER'S. The WE'S occupy the left half of the stage with a radio studio in the back, and the THER'S occupy the right half of the stage with a similar setting. A barbed-wire fence down the center symbolizes the border between the two countries.

I 'M NOT READY YET

Play in one act by Roderich Graeff

Madison, Wisconsin, USA
1952, during the Korean war

This play takes place between two countries -- the WE'S and the OTHER'S. The WE'S occupy the left half of the stage with a radio studio in the back, and the THER'S occupy the right half of the stage with a similar setting. A barbed-wire fence down the center symbolizes the border between the two countries.

Cast of characters:

The WE'S
- Paul - announcer and reporter for a radio station
- Ann - an assistant in the studio
- Radio engineer
- Studio director
- Messenger boy
- Mr. Advertiser
- Businessman
- Farm woman with a child
- Crippled man
- Chief engineer
- Three men

The Other's
- Officer - Army officer and news analyst
- Interpreter
- Soldiers
- An old woman

This play is a fantasy taking place in another world. Any similarity to living persons or existing countries is unintentional.

(Lights turn out completely; curtain rises; martial music is softly playing. Left half of stage slowly brightens. Paul is sitting in a chair, looking thoughtful. Ann and the radio engineer are working around the studio. Mr. Advertiser is pacing the floor).

ENGINEER: I think it's time again.

ANN: Paul,...Paul! Do I dare ask your royal highness to come to the microphone? (Paul slowly gets up, moves to the studio in the rear and picks up the microphone)

PAUL: Attention! Attention, please! We are expecting a very important announcement! Attention!! Attention, please! Keep your radio tuned to this station for a very important message from the government. (Soft martial music begins again an Paul returns to his chair) Indeed! (sarcastically) We are expecting a very important announcement.

ANN: Come on, what's the matter with you, Paul? You're getting to be a regular old grouch.

PAUL: Excuse me, Ann, but I'm sick and tired of it! An important message -- and what will, that be?
ANN: How should I know?

PAUL: (getting up and helping her rewind the tape on a wire recorder) I'll tell you. (in his announcer's voice) Attention, please! Here is a very important message from the government: On its peace voyage around the world, our fleet this morning entered the harbor of the capital city of Debterland. The town was bedecked with flags. Shortly after the arrival, King Giveme was welcomed by Admiral Sinkem on board the battleship Freedom.

"This friendship voyage by our fleet will show the entire world that we want peace" said our admiral. On the tour of the ship, King Giveme was much impressed by the new 21cm cannons. "We can destroy any ship within five miles of us with five salvos from these guns, and we hope to reduce that very shortly to three" said our admiral.

Ann: With only three salvos?

Paul: Yes, with only three salvos. Highlight of the day was the signing of a friendship pact between the two countries. "We are proud to form an alliance with the most powerful nation in the world, with that nation which shows, not with mere words, but with action that it is for peace" said King Giveme at the conclusion of the negotiations. Officers from the fleet were entertained at a royal banquet given by His Majesty in the courtyard of the palace. Shortly before dusk the fleet left the harbor. The flags were put away again.

ANN: And?

PAUL: And? Day by day I tell thousands of people anxious for peace: "We are fighting for peace, and the best way to do that is to show the Others what big cannons we have."

Ann: I hear you talking, but you yourself took part in the last war, did you not?

Paul: Yes, but not as a soldier. I had a deferment as a technical specialist and worked in research. I even made quit a neat invention.

Radio engineer: An invention?

Paul: Yes, a siren made out of plastic.

Ann: What was that good for?

Paul: In detail I learned about it after the war. The siren was mounted in one of those huge bombs called air mines, containing 1 ton of explosive. We dropped those over the cities of the OTHER'S to prevent them to put out fires.

Radio Engineer: How did that work?

Paul: One was not able to hit the armament factories accurately enough. Therefore it was decided to burn down the homes of the people working to prevent them to work in these plants. Our airplanes dumped thousands, tens of thousands of little incendiary bombs but the success was limited.

Radio Engineer: The success was limited?

Paul: Yes, women and boys came out of their cellars once they noticed these incendiary bombs hitting their homes and put out the fires with sand filled paper bags which stood in all rooms for that purpose.

Ann: Women and boys? Why not the men?

Paul: The men were fighting at the front.

Radio Engineer: Where is the connection with your siren?

Paul: One wanted to increase the scary and intimidating effect of the bomb. Up to now their effect was limited to an area of about 200 meters around the point of impact. So I invented a little propeller made out of plastic which mounted in each bomb create a terrifying noise, which got louder and louder when approaching the point of impact.

Ann: What effect did this have on the people. who wanted to extinguish those fires?

Paul: The people in their cellars heard a ton of a siren which became louder and louder. It is like standing at a railroad crossing and hearing a train approaching, blowing its whistle. Now they realized, a air mine is approaching. How close will it hit? The noise of the siren kept increasing. It seemed to be like eternity. One is scared, terribly scared and would like to crawl into the earth. One pulls in ones head, the heart tightens and one stops breathing. And then comes the explosion, a terrifying noise, the earth shakes, the house seems to be swinging back and forth, a pressure wave arrives, the candle is blown out, the air drums seem to burst, total darkness is around you, plaster falls from the ceiling and dust settles in your eyes.

In the next moment you start breathing again and you realize, I am still alive, and the people around me also. And than you think of the people 400 or ~0O meter away in those buildings, where the bomb exploded. And a warm~ feeling flows through you, I am still alive. But you where deeply scared, you stayed in your cellar, you would not clime up the stairs to search for incendiary bombs.
Ann~ And then?

Radio Engineer: .. . .And then???

Paul: Then fire started in more and more homes, whole areas where covered by flames growing together to a huge inferno, a powerful fire storm started swallowing up whole sections of a city.

Ann: And the people, the woman and the children, Paul, what happened with the people?

When it got to hot in the cellars or the breathing became to difficult, they tried to escape through the streets. They burned to death in the fire storm, thousands, ten thousands in one night.

Ann: Did the airmen in the planes knew what they had done?

Paul: I have no idea.

Ann: This is terrible. I can not believe, that our people planned and proceeded with such an undertaking. And you have assisted it with your siren!
Paul: Ann, you know I love you, I did not know that!

(Light changes from the left to the right side of the stage. Interpreter, with earphones on, is seated by receiving set. Officer is seated near by.)

OFFICER: What was that ?

INTERPRETER: Still the same. They're expecting an important message.

OFFICER: I must say you have a nice job. Whenever anything interesting happens, you're the first to hear it. On the other hand, you always have to listen to their stupid lies. It's a good thing that our people are protected from hearing that terrible propaganda. Our interpretation gives them the truth!

INTERPRETER: Truth! How difficult to find the truth, to distinguish what is true from what is untrue, to recognize what is emphasized or colored.

OFFICER: Oh, yes, it's difficult since most people don't have the time or the desire to study all the sources of information in order to make an objective judgement. That's why we have specialists like me to take the news from all the countries, and with our knowledge of the history and language,~ to interpret the facts for our people.

INTERPRETER: Certainly, (sarcastically) that is well known. (The Officer looks up startled). 1 mean. ..it's the best way to inform our people.

OFFICER: The situation on The Other side, with each citizen having access to all sources of information, sounds good along with the theory that each man is capable of finding the truth if he has all facts. But how does this theory look in practice? The normal citizen reads only one newspaper and usually forms his opinion from just this one. Is it possible to be more one-sided?

(pause) I wonder when this important message will come through? I can't make my comment for the day until it does.

(Left side lights as right side grows dark)

ADVERTISER: I can't stand it any longer. Tell me, who's paying you? - - these silly people with their important announcement or my company? I refuse to wait any longer!

RADIO ENGINEER: You elected these "silly people." In fact, it's your government. You can read your commercial right after the message. I'm sorry I can't let you start now.

PAUL: Mr. Advertiser, I agree with you completely; in the next election we will vote them out of office.

MESSENGER BOY: (comes running in waving some paper) The message!

RADIO ENGINEER: The message! Just glance through it quickly, Paul. We have to broadcast it immediately. Are you ready?

PAUL: (looking at the message in amazement) I.. .1 can't believe it! Isn't this a mistake?

RADIO ENGINEER: Ready? Go ahead.

PAUL: Attention, attention, please! I am going to read an important message from the government. Please listen carefully! This message vitally concerns every citizen.

"For several years the great powers of this earth have been arming themselves on a rapidly growing scale. Day by day more scientific ingenuity and productive capacity is being devoted to the development and building of weapons of war; and an ever - increasing part of life is being spent in learning to use these weapons. Each nation gives as its reason for this arms race the need for a strong defense force to repel threatened aggression.

"We believe that modern warfare is so devastating that an army is no longer a protection for life and freedom. In a future war, a great part of the values which make life worthwhile will vanish. The few who would survive another war would be unable to think of themselves as victors in a world destroyed.

We have viewed these developments for a long time with the deepest concern. But we believe that war is only one of many inhuman consequences of the basic attitude in life which overemphasizes the value of external forces. Our past attempts to find a solution were bound to fail because we weren't able to touch this fundamental cause. We came to the conclusion that the first steps in development of a new attitude will be the abolition of some of the external forces, and the attempt to help more and more people to develop their own inner values.

"Therefore, we, the elected representatives of the people, with the fullest consciousness of our responsibility to our nation and to the world, have decided to make the following declaration:

All development and production of arms will cease immediately. Excluding those materials which can be used for peaceful purposes, all existing weapons and military materials will be completely destroyed. To replace military training a voluntary work force of young men and women will work in those countries which want their services. We will equipped them with all necessary training and tools to accomplish their work. In exchange, we will welcome all those from other countries who which to live and work with us for a period of time.

"For this program we will need only a small part of our present military budget. Another part will be used for converting our war factories to peace production. The goods manufactured in these factories will be at the disposal of all people who need them. In addition, we will be able to increase our expenditures for cultural and educational purposes.

"We are convinced that this program will have a great influence on human relationships throughout the world. We recognize what great individual sacrifices it will require, but we believe that our people have the moral strength to make this idea succeed (left side becomes dark).

(Right side, still darkened)

OFFICER: Ha,ha,ha,ha,ha,ha! (stage slowly lightens) Is that really a literal translation?

INTERPRETER: Certainly! I don't understand it.

OFFICER: I don't completely understand it either. But what does it matter? It can't bother us anyway!

INTERPRETER: It doesn't sound like their usual propaganda.

OFFICER: No, even more obscure than usual!

INTERPRETER: Obscure? I have an ominous feeling about it. This is the first time that they've come up with a new idea. How could you fight this?

OFFICER: (thoughtfully) Yes, I know it's difficult to fight a real idea, but anything they produce out of their superficial lives, we'll never have to fear. What will it be? . . .only a new propaganda trick, a new trap; . . .but we'll not be deceived! (right side grows dark)

(Left side is lightened)

PAUL: This is incredible! What does it all mean?

RADIO ENGINEER: Mr. Advertiser...

PAUL: But. . .didn't you hear the message?

RADIO ENGINEER: Ready, Mr. Advertiser? . . .Now!

ADVERTISER: (into microphone) And again we have the daily Anti-Itching Hour. Today we are on board the destroyer Enterprise and... ah... just a minute. here is a seaman coming along the deck. Seaman, do you have a few words for our radio audience? (Advertiser changes his voice to sound like a seaman) "Yeah, sure. What do you want to know?" (resumes his own voice) Tell me, Seaman, you men have looked more comfortable lately. Would you please tell our listeners why this (voice fades away).. .has...happened?

ANN: What are you saying now?

PAUL: (bewildered) I just can't fully grasp it. Until today we rearmed with all our energy and believed that our very life depended on that,...and now, after tomorrow, everything will be different. Instead of cannons we will build farm machinery; instead of tanks, tractors and cars; and instead of air fields we'll be building roads and swimming pools; in place of barracks we'll have new schools; and instead of military service, we will have a world exchange program of youn people. It's incredible! Simply incredible!

ANN: Oh, Paul. Do you think it will succeed?

DIRECTOR: Paul.. .Paul! Why are you standing around? Don't you understand what's going on? A new idea has been born, and we're the ones to nurture it and make it grQw. Other peoples have given the world their music, their art, their philosophies --but we show the way to peace. (looking toward Mr. Advertiser) Listen! Is this the first commentary?

ADVERTISER: (voice can be heard again) So send anti-itch cream to your husbands and sons who are dedicating their lives fighting for your freedom. Anti-itch cream.. . anti-itch cream in the fight for peace and freedom. Don't forget the special sale coming soon to (voice fades out) . . .your local stores.

DIRECTOR: Good Lord! How is it possible? . . .But you, Paul, go out into the streets and find out how the public is reacting to all this.

PAUL: Yes, out on the street! I will ask the people.. .all people. . .everybody. This idea can succeed only if everyone is convinced. But do they understand the full implications of this idea, how it will and must affect every life, mine (coming to the front of the stage and speaking to the audience) and yours? What do you think about it?.. .What do you feel about it?.. .Did you really understand it?

(business man comes from the auditorium onto the stage)

BUSINESS MAN: Hello, is this the way to the airport?

PAUL: To the airport? What do you want there?

BUSINESSMAN: (excited) I'm going home! I'm going home as quickly as possible. You know, now that I am allowed to tell it, I have a little war plant over there, and I've wanted to reconvert it for a long time. I already know what I will produce. Wind motors! I think the OTHER'S will need them in those areas where they don't have much electricity.

PAUL: Oh, but listen! Until now we're the only ones who have announced that we will disarm.

BUSINESSMAN: (going away of f stage) Yes, but do you really think there would be any sense in our little country keeping an army if you don't have one any more? Do you think we could stand a war without you? Ho, ho,... I guess not. (a farm woman enters with a child)

FARM WOMAN: I can't believe it... there is still someone who speaks about war?

PAUL: Oh, no! He didn't mean it the way you think.

FARM WOMAN: Young man, you just don't understand that yet. Haven't you heard many people saying one thing about war but meaning something quite different?

PAUL: So you don't believe that our government is serious with their announcement to completely disarm?

FARM WOMAN: I'll knock their heads together if they don't do what they announced. My son will stay at home! He won't be sent away to foreign countries to shoot other people or be shot. Oh, no, of course he will go to foreign countries, but as a farmer! And for the same length of time we will take a boy from the OTHER'S. He will have a good life in our home, and no one will hurt him. (a cripple comes in on crutches) Come on, my son, may I help you?

PAUL: Were you wounded in the last war?

CRIPPLE: Yes, but not so bad that I couldn't join the next one. I will show the OTHR'S what it means to cripple me!!

PAUL: You are speaking about the next war as though you wanted one.

CRIPPLE: What do you mean "as if you wanted one"? Have you ever been asked whether or not you wanted a war? Of course not! But we will have one sooner than most people think.

FARM WOMAN: You haven't heard the announcement, have you?

CRIPPLE: Sure, I heard it, and that's just why we'll have a war. Do you believe the OTHER'S will destroy their weapons if they see us doing it?

FARM WOMAN: Why shouldn't they? They will be glad if they don't have to go to war anymore and don't feel threatened. They want to spend their money for more useful things, and this with our help, machines, materials, skilled workers..

CRIPPLE: Ha, ha, ha. (cracked up, the Chief' Engineer comes from the auditorium) Hey, Dick! Come here a minute! Here is someone who believes in that big fairy tale!

CHIEF ENGINEER: Fairy tale! It's national suicide! This is just what the Others have been waiting for. They've been arming for years and years, always hoping we would expose a weak side that they could attack. And we're making it so easy for them! Just wait until we've destroyed all our weapons, our only protection, and the life's work of' many. Day and night we've worked on defence. Very soon we would have been able to produce the new double V2 in mass. Then we would have been at least a year and perhaps a year and a half ahead of The Others. In the event of war we could have destroyed their principal cities in half an hour. That's what I call protection! That's what I call real defence!

PAUL: And the OTHER'S, couldn't they do the same?

CHIEF ENGINEERS Tomorrow they will come, plundering, robbing and murdering, through the whole world! And we will meet them, completely defenceless! Those who survive will either be deported or remain as forced labour under their lash.

CRIPPLE: Now you kow what will happen.

PAUL: (in a skeptical tone) So that's what will happen? How can you predict what will happen?

CRIPPLE: Don't you keep yourself informed? Don't you read the newspaper?

CHIEF ENGINEER: Now don't joke about it. You know past history has shown what people are able to do.

PAUL: Yes, but that happened during wars against people who offered armed resistance, and it was simple for the leaders to convince their citizens that they had to defend themselves. How can you continue to fight someone who doesn't fight back and who only wants to help you?

FARM WOMAN: Why should we try to predict what the OTHER'S will do? Why don't we ask them?

CHIEF ENGINEER: As if it were as simple as that. . .just to speak to the OTER'S.

PAUL: Why don't we try? Did we ever really try? Let's ask them. (goes to the border) Hello., you Others.. .do you hear me? Did you hear our message?

(Interpreter and Officer are looking at each other in surprise as the right side of the stage is lightened)

PAUL: Hello, did you hear our announcement?

OFFICER: Who are they? What do they want?

INTERPRETER: Who are you?

PAUL: We are the WE's.

INTERPRETER: They say they are the WE's.

OFFICER: The WE's? Those people don't belong to us. They speak a different language. They must be the OTHER'S.

INTERPRETER: You have made a mistake. We are the WE'S. You are the OTHER'S!

CRIPPLE: (to his companions) There, you can see it! (calling over the border) No!! We are the WE'S! You are the OTHER'S!
CHIEF ENGINEER: What did I tell you? They want to fight! Any talking with them is nonsense, wasted time and energy.

CHILD: Mother, is it the same with children? Are there the WE children and the OTHER children? Or do only grown-ups have that? (pause as the farm woman looks at her friends)

PAUL: (calling over the border) Did you hear our message? What did you think about it?

INTERPRETER: (to Officer) They are asking what we think about their message.

OFFICER: Since when have they been asking our opinion about anything? This whole thing is becoming more and more cloudy -- or -- (as he thinks he sees through some ulterior motive) or -- or clearer and clearer. Yes, I will answer them. (to them) I will answer you. I will tell you what I think of you and your idea as I have already told my people. (to the Interpreter) Did you translate that?

INTERPRETER: Our specialist in foreign news will tell you what our people think about you and your message.

FARM WOMAN: His language is different from ours, but you can understand him if you really want to.

OFFICER: Never before, in all my years of commenting on and interpreting the news, has such a thing happened to me. There you stand, after arming yourself for years, just waiting for us to show some weak side you can attack, and now you expect us to believe you are throwing all your weapons into the sea. What kind of fools do you take us for? Can't you think up a better story than that? You just want to lull us into a false security. You know our people are peace-loving and would like to destroy their weapons. But then you would come.. .No, I see through your trick! But you will not be successful! We will double our defense efforts and repel any aggression, no matter where it may comes from.

FARM WOMAN: They don't believe us! (calling) But we really will destráy our arms.

OFFICER: Lie!!

PAUL: We really are disarming!

OFFICER: Fraud!

FARM WOMAN: We are going to bring you help, material, tractors. (to her companions) How can we show them that we really will do it?

OFFICER: All right, bring us tractors. I myself will receive you. And I will expose your real intentions to the entire world. I will be expecting you.

PAUL: (quietly) Let's go. Let's show them what we believe. Let's convince them. (calling) Yes, we will come!

(light goes out on both sides, and after a few minutes the sound of moving tractors can be heard; it becomes quiet again; whole stage lightens, and one tractor can be seen in the rear of left stage)

PAUL: Here we are. We will meet them here! (a group of men along with the farm woman follow Paul on stage)

FARM WOMAN: A beautiful day!

FIRST MAN: It is so quiet! I'm terribly anxious to know what is going to happen.

SECOND MAN: I'm going to write home as soon as I see how it is over there. My family asked to be sure to do that.

FIRST MAN: Is it true that the OTHER'S crossed the border and harmed some people?

PAUL: You mean our friends? Why should they? And even if a few did, that will end very soon.

THIRD MAN: I heard they shot a woman.

PAUL: Shot a woman! That's bad, very bad. But how many millions were killed in the last war? How many would it be in a future war? You were told they didn't die in vain.. .until you got into another war. But this time it is true! They will not die in vain. We will have to sacrifice, but for a much higher goal. Not only will we no longer have these devastating wars, but external power, with its other evil results, will lose the mighty hold it has over us today. With inner strength and firm belief in our idea, we have nothing to fear.

FIRST MAN: I wonder how they will receive us. I hope they show up soon.
(from right side)

OFFICER: Hands up! Any resistance is useless! (Officer and some soldiers come in from the right)

Officer: What did I tell you, men? Didn't I tell you they would come? And here they are! Search them! Take away all weapons! Well, well, look over there. They really brought their tractors! Search them also, but be careful; they may be loaded with explosives! When you've finished, we'll have a little talk with them. (soldiers look for weapons) Bring all the weapons you found to me.

FIRST SOLDIER: I didn't find any.

SECOND SOLDIER: Neither did I.

THIRD SOLDIER: Not a single one.

FOURTH SOLDIER: The tractors are good. . .and with new tires.

FIFTH SOLDIER: They want to help us they say.

SIXTH SOLDIER: We can keep the tractors.

SEVENTH SOLDIER: They will bring more.

OFFICER: Quiet! What nonsense! No weapons?.. .no weapons?

Why did you come? What do you want here with your tractors?

PAUL: We don't have weapons any more. We want to help you.

SOLDIERS: They don't have weapons anymore. They want to help us. (Messenger comes from the right)

MESSENGER: Soldiers from the third army have shot one of' their own officers. They've thrown their weapons away and are driving the new tractors. They're talking about a new idea, don't obey anyone, and say they are going to destroy all arms.

OFFICER: (stamping on the ground) Damn it! That means I must act quickly. (loud) No weapons, you say? None at all?

SOLDIERS: No! Not a single one!

OFFICER: So they are spies! I suspected it immediately.

Spies and agitators.

SOLDIERS: They want to help us they say.

OFFICER: Quiet! Who gives the orders here? You or I? You don't believe me? Since when do I have to convince you? Well, I will do it. They say they want to give us tractors, don't they? But how do ~ tractors look?

SECOND SOLDIER: They're painted red.

OFFICER: Of course their color is red. And what is written on them?

THIRD SOLDIER: "Our farmers, our army -- guarantees for peace."

OFFICER: (to Paul) You understand that we can't accept any tractors which look different from ours. But I will give you time right now to change that.

FIRST MAN: How can you ask such a thing?

OFFICER: Do you hear me? I will give you the choice. Will you start to paint?

FIRST and SECOND MEN: That is too much! You can't demand that! How could we write that the army is the guarantee for peace?

OFFICER: (to his soldiers) Did you hear that?

ALL MEN: No, no! We won't do it! That is asking too much! Never!

OFFICER: SeIze them! Take them away! They will have to face a people's trial. (soldiers and men go, but do not see Paul who has moved to the tractor)

OFFICER: (speaking to himself) These people are dangerous! They really don't have any weapons. They believe in thei~r message! (to a soldier) Lieutenant! If more should come, shoot immediately!
(sees Paul painting one of the tractors) What's going on over there? Stop it! What are you doing there?

PAUL: I'm painting the tractor red, and will paint on the sign as you asked!

OFFICER: Stop! I say stop!
(takes a revolver, approaches him and shoots)
Dead! Dead! (stamps his foot, paces around, walks away but returns to the body;

an old woman comes in from the right side; officer runs to her and points at Paul's body after dropping the revolver) I told him to stop. He's dead, mother!

OLD WOMAN: What harm did he do?

OFFICER: What harm did he do? He wanted to bring a new idea.

OLD WOMAN: And you rejected it by doing that?

OFFICER: I shothim, mother, but I couldn't hit their idea.
The weapons I have don't work any longer.

OLD WOMAN: What have you done, my son?

OFFICER: I killed him, mother but I am defeated! (goes to the body)

PAUL: (turning and slowly rising) Don't touch me. I tell you, don't touch me! No, I'm not going to stay here. I don't recognize my death.
(Officer in astonishment, retreats a few steps. Paul is getting up:)
I believed in this idea. I devoted my thoughts, my work, my strength in it -- my soul, and even my life. But did I accomplish anything? I could have gone on living, and I'm going to do that, even if it might be for only a few more years. Who knows? Then I will probably have go to war. I will be ordered to duty on a big ship, and I will fight there as one link in a big chain, until we blow up.. .or you.. .or both together. But at least, I will not have to decide anything now. That's what I will do right now, I'm not going to make a decision. You can't ask a man to die by his own conviction just for an idea.

(Paul turns to the audience and calls out:)

I'M NOT READY YET!

(while Paul is speaking, Officer listened with astonishment, he picks his revolver up again and pushing the old woman aside orders one of the soldiers:)

OFFICER: Take him to the other prisoners!
(The curtain falls.)

Fragment, where does it belong?

(Shouting) Shoot them all! This time I will hit more than just a person!

(Lights turn out, martial music starts to play, lights slowly come to the left side of the stage, showing the studio as it was in the beginning)

MESSENGER: The announcement!

ANN: Paul, the announcement! Hurry up!

RADIO ENGINEER: Can you read it right away? Ready

PAUL: Attention, please! Here is a very important message from the government: On its peace voyage around the world, our fleet this morning entered the harbor of the capital city of Debterland. The town was bedecked with flags

(His voice fades away, the lights get dimmed, the curtain do ~ ses.)

My Path to Peaceful Energy
Appendix 2

Net Journal: Produktion von nutzbarer Energie aus einem Wärmebad

Roderich W. Graeff

International Licensing
Prof.-Domagk-Weg 7, D-78126 Königsfeld, Germany
102 Savage Farm Drive, Ithaca, NY 14850-6500, USA
E-mail rwgraeff@yahoo.com

December 2007

Zusammenfassung

.

Messungen der Temperaturverteilung in senkrechten Gas- und Flüssigkeitssäulen in isolierten Systemen zeigen einen Temperaturgradienten, oben kalt und unten warm. Dies ist erklärbar durch den Einfluss der Schwerkraft auf die Bewegung von Molekülen.

Eine derartige Temperaturschichtung macht die Erzeugung nutzbarer Energie ohne Zuführung zusätzlicher Energie von außen aus einem Wärmebad möglich. Dies steht im Gegensatz zu der Formulierung des zweiten Hauptsatzes der Thermodynamik, die besagt, dass ein Perpetuum Mobile der zweiten Art nicht möglich ist.

Produktion von nutzbarer Energie aus einem Wärmebad

Roderich W. Graeff

International Licensing
Prof.-Domagk-Weg 7, D-78126 Königsfeld, Germany
102 Savage Farm Drive, Ithaca, NY 14850-6500, USA
E-mail rwgraeff@yahoo.com

December 2007

Einleitung

In einer Sitzung der Kaiserlichen Akademie der Wissenschaften in Wien im Jahre 1876 führte J. Loschmidt aus, dass seiner Ansicht nach Festkörper ebenso wie Gase unter dem Einfluss der Schwerkraft eine Temperaturverteilung zeigen sollten, oben kalt und unten warm [3]. Er erklärte:

Damit wäre auch der terroristische Nimbus des zweiten Hauptsatzes zerstört, welcher ihn als vernichtendes Prinzip des gesamten Lebens des Universums erscheinen lässt, und zugleich würde die tröstliche Perspektive eröffnet, dass das Menschengeschlecht betreffs der Umsetzung von Wärme in Arbeit nicht einzig auf die Intervention der Steinkohle oder der Sonne angewiesen ist, sondern für alle Zeiten einen unerschöpflichen Vorrat verwandelbarer Wärme zur Verfügung haben werde. ...

Damals waren die Messmöglichkeiten noch nicht vorhanden, die eine Überprüfung dieser Auffassung durch Versuche zugelassen hätten. Nur theoretische Überlegungen, insbesondere die Temperaturverteilung in senkrechten Gassäulen betreffend, waren möglich. Diese wurden durch L. Boltzmann [1] und J.C. Maxwell [2] angestellt. Im Gegensatz zu J. Loschmidt vertraten sie die Auffassung, dass die Temperatur über die Höhe konstant sein müsse. Nur so sei der zweite Hauptsatz erfüllt. A. Trupp [4] gibt eine gute Zusammenfassung dieser über Jahre andauernden Diskussion.

Nur Versuche können wohl eine klare Antwort auf diese unterschiedlichen Auffassungen bringen. Dies wird seit einigen Jahren durch die Entwicklung von Spannungsmessgeräten möglich, die zusammen mit Thermoelementen die Messung von Temperaturdifferenzen im Bereich von 0,001 °K erlauben. Da ein isoliertes System nicht herstellbar ist, das die immer vorhandenen Temperaturschwankungen der Umgebung vom Versuchsobjekt völlig fernhält, sind Messungen über längere Zeiträume für die Berechnung von langfristigen Durchschnittswerten nötig. Auch solche Durchschnittsmessungen wurden erst mit modernen Datenerfassungssystemen durchführbar.

Im Folgenden wird über Langzeitmessungen an senkrechten Gas- und Flüssigkeitssäulen berichtet. Eine theoretische Erklärung der Messergebnisse wird vorgestellt, die auch quantitative Aussagen erlaubt. Die theoretischen Werte stimmen mit den Messergebnissen gut überein.

Typische Versuchsanordnung

Die Abbildung auf der nächsten Seite zeigt eine typische Versuchseinrichtung, wie sie z.B. für den Versuch B74 eingesetzt wurde. Der Glaseinsatz (Dewargefäß) einer normalen Thermosflasche (1) ist in einem weiteren, 1 Liter großen Dewargefäß (2) angeordnet. Auf dieser ruht eine dritte, 1/2 Liter große Dewarflasche (3). Der Zwischenraum (4) zwischen den Dewargefäßen ist mit feinen PET-Fasern ausgefüllt.

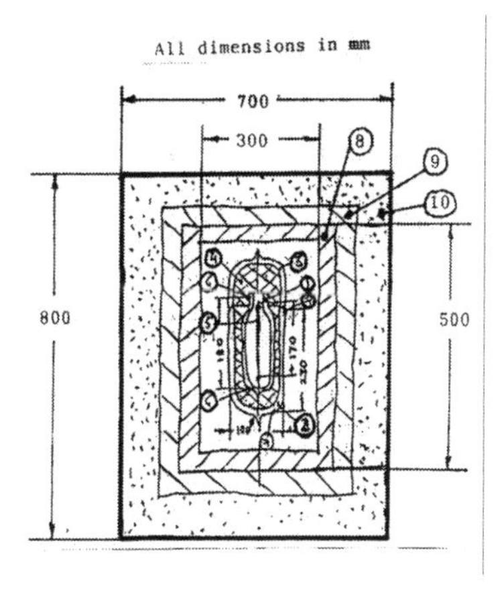

Der Innenraum der inneren Dewarflasche (1) ist der eigentliche Versuchsraum. In ihm befindet sich das zu messende Gas oder eine Flüssigkeit. Um Konvektionsströme und Wärmestrahlung zwischen oben und unten zu verhindern, ist der Innenraum zusätzlich mit einem feinen Pulver, z.B. einem Glaspulver, angefüllt. In der mittleren Achse ist ein Thermoelement (5) angeordnet. Ein zweites Thermoelement (6) ist außen auf dem Dewarglas (1) angebracht. Ein drittes Thermoelement (7) befindet sich außen auf dem Dewarglas 3.

Feine PET-Fasern fixieren die Dewargläser innerhalb eines Gehäuses (8), das aus 40 mm dicken Styroporplatten hergestellt ist. Dieses Gehäuse ist von 50 mm dicken Kupferplatten umgeben, die aus zusammengepressten Kupferdrähten bestehen. Der gesamte Versuchsaufbau ist schließlich von 100 mm dicken Polystyrolschaumplatten (10) umgeben [6].

Der Durchmesser der Thermoelementdrähte wurde möglichst klein gewählt, um die nicht vermeidbaren Wärmeleitungsverluste durch die Drähte klein zu halten. Das Voltmeter misst die Spannung der Thermoelemente mit einer Auflösung von 1×10^{-7} Volt entsprechend einer Temperatur von 0,0003 °K.

Versuchsumfeld.

„Isoliertes System" bedeutet, dass das System kein Material und keine Energie mit dem Umfeld austauscht. Während ersteres praktisch einzuhalten ist, kann ein Energieaustausch nicht völlig unterbunden werden. Selbst eine optimale Isolation kann nicht verhindern, dass wegen des Temperaturunterschieds zwischen innen und außen ein wenn auch sehr kleiner Energieaustausch mit der Umgebung stattfindet.

Einen typischen Versuchsaufbau, wie er in ähnlicher Form für den

Versuch B74 benutzt wurde, zeigt die obige Abbildung: Drei ineinander steckende Dewargläser, die die Hälfte eines Versuchsaufbaus darstellen. Dieser Aufbau wird in einem der beiden Aluminumrohre montiert, wie in der nächsten Abbildung dargestellt. Man erkennt den inneren und äußeren Styropor-Formkörper mit den dazwischen angeordneten Kupferplatten.

Die Abbildung unten zeigt eine Variante des Versuchsumfeldes [5]. Eine innere feststehende Trommel ist von einer kontinuierlich kreisenden äußeren Trommel umgeben. Beide Trommeln sind aus 25 mm dickem Aluminum hergestellt. Dies hat eine Temperaturdifferenz in der inneren Trommel, oben warm und unten kalt, von nur 0,002 °K pro 500 mm Höhe der inneren Trommel zur Folge. Ein weiterer Vorteil dieser Anordnung besteht darin, dass die Innere Trommel um 180 Grad um die horizontale Achse gedreht werden kann, ohne die Temperaturmessungen zu unterbrechen oder die Isolation negativ zu beeinflussen. Hierdurch wird die innere Versuchsanordnung auf den Kopf gestellt. [6]

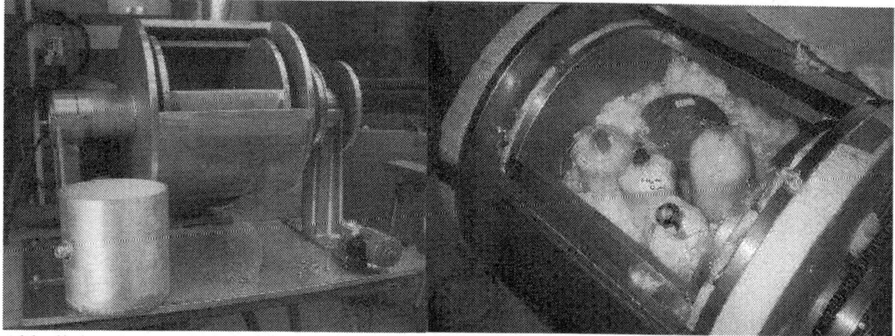

Abb. links: Apparat mit äußerer rotierender und innerer feststehender Trommel
Abb. rechts: innere Trommel mit 5 Versuchskörpern

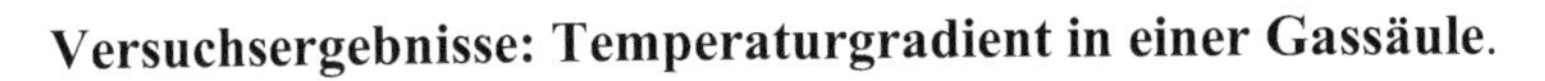

Versuchsergebnisse: Temperaturgradient in einer Gassäule.

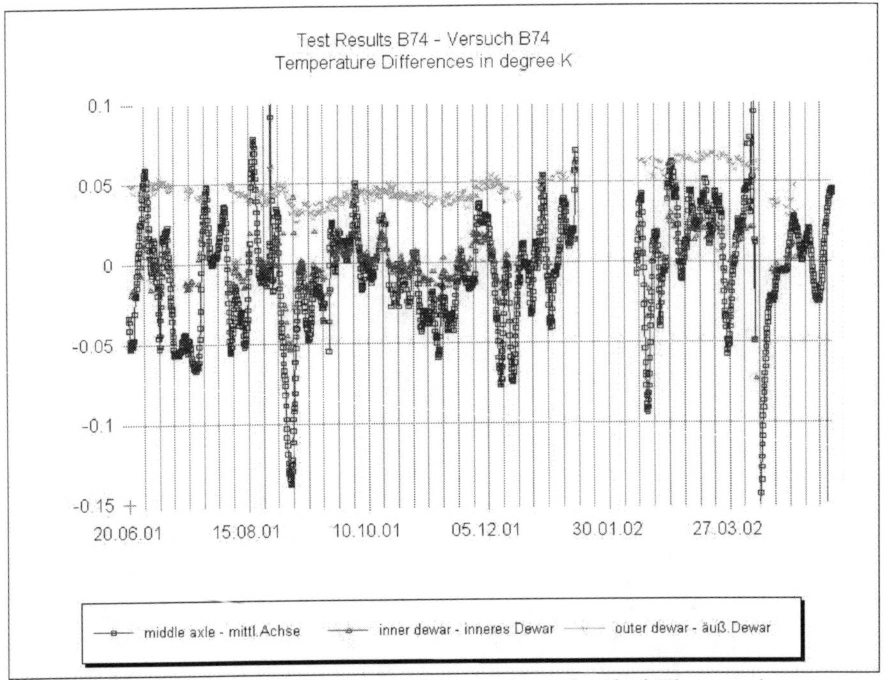

Die obige Grafik zeigt die Temperaturmessungen der drei Thermoelemente von Versuch B74, dargestellt für ein halbes Jahr, worüber schon auf der Webseite [5] und auf der Konferenz *Quantum Limits to the Second Law* in San Diego im Juni 2002 berichtet wurde [8]. Jeder Punkt der Kurve ist der Mittelwert von 60 Temperaturwerten, die alle vier Minuten innerhalb von 4 Stunden gemessen wurden.

Jeder Punkt oberhalb der Nulllinie bedeutet, dass die obere Messstelle des Thermoelements wärmer war als die untere.

Man erkennt, dass das äußere Dewarglas 2 oben immer um etwa 0,04 °K wärmer war als unten. Dies gibt den Einfluss der umgebenden Raumluft wieder, die typisch oben um etwa 1 °K pro Meter wärmer ist als unten. Die gute Isolation um den Versuchsraum herum und insbesondere die wärmeleitenden Kupferplatten sind der Grund für den stark reduzierten Temperaturabfall, der sich außen am Dewarglas 2 von oben nach unten zeigt.

Im Gegensatz hierzu zeigt die innere Achse meistens einen umgekehrten Temperaturverlauf, nämlich oben kalt und unten warm.

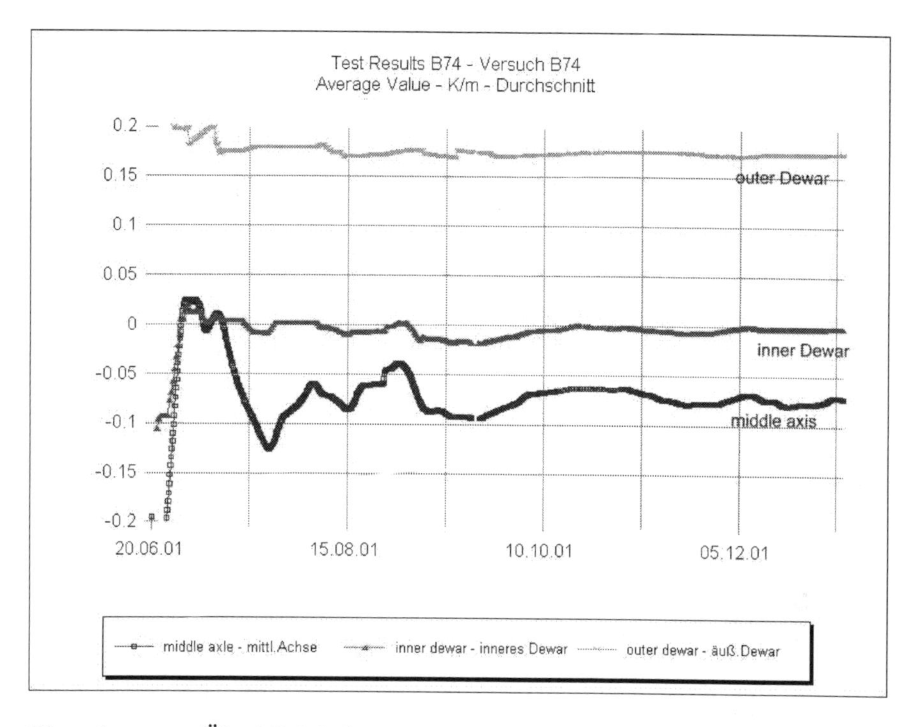

Einen besseren Überblick bekommt man durch den Durchschnitt aller Messwerte vom Anfang des Versuchs bis zum jeweiligen Zeitpunkt der obigen Grafik. Alle Messwerte wurden auf eine Höhe von 1 Meter umgerechnet.

Tabelle 1: Durchschnitt der Temperaturen über 6 Monate:

	aus Abb. 5:	Abstand zwischen Thermoelementperlen:	bezogen auf 1m:
Dewarglas 2:	+0,040 °K	0,23 m	+0,174 °K
Dewarglas 1:	-0,001 °K	0,18 m	-0,0055 °K
Mittelachse:	-0,012 °K	0,17 m-0,0705	°K

Der Durchschnittswert über ein halbes Jahr für Dewarglas 2 zeigt einen deutlich positiven Wert; das Glas ist oben warm und unten kalt. Die Temperaturdifferenz für das Dewarglas 1 war praktisch gleich 0. Der Durchschnittswert für die Mittelachse im Innenraum von Dewarglas 1 zeigt einen negativen Wert, was bedeutet, dass es im Durchschnitt oben kälter war als unten, also genau das Gegenteil zu den Werten am äußeren Dewarglas 2.

Versuchsergebnisse: Temperaturgradient in einer Flüssigkeitssäule

Die nächste Abbildung zeigt die Ergebnisse der Temperaturmessungen von zwei Thermoelementen aus Versuch B890 über fünf Wochen. Der Versuch war wie in Abb. 1 dargestellt angeordnet, jedoch mit dem Unterschied, dass der innerste Versuchskörper aus einer Glasflasche bestand, die nicht in zwei Dewargefäßen, sondern in zwei Aluminumrohren montiert war. Die Glasflasche mit einem Inhalt von 1/2 Liter war mit Leitungswasser und einem feinen Glaspulver angefüllt. Gemessen wurden die Temperaturdifferenzen an der inneren Achse und außen am inneren Aluminumrohr.

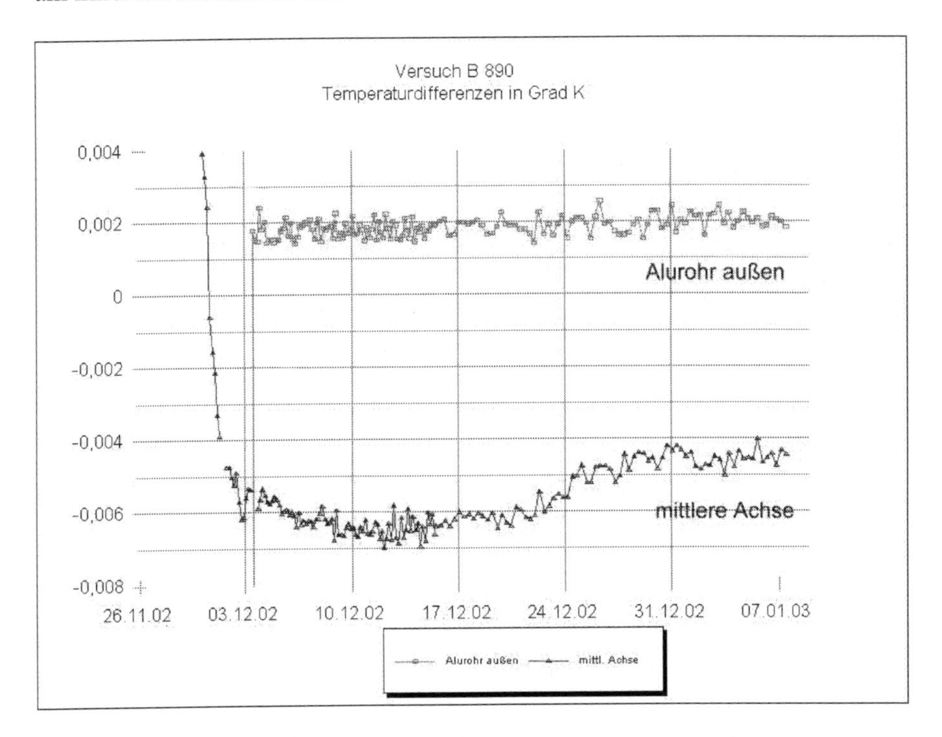

Jeder Kurvenpunkt in der Grafik ist wieder der Mittelwert von 60 Temperaturmesswerten, die, 4 Stunden lang im Abstand von 4 Minuten aufgenommen, für einen Zeitraum von 5 Wochen aufgezeichnet wurden. Jeder

Wert oberhalb der Nulllinie bedeutet, dass die obere Messperle des Thermoelementes wärmer war als die untere.

Man erkennt, dass die Temperatur außen am Aluminumrohr oben immer um etwa 0,002 °K höher war als unten. Im Gegensatz hierzu zeigt die innere Achse für den ganzen Zeitraum einen umgekehrten Temperaturverlauf, nämlich oben kalt und unten warm.

Die Tabelle 2 gibt die Durchschnittswerte über 4 Stunden wieder, umgerechnet auf eine Höhe von 1 Meter zwischen den Thermoelementperlen:

Tabelle2:

	aus Grafik:	Abstand zwischen Thermoelementperlen:	bezogen auf 1m:
Alu-Rohr:	+0,002 °K	0,21 m	+0,01 °K
Mittelachse:	-0,0046 °K	0,18 m	-0,026 °K

Obwohl Abb. 6, Versuch B890, vorerst nur Messwerte für 5 Wochen wiedergibt; macht ein Vergleich mit Abb. 4 deutlich, dass die Messwerte in der Flüssigkeitssäule viel stabiler sind als in der Gassäule von Versuch B74.

Genauigkeit der Messergebnisse

Es ist schwierig, eine umfassende Aussage über die Genauigkeit der Temperaturmessungen zu machen. Ich glaube, dass diese bei +/- 0,002 °K liegt. Ein derartig niedriger Wert ist möglich, da wir nicht absolute Temperaturen, sondern Temperaturdifferenzen messen. Die Genauigkeit wird auch dadurch erhöht , dass wir für die Messung kleiner Differenzen 5 in Reihe geschaltete Thermoelemente einsetzen. Bei den Messungen, bei denen das Vorzeichen des Temperaturgradienten wichtiger ist als der absolute Wert, wird jeder Wert mit umgedrehter Polarität ein zweites Mal gemessen.

Interpretation der Messergebnisse

Die Temperaturdifferenzen, gemessen in den Mittelachsen, oben kalt und unten warm, können nicht mit den normalen Vorstellungen über den Wärmetransport erklärt werden. In einer Umgebung, die oben warm und unten kalt ist, kann nicht von alleine, ohne Zuführung von Energie, in den Mittelachsen ein umgekehrter Temperaturgradient, oben kalt und unten warm, entstehen. Aber die

Versuchsergebnisse können durch den Einfluss der Schwerkraft auf die senkrechte Bewegung der Moleküle erklärt werden.

Die Schwerkraft beschleunigt ein Molekül auf dessen Weg von oben nach unten und bremst es auf seinem umgekehrten Weg von unten nach oben ab, wie in Abbildung 7 schematisch dargestelt. Im Durchschnitt treffen die Moleküle daher auf die obere Wand des Behälters mit einer niedrigeren Geschwindigkeit auf als auf die untere. Die Moleküle nehmen bei ihrem Auftreffen auf die Wände deren Temperatur an. An der oberen Wand werden sie beschleunigt und nehmen daher Wärmeenergie auf, die sie durch die Abbremsung an der unteren Wand dort wieder abgeben. Sie transportieren Wärmeenergie von oben nach unten, bis sich schließlich in dem isolierten System ein Gleichgewichtszustand einstellt mit einer niedrigeren Temperatur in der oberen Wand als in der unteren. Diese Temperaturdifferenz, die sich durch die Wirkung der Schwerkraft oder Gravitation einstellt, nenne ich T(Gr).

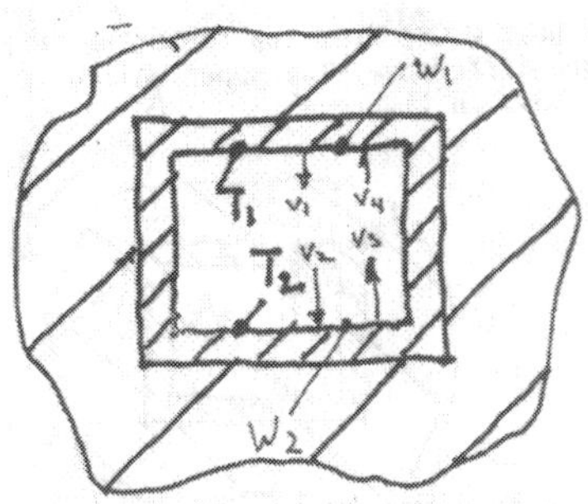

T(Gr) kann ausgerechnet werden, wenn man die potentielle Energie der Moleküle gleichsetzt mit der Vergrößerung ihrer vertikalen Geschwindigkeit. Dies führt zu der Beziehung

$$T(Gr) = -g \times H / c$$

mit g = Konstante der Erdbeschleunigung
H = Höhenunterschied
c = spezifische Wärme

Da sich die Schwerkraft nur auf den vertikalen Anteil der Molekülbewegung auswirkt, muss die spezifische Wärme c - bei Gasen c_v - durch die Anzahl der Freiheitsgrade dividiert werden, so dass sich ergibt

$$T(Gr) = -g \times H / c(Gr)$$

mit $c(Gr) = c / n$ mit n = Anzahl der Freiheitsgrade.

Für einige Gase und Flüssigkeiten ergeben sich die Werte der Tabelle 3:

Tabelle 3: Berechnung von T(Gr):

	Luft	Xenon	Wasser
Freiheitsgrade	5	3	16
T(Gr) °K/m	-0,07	-0,311	-0,04

Ein Vergleich der gemessenen Werte von Tabelle 1 und 2 mit den berechneten von Tabelle 3 zeigt eine sehr gute Übereinstimmung für Luft. Selbst für Wasser ist die Übereinstimmung recht gut wenn man bedenkt, dass vorerst nur Werte über 5 Wochen vorliegen.

Ausblick

Messungen des Temperaturverlaufs in Gas- und Flüssigkeitssäulen ergeben eine vertikale Temperaturschichtung in Übereinstimmung mit der Auffassung von J. Loschmidt [3]. Es zeigt sich eine gute Übereinstimmung mit berechneten Werten. Weitere Versuche müssen klären, ob es Materialien und Konstruktionen gibt, die eine wesentlich größere Temperaturdifferenz erzeugen, um eine wirtschaftliche Energieproduktion zu ermöglichen.

Dieses Resultat ist ein Widerspruch zum 2. Hauptsatz der Wärmelehre. Falls es sich bestätigt, wäre die Erzeugung brauchbarer Energie aus einem Wärmebad möglich, was ein Perpetuum Mobile der 2. Art darstellen würde. Es wäre sehr wünschenswert, wenn die vorgestellten Messergebnisse von dritter Seite überprüft würden.

Literatur

[1] Boltzmann, L., Wissenschaftliche Abhandlungen, Bd.2, Hrsg. F.Hasenöhrl, Leipzig, 1909
[2] Maxwell, J.C., The London, Edinburgh, and Dublin Philosophical Magazine of Science, 35, S. 215 (1868)
[3] Loschmidt, L., Über den Zustand des Wärmegleichgewichts eines Systems von Körpern mit Rücksicht auf dieSchwerkraft, Sitzungsberichte der Mathematisch-Naturwissenschaftl. Klasse der Kaiserlichen Akademie der Wissenschaften 73.2 (1876), S. 135
[4].Trupp, A., Physics Essays, 12, Nr. 4, 1999
[5] Gräff, R.W., Gravity Machine, Internet <firstgravitymachine.com>
[6] In- und ausländische Patente beantragt

[7] Gräff, R.W., Gravity Machine, my Search for Peaceful Energy, erscheint voraussichtlich im Sommer 2003
[8] Gräff, R.W., Measuring the Temperature Distribution in Gas Columns, CP 643, Quantum Limits to the Second Law; First International Conference, Melville, New York, 2002, American Institute of Physics, AIP Conference Proceedings, Vol. 643

My Path to Peaceful Energy
Appendix 3

US Patent Application No.: US 2003/0145883 A1

Roderich W. Graeff

International Licensing
Prof.-Domagk-Weg 7, D-78126 Königsfeld, Germany
102 Savage Farm Drive, Ithaca, NY 14850-6500, USA
E-mail rwgraeff@yahoo.com

Abstract

A method and apparatus for creating temperature differences in columns of gases, liquids or solids in a closed system under the influence of gravity is used to provide energy in the form of electricity or heat. A temperature differential element, optionally a solid, liquid or gas, is suspended vertically in a chamber inside an enclosure. The chamber optionally is evacuated, filled with fibres, powder or small spheres, or otherwise arranged to minimize the effects of convection currents and radiation. Under the effect of gravity, the upper end of the temperature differential element becomes cooler than the lower end. A thermocouple can be used to generate electrical energy from the temperature difference between a vertical segment, for example the upper and lower ends, of the temperature differential element, or heat exchangers used to extract heat.

US patent application no. 20030145883

Roderich W. Graeff

United States Patent Application	**20030145883**
Kind Code	**A1**
***Graeff,* Roderich W.**	**August 7, 2003**

Gravity induced temperature difference device

Abstract

A method and apparatus for creating temperature differences in columns of gases, liquids or solids in a closed system under the influence of gravity is used to provide energy in the form of electricity or heat. A temperature differential element, optionally a solid, liquid or gas, is suspended vertically in a chamber inside an enclosure. The chamber optionally is evacuated, filled with fibres, powder or small spheres, or otherwise arranged to minimize the effects of convection currents and radiation. Under the effect of gravity, the upper end of the temperature differential element becomes cooler than the lower end. A thermocouple can be used to generate electrical energy from the temperature difference between a vertical segment, for example the upper and lower ends, of the temperature differential element, or heat exchangers used to extract heat.

Inventors:	***Graeff,* Roderich W.**; *(Ithaca, NY)*
Correspondence Name and Address:	**BROWN & MICHAELS, PC 400 M & T BANK BUILDING 118 NORTH TIOGA ST ITHACA NY 14850 US**

Serial No.: **351946**

Series Code: **10**

Filed: **January 27, 2003**

U.S. Current Class: **136/201**; 165/138

U.S. Class at Publication: **136/201**; 165/138

Intern'l Class: F28F 007/00; H01L 035/34

Claims

What is claimed is:

1. A method to create a temperature difference within a mass of solid, liquid or gas, comprising the steps of: a) providing a closed system in the form of an elongated insulated container having an interior; and b) enclosing the mass, comprising a temperature difference element, within the interior of the container in a vertical arrangement.

2. The method of claim 1, further comprising the step of extracting energy from the temperature difference element.

3. The method of claim 2, in which the step of extracting energy comprises providing a thermocouple along a vertical segment of the temperature difference element, and energy is extracted in the form of electricity generated by the thermocouple from a temperature difference along the vertical segment.

4. The method of claim 2, in which the step of extracting energy comprises the steps of providing a heat exchanger in the interior and passing a heat-exchange fluid through the exchanger, and energy is extracted in the form of heat.

5. The method of claim 1, further comprising the step of filling the interior of the container with a permeable material, such that the effects of convection and radiation in the mass are reduced.

6. The method of claim 5, in which the permeable material is selected from a group comprising solid glass fibres, hollow glass fibres, powder, solid spheres, hollow spheres, sand and foam.

7. The method of claim 1, in which the mass is a solid, further comprising the step of evacuating the interior of the container.

8. The method of claim 1, further comprising the steps of surrounding the container with a rotatable drum, and rotating the drum.

9. The method of claim 1, further comprising the step of mounting the container on a pivot, such that the container can be inverted.

10. An apparatus to create a temperature difference within a mass of solid, liquid or gas, comprising: an elongated container comprising a closed insulated chamber with an interior; a mass comprising a temperature difference element disposed vertically inside the interior of the closed chamber.

11. The apparatus of claim 10, further comprising a thermocouple arranged along a vertical segment of the temperature difference element, such that energy is extracted in the form of electricity generated by the thermocouple from a temperature difference along the vertical segment.

12. The apparatus of claim 10, further comprising a heat exchanger in the interior, such that energy is extracted in the form of heat.

13. The apparatus of claim 10, in which the interior of the container is filled with a permeable material, such that the effects of convection and radiation in the mass are reduced.

14. The apparatus of claim 13, in which the permeable material is selected from a group comprising solid glass fibres, hollow glass fibres, powder, solid spheres, hollow spheres, sand and foam.

15. The apparatus of claim 10, in which the mass is a solid and the interior of the container is evacuated.

16. The apparatus of claim 10, further comprising a rotatable drum surrounding the elongated insulated chamber.

17. The apparatus of claim 10, in which the elongated insulated container is mounted on a pivot, such that the container can be inverted.

18. The apparatus of claim 10, further comprising a housing surrounding the elongated container.

Description

REFERENCE TO RELATED APPLICATIONS

[0001] This application claims an invention which was disclosed in Provisional Application No. 60/353,307, filed Feb. 1, 2002, entitled "GRAVITY INDUCED TEMPERATURE DIFFERENCE DEVICE". The benefit under 35 USC .sctn.119(e) of the U.S. provisional application is hereby claimed, and the aforementioned application is hereby incorporated herein by reference. BACKGROUND OF THE INVENTION

[0002] 1. Field of the Invention

[0003] The present invention pertains to the field of energy production. More particularly, the invention pertains to a method and apparatus for creating temperature differences in columns of gases, liquids or solids in a closed system under the influence of gravity, and using the temperature differential to extract useful energy.

[0004] 2. Description of Related Art

[0005] Of all processes that are permitted by the first law of thermodynamics, such as energy conversions, only certain types of processes actually occur in nature. It is the Second Law of thermodynamics that determines whether a process will or will not occur. Examples of such processes include, but are not limited to heat flow from a warmer to a cooler body, the spontaneous dissolution of salt in water, and the decrease in amplitude of the oscillations of a pendulum over time. These are examples of "irreversible" processes, or processes that occur naturally in one direction only, otherwise requiring the input of energy or work. In general, a process is irreversible if the system and its surroundings cannot return spontaneously to their initial states. Such processes can only be returned to their initial states by changing the surroundings, or doing external work. [0006] The Second Law of thermodynamics can be stated in numerous ways. The Kelvin-Planck form of the Second Law states that no heat engine operating in a cycle can absorb thermal energy from a reservoir and perform an equal amount of work. Alternatively, in the words of Rudolf Clausius, it is impossible to construct a cyclical machine that produces no other effect than to transfer heat continuously from one body to another body at a higher temperature. Although these statements may appear to be unrelated, they are in fact equivalent. In essence, the Second Law states that a device capable of converting thermal energy into other forms of energy at 100% efficiency cannot be constructed.

[0007] A closed system in the field of thermodynamics is defined as an enclosed region of constant volume, wherein neither mass nor energy crosses the boundary. Inside the enclosed region, in accordance with the Second Law of thermodynamics:

[0008] initial temperature differences will become smaller and temperatures will eventually equalize, meaning that the entropy will increase;

[0009] heat will not flow from a cooler to a hotter object; and

[0010] no process can take place where the entropy would decrease.

[0011] In other words, the Second Law of thermodynamics states that, in a closed system, real processes occur in only one preferred direction. Thus, heat flows spontaneously from a warmer object to a cooler one, while the reverse reaction does not occur spontaneously, but rather requires the input of work or energy.

SUMMARY OF THE INVENTION

[0012] In accordance with the invention, it was found that it is possible within the right apparatus to create temperature differences in a closed system without introducing work and thereby to decrease the entropy. The temperature differences so produced can be used to perform work outside the closed system.

[0013] The present invention is a method and apparatus for creating temperature differences in columns of gases, liquids or solids in a closed system under the influence of gravity. The temperature differences can be used to provide energy in the form of electricity or heat. The invention also pertains to apparatus for reducing temperature influences from outside the closed system.

[0014] In an embodiment of the invention, a temperature differential element, optionally a solid, liquid or gas, is suspended vertically in a chamber inside an enclosure. The chamber optionally is either evacuated, filled with fibres, powder or small spheres, or otherwise arranged to minimize the effects of convection currents. Under the effect of gravity, the upper end of the temperature differential element becomes cooler than the lower end. A thermocouple can be used to generate electrical energy from the temperature difference between a vertical segment, for example the upper and lower ends, of the temperature differential element.

BRIEF DESCRIPTION OF THE DRAWINGS

[0015] FIG. 1 shows an apparatus demonstrating the method of the invention.

[0016] FIG. 2 shows another embodiment of the invention, using a metal rod as the temperature differential element.

[0017] FIG. 3 shows still another embodiment of the invention, in which the chamber is in a vacuum instead of being gas-filled.

[0018] FIG. 4 shows another embodiment of the invention, in which the chamber is divided by a number of horizontally arranged thin films.

[0019] FIG. 5 shows an optional apparatus for reducing the temperature influence from outside the closed system, according to an embodiment of the invention.

[0020] FIG. 6 shows another optional apparatus, wherein a Dewar glass is used for reducing the temperature influence from outside the closed system and heat exchangers extract energy, according to an embodiment of the invention.

[0021] FIG. 7 shows an arrangement of apparatus to reach higher temperature differences.

[0022] FIG. 8 is a section along the line 8-8 in FIG. 5.

DETAILED DESCRIPTION OF THE INVENTION

[0023] The method of the invention comprises the step of exposing atoms or molecules to the effects of gravity in a closed system in a vertical arrangement.

[0024] Due to the tendency of warm air to rise and of cold air to flow downwards, a room filled with air, either heated or cooled, typically has a higher temperature in the upper parts than in the lower parts. Typically, the temperature distribution over the height is on the order of about 1 degree C. per meter. One would therefore assume that, even when very well insulated from the surroundings, lead rod 1, as shown in FIG. 2, would have a higher temperature at the top than at the bottom. However, to the contrary, once equilibrium is reached, lead rod 1 actually has a colder temperature at the top than at the bottom. The method of the present invention is based upon this unexpected observation.

[0025] Referring to FIG. 1, when an elongated container 2 containing a gas (such as, e.g., air) is placed in a vertical position, the air molecules move randomly within the container in a movement called "Brownian Motion". However, even a closed system is exposed to the effects of gravity. Therefore, any molecule moving in an upward direction will be slowed down, and any molecule moving downward will be accelerated by the force of gravity. This means that molecules in the upper part of the container will have, on average, a slower speed (i.e., less kinetic energy) than molecules in the lower part. The temperature of a number of molecules (e.g., a gas) is proportional to the average velocity or kinetic energy of the molecules. Therefore, the temperature in the upper part of the container is lower than the temperature in the lower part. By hitting the upper and lower wall, the molecules attain the temperatures of the walls, cooling the upper walls and heating the lower walls. Thus, energy is transported from the upper walls to the lower walls, thereby creating a lower temperature in the upper walls and a higher temperature in the lower walls. This temperature difference is used in the present invention to perform work inside or outside the closed system.

[0026] In fluids, whether liquids or, in particular, gases, warmer molecules tend to rise and colder molecules tend to fall within the container, due to their different densities, which are proportional to their temperature. This fluid flow is called "convection". The convection effect tends to negate the effect of energy transport from the top to the bottom due to gravity. In accordance with the invention, it is therefore helpful optionally to fill the container with, for example, a mass of very fine fibres 6, such as, for example, glass wool (Fiberglas.RTM.), natural or plastic fibres (solid or hollow), powder, foam, or small solid or hollow spheres (e.g., made of plastic, metal, or glass), in order to reduce convection currents and radiation to a minimum. This might not be necessary when using highly viscous liquids like heavy oils, petroleum jelly or gelatine.

[0027] When using a gas, it is helpful also optionally to reduce the pressure in container 2 to such a degree that the free path length of the molecules therein is greater than the inner dimensions of the container. Thus, the molecules fly from wall to wall, without hitting any other molecule. For example, if the pressure is reduced, e.g., to 0.00001 mbar, the free path for air at 20 degrees C. increases to 6 meters, and thus the molecules fly directly from one wall to the other in a container with a height of 1 meter.

[0028] In another embodiment of the invention, container 2 containing a gas with a lower pressure, also called a rarified gas, is divided by a number of horizontally arranged thin films (such as, e.g., films comprising thin metal or plastic). Using, for example, ten evenly spaced apart films within the container, allows one to increase the pressure of the gas to 0.0001 mbar, thereby reducing the free path to 0.6 meters. The maximum vertical length of the free space is thus only 0.1 meters, and the molecules therein then fly directly from wall to wall. The higher pressure also has the advantage that the amount of energy transported from the upper to the lower wall increases in proportion with the pressure.

[0029] The opposite, namely to increase the gas pressure can have a beneficial effect, as the amount of energy transferred from the top of the container to its bottom increases proportional to the gas pressure.

[0030] The temperature difference between the top and the bottom can be calculated, as it is proportional to the difference of the potential energy of the molecules at the top and at the bottom, as follows:

[0031] E=W.times.H where E is potential energy, W is weight, and H is vertical distance (i.e. height).

[0032] This amount of energy equals the temperature increase of the mass in question:

[0033] E=M.times.c(Gr).times..DELTA.T, where c(Gr) is the specific heat c(for gases cv) divided by n where n is the number of freedom of degrees, .DELTA.T is the temperature difference between the two ends of the column, and M is the mass of the molecules.

[0034] Therefore:

W.times.H=M.times.c(Gr).times..DELTA.T or

[0035] 1 T = W M .times. H c (G r) with W M = g 2 T = - g .times. H c (G r)

[0036] (minus as gravity acts against direction of H)

[0037] Using the above formula, one can calculate the temperature differences for various atoms and molecules. Table 1, below, gives the results for a height of one meter for various materials.

1TABLE 1 Material Aluminum Lead Water Glass Mercury Xenon Iodine Gas Air Specific heat 900 130 4185 800 140 160 140 718 (m.sup.2/sec.sup.2, .degree. C.) .DELTA.T (.degree. C.) -0.009 -0.08 -0.04 -0.009 -0.06 -0.311 -0.355 -0.07

[0038] FIG. 1 shows the simplest form of the invention. Enclosure 2, preferably comprising aluminum with a thickness of about 2 cm, has an interior or central chamber 12, which is filled with a gas or liquid. This massive aluminum enclosure helps to equalize the temperature of the inside surface of the enclosure as much as possible. The enclosure is insulated from its surroundings by an insulating material 3, such as, for example, polystyrene foam in a thickness of about 50 cm. Preferably, chamber 12 is filled with a mass of very fine solid or hollow fibres 6, such as, for example, glass wool (e.g., Fiberglas.RTM.), natural fibres, or other materials such as foam, powder, or small solid or hollow spheres (e.g., comprising plastic, glass or metal), in order to reduce convection currents and radiation effects to a minimum. The fibres are optionally used to help avoid convection currents in the enclosed gas or liquid, as this tends to decrease the creation of the temperature difference, upon which the present invention is based.

[0039] Referring again to FIG. 1, thermocouple 4 measures the temperature difference between the top and the bottom (or some other vertical segment) of the thermal differential element (i.e., gas in this example). Wires 5 transfer the voltage output (V+/-) from thermocouple 4 to the outside of chamber 12. For example, using a Type E thermocouple, in which one leg of the thermocouple is made from copper/nickel and the second leg is made from nickel/chromium, the system of the invention yields a voltage of about 70 mV per 100 degrees C. In order to increase the usefulness of this voltage, one optionally employs a large number of thermocouples arranged in sequence, such as in a thermopile, as shown schematically in FIGS. 2 and 3. Thus, using 100 thermocouples in sequence, the voltage in example 1 increases to 7V per 100 degrees C.

[0040] As predicted by Table 1, one measures a temperature difference between the top and the bottom for the enclosure with a vertical height of 1 meter of about 0.01 to 0.3 degrees C. In this case, the temperature differential depends on the relative weights and weight distribution of the gas molecules, the glass wool, and the material of the enclosure. The choice of such materials is within the skill of one of ordinary skill in the art, based on the teachings herein, and determines the extent of the temperature difference obtained in the system of the present invention.

[0041] Referring to FIG. 2, a lead rod 1 with a diameter of about 5 cm and a length of about 100 cm is arranged in a vertical position within enclosure 2, thereby forming the temperature differential element of the invention. The enclosure of this example is built with a diameter of about 20 cm and a height of about 120 cm, and the space between rod 1 and enclosure 2 is filled with loose glass fibres 6.

[0042] Although the height of the chamber in this example is shown as being much larger than the width or depth, such a limitation is not required of the invention. The height of the container simply determines the maximum possible temperature differential, whereas the total mass of the temperature difference element determines the maximum possible energy transfer from the top to the bottom.

[0043] The use of lead rod 1 in FIG. 2, as opposed to a gas or liquid as shown in FIG. 1, can be predicted from Table 1 to produce a larger temperature differential, due to the smaller specific heat of lead as compared to a typical gas.

[0044] A further increase in the usefulness of the invention optionally is obtained by increasing the length of rod 1 and by combining a number of enclosures 2 in one installation.

[0045] Alternatively, if the mass is a solid as shown in FIG. 3, the space inside enclosure 2 can be evacuated, and the inside surfaces 7 are optionally made highly reflective to radiation, in order to insulate rod 1 from the surrounding enclosure and thereby avoid heat transfer between them as much as possible.

[0046] As shown in FIG. 3, the apparatus of the invention optionally is surrounded by a housing 8, which preferably is kept at a relatively constant temperature (e.g., about +/-0.5 degrees C.) through a thermostatically controlled heating system 9, which optionally is powered from outside the surrounding area by electrical connections 10, liquid flow, or like means. Improved thermal insulation also is obtained optionally by evacuating the space or by dividing the space in two parts, as shown in FIG. 3, one being filled with insulation 3 and one being evacuated 11.

[0047] It is also within the scope of the invention to reduce the pressure in chamber 12 to, for example, 0.0001 mbar, using a rarified gas or gases with a free path greater than the inner dimensions of the container. Referring to FIG. 4, container 2 containing a gas with a lower pressure, also called a rarified gas, is divided by a number of horizontally arranged thin films 13 (such as, e.g., films comprising thin metal or plastic). Using, for example, ten evenly spaced apart films within the container, allows one to increase the pressure of the gas to 0.0001 mbar, thereby reducing the free path to 0.6 meters. The maximum vertical length of the free space is thus only 0.1 meters, and the molecules therein then fly directly from wall to wall. The higher pressure also has the advantage that the amount of energy transported from the upper to the lower wall increases in proportion with the pressure.

[0048] It is further within the scope of the present invention optionally to expose the enclosure to any constant field, be it gravitational, electrical or electromagnetic. Any known solid, liquid or gaseous material optionally is used for creating these temperature differences, but those with a low specific heat and/or a high number of degrees of freedom are most advantageous.

[0049] In order to use the method of the invention in accordance with the preferred embodiment, the surroundings of the apparatus of the invention preferably have a temperature as constant as possible. This generally is difficult to attain, as in any space or room, there is typically a temperature distribution over height of about 1 degree C. per meter. In order to alleviate this problem, in a further embodiment of the invention, an apparatus in accordance with FIGS. 5 and 8 optionally is used. Referring to FIGS. 5 and 8, an embodiment of the present invention comprises a drum-like enclosure 14 that has a horizontal axle 15. The axle is held in bearings 16 supported by stand 17. A motor 18 (or similar suitable means) is connected through chain drive 19 with axle 15, and turns drum 14 around its horizontal axle at a rate of about 1-10 times per minute. Drum 14 preferably comprises a metal with a high heat conductivity, such as, for example, aluminum, and preferably is relatively thick, such as, for example, about 3 cm, such that drum 14 has a relatively small temperature difference within its body. The rotation of drum 14 tends to equalize the temperature difference of its surroundings, typically from about 1 degree C. per meter of height to a very small temperature difference, thereby increasing the efficiency of the apparatus.

[0050] Drum 14 typically has a cover 20 (or similar suitable means) for allowing access to container 21. Container 21, in turn, has its own horizontal axle 22, which is mounted inside axle 15 in such a way that it is able to remain stationary, while drum 14 rotates around it. Container 21 is held in a stationary vertical position via arm 23, which optionally is fixed in its position through a removable bolt 24 (or similar suitable means) bolted to stand 25. The apparatus in accordance with FIGS. 1-4 optionally is installed inside container 21, through covers 20 and 26. Due to the rotation of drum 14, container 21 shows a temperature difference of only about 0.001 degrees C., as measured over a height of about 1 meter between the top and the bottom of container 21. As a result, the temperature difference due to the effect of gravity inside container 2, in accordance with FIGS. 1-4, is much more pronounced, as the temperature differential is thus not negatively affected by a temperature differential within the room in which the apparatus of the invention is located. The axle on one side of container 21 is optionally hollow, such that any thermocouple wires coming from container 2 can be led to the outside thereof, to be connected to a voltmeter, generator, or similar suitable device.

[0051] FIG. 6 shows the optional use of Dewar glasses 27 and 28. Dewar glasses are two-walled glass containers, which are very well known in the prior art. The space between the two glass walls typically is evacuated to a pressure of about 0.001 mbar. This results in a very low heat conductivity between the two walls. Therefore Dewar glasses typically are used to store liquids that would normally exist in the gaseous state at ambient temperature. In accordance with the invention, the Dewar glasses form the elongated enclosure, and enclose the interior chamber 29, which is extremely well insulated from the area outside the enclosed system. The Dewars can be placed in a housing 62, preferably covered in insulation 63, and the space 72 between the outer Dewar 27 and the housing 62 can be evacuated or filled with foam, solid or hollow fibers or small particles, such as, for example, solid or hollow spheres or sand.

0052] FIG. 6 shows the use of heat exchangers 60 and 61, in place of the thermocouples shown in other figures. This is an alternative method of extracting energy from the mass, in the form of heat or cold, by passing a heat-exchange fluid through the exchanger. Heat exchanger 60 is the colder, heat exchanger 61 the warmer exchanger. The energy from exchanger 60 or 61 could be used directly to warm or cool something, or used to power a refrigerator, or provide heat to drive a heat-powered device such as a Stirling cycle engine. In this arrangement the temperature difference element would act like the pump in a heat pump, transferring energy from the heat exchanger 60 to the heat exchanger 61.

[0053] To use the temperature difference to create electrical energy or to use it through a heat exchanger producing a warm and/or cold liquid, it is advantageous to have a temperature difference as high as possible. To reach a higher temperature difference by increasing the height of the temperature differential element is often not possible due to space limitations. In accordance with the invention a number of elements can be combined in one housing 3 as shown in FIG. 7 for three elements. They are arranged in sequence to each other by connecting the cold top of element 31 with the warm bottom of element 32 and the warm top of this element with the bottom of element 33. The connecting element 30 can be a metal rod with a high heat conductivity, e. g. made from copper or silver, or it can be a temperature differential element filled with a gas having a smaller temperature difference per meter of height than the elements 31, 32, and 33. Through this temperature difference between the bottom of element 31 and the top of element 33 is greater than the temperature difference of one element alone.

[0054] Accordingly, it is to be understood that the embodiments of the invention herein described are merely illustrative of the application of the principles of the invention. Reference herein to details of the illustrated embodiments are not intended to limit the scope of the claims, which themselves recite those features regarded as essential to the invention.

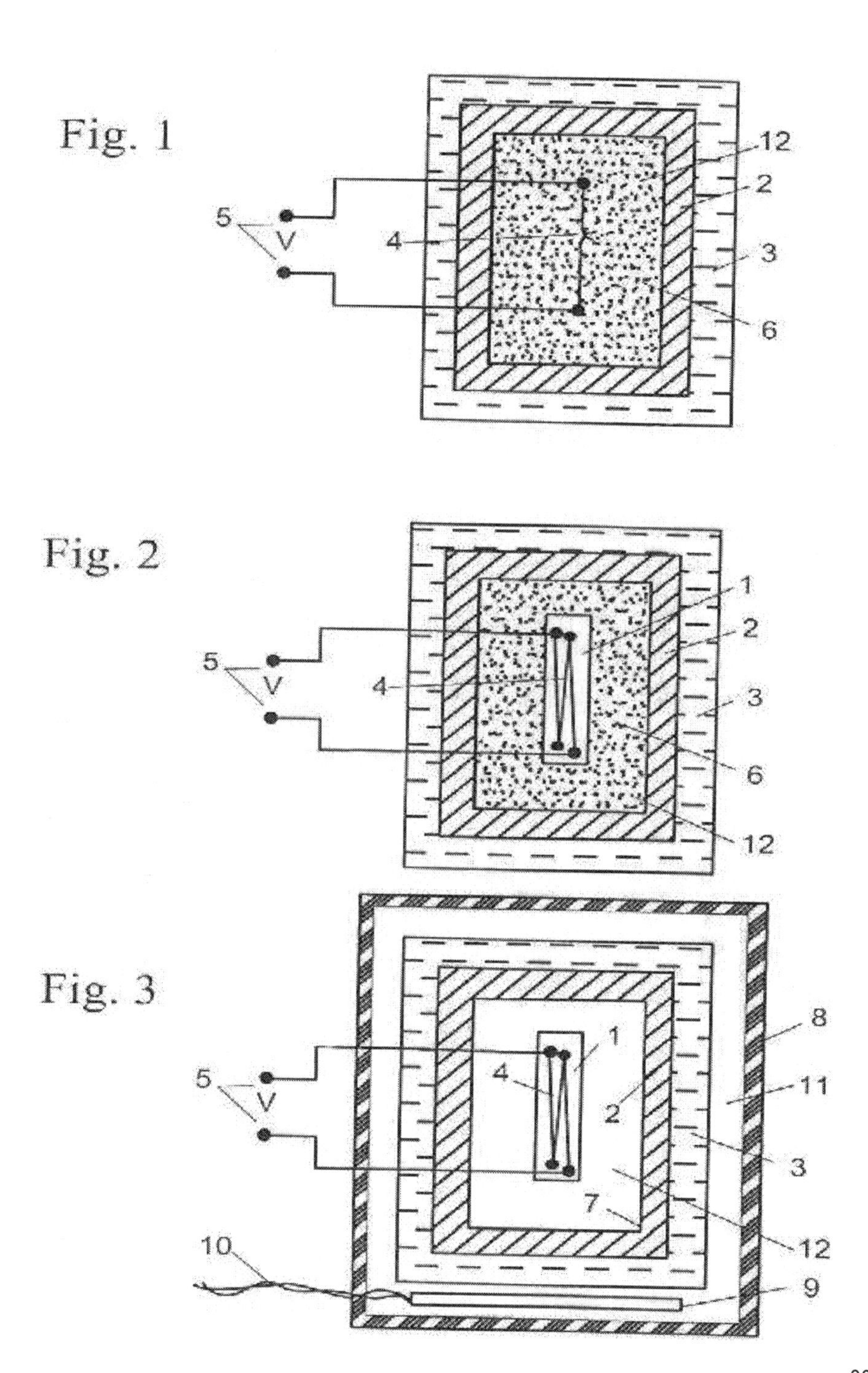
Fig. 1
12
2
5
V
4
3
6
Fig. 2
1
2
5
V
4
3
6
12
Fig. 3
8
1
4
5
V
2
11
3
7
12
10
9

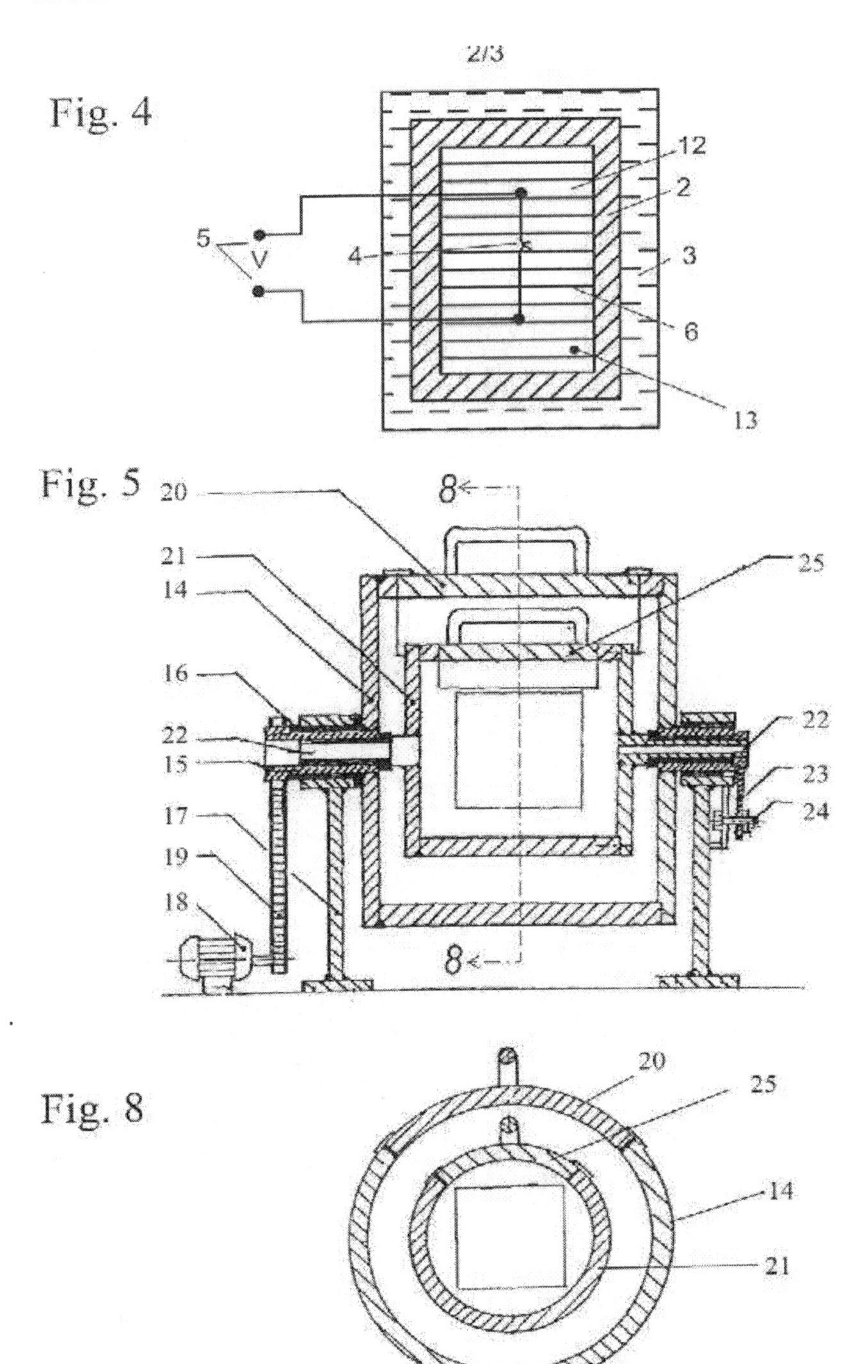
2/3
Fig. 4
12
2
5
V
4
3
6
13
Fig. 5
20
8
21
14
25
16
22
15
22
23
24
17
19
18
8
Fig. 8
20
25
14
21

3/3

Fig. 6

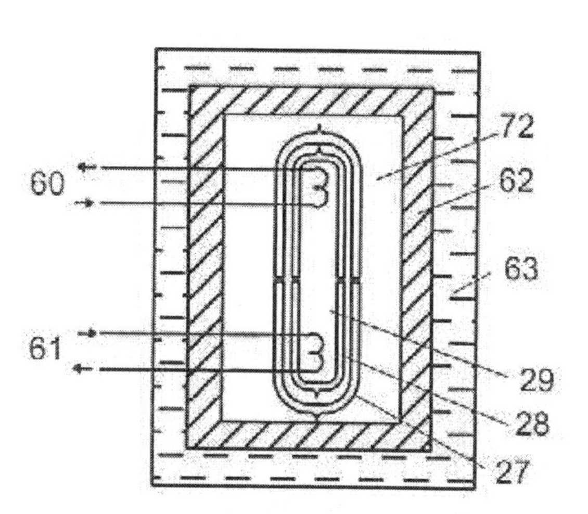

Fig. 7

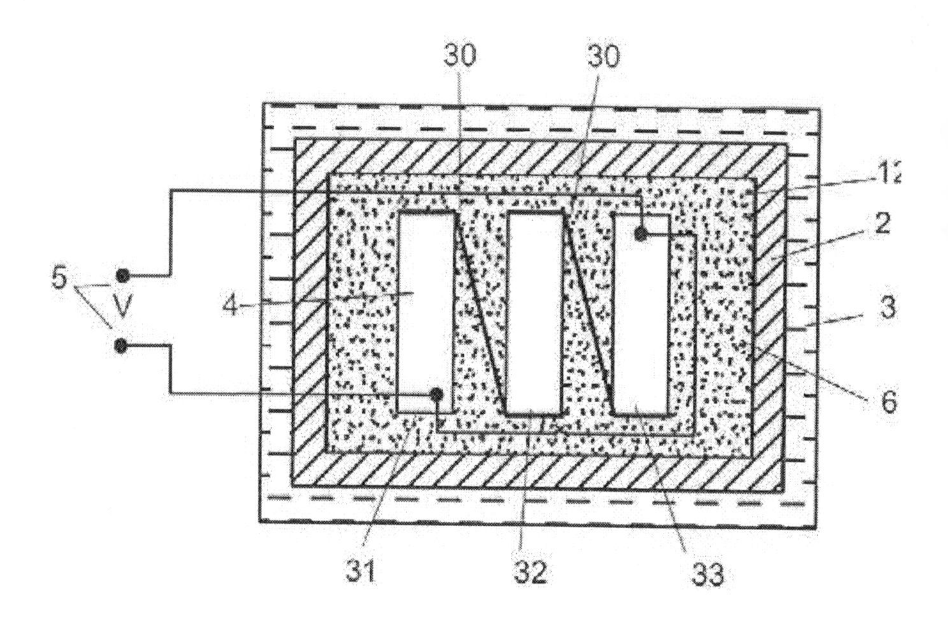

My Path to Peaceful Energy
Appendix 4

B76 San Diego: MEASURING THE TEMPERATURE DISTRIBUTION IN GAS COLUMNS

Roderich W. Graeff

International Licensing
Prof.-Domagk-Weg 7, D-78126 Königsfeld, Germany
102 Savage Farm Drive, Ithaca, NY 14850-6500, USA
E-mail rwgraeff@yahoo.com

Abstract. Late in the 19th century J. Loschmidt believed that a vertical column of gas in an isolated system would show a temperature gradient under the influence of gravity, cold at the top and warm at the bottom. L.Boltzmann and J.C. Maxwell disagreed. Their theories tried to prove an equal temperature over height.
Experiments with various test setups are being presented which seem to strenghten the position of Loschmidt. Longterm measurements at room temperature show average temperature gradients of up to 0,07 °K/m in the walls of the enclosure, cold at the top and warm at the bottom.
The measured values can be explained by the conversion of the potential energy of the molecules into an increase of their average speed through gravity.

This paper was presented on July 30, 2002 in the conference *Quantum Limits to the 2nd Law: First international conference*
and published by American Institute of Physics, New York, AIP Conference Proceedings, 2002, vol. 643
In the original the graphs, especially Figure 4 and 5, are printed in color with the result that they are much better understandable. The original paper in color can be downloaded from the webpage www.firstgravitymachine.com.

MEASURING THE TEMPERATURE DISTRIBUTION IN GAS COLUMNS

Roderich W. Graeff

International Licensing
Prof.-Domagk-Weg 7, D-78126 Königsfeld, Germany
102 Savage Farm Drive, Ithaca, NY 14850-6500, USA
E-mail rwgraeff@yahoo.com

INTRODUCTION

In the late 19th century an animated discussion went on between L. Boltzmann [1], J.C.Maxwell [2], and J. Loschmidt [3]: Would a vertical column of gas under equilibrium conditions in a closed system show equal temperatures over height under the influence of gravity? The 2nd law of thermodynamics seemed to demand this. Only J. Loschmidt presented a different position. He argued for a temperature gradient over height, cold at the top and warm at the bottom. A. Trupp [4] gives a good overview over this debate. In a meeting of the Austrian Imperial Academy of Science in February 1876 Loschmidt declared [3]:

... *With this the terroristic nimbus of the Second Law which makes it look like the destructive principle of any life in the universe would be destroyed. On the other side it opens up the comforting perspective that humanity is not solely dependent upon using coal or the sun to produce work out of heat, but that for all times an inexhaustible reserve of changeable heat will be available*

The author reports on experiments measuring the temperature distribution in vertical columns of rarified and dense gases, primarily air and Xenon. The height of the columns varies from 0,2 m to 2 m. Special efforts were taken to create an isolated system as close to a "closed system" as possible. Thermocouples were used to measure temperatures in order to avoid any introduction of energy into the systems.

TYPICAL TEST ASSEMBLY

Figure 1 shows a typical experimental setup. A Dewar insert of a commercial Thermos bottle (1) is mounted within a wide mouth Dewar insert of 1 liter size (2) which is covered by a similar Dewar insert of ½ liter (3). The space (4) between the innermost Dewar and the two outside Dewar inserts is filled with fine PET fibers. The innermost Dewar (1) of ½ liter is filled with a fine powder in order to eliminate convection currents and radiation between the inner wall surfaces. A thermocouple (5) is arranged in the middle axle with a height distance between junctions of 170 mm. A second thermocouple (6) is mounted on the outside of the Dewar insert (1) with a vertical distance of 180 mm. A third thermocouple (7) is mounted on the outside of Dewar insert 2 with a vertical distance of 230 mm.

The Dewar inserts are held in place by fine PET fibers within a box (8) fabricated out of 40 mm thick polystyrene foam panels. This box is surrounded by 50 mm thick panels (9) consisting of pressed copper wires. The whole setup is insulated against the room air by 100 mm thick polystyrene foam panels (10).

The diameter of the thermocouple wire is kept as small as possible in order to reduce any heat conduction through these wires. The measuring instrument measures the voltage of the thermocouples with a resolution of 1×10^{-7} V corresponding to 0,0003 °K.

FIGURE 1

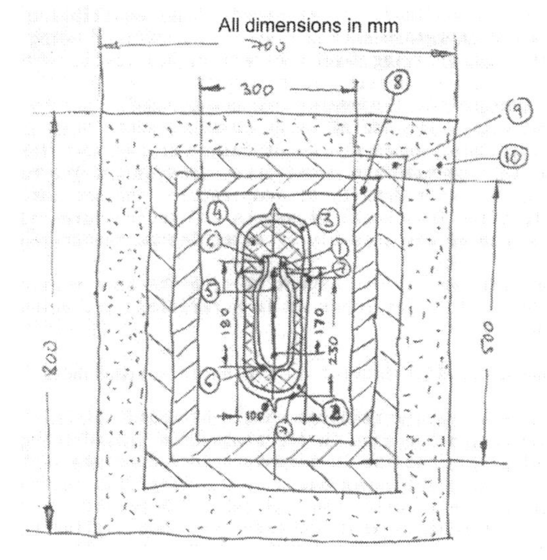

Typical test setup of Dewar glasses arranged in an isolated box

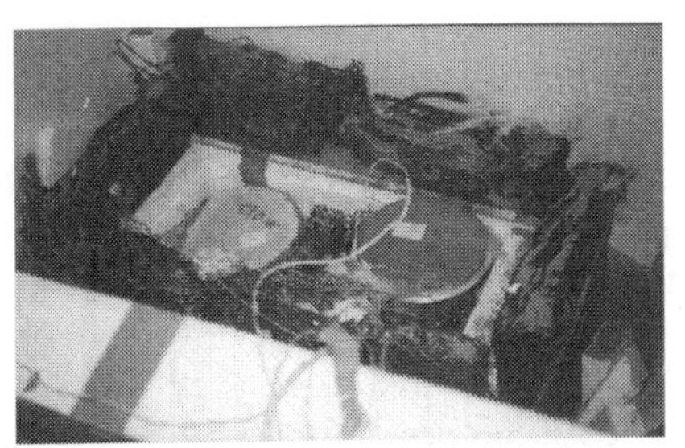

FIGURE 2: Assembly of Dewar inserts and highly insulated box with 3 test columns

TEST ENVIRONMENT

An isolated system demands that there is no material exchange between the system and the outside and no energy exchange between the gas and the surrounding area. While it is practically possible to fulfill the first condition it is impossible to exclude any exchange of energy. Even the best insulation will always transmit some energy based on the temperature difference between the inside and the outside.

A typical test assembly of a number of Dewar glasses is shown in figure 2. Half of one assembly consisting of three Dewar glasses is shown. They would be mounted inside aluminum containers two of which are shown on the right side of figure 2. They are insulated from the outside by 40 mm thick polystyrene foam panels, 50 mm thick copper blocks made out of copper wires pressed together, and finally by 100 mm thick polystyrene foam panels.

Figure 3 shows a double drum apparatus [5]. While the inner drum remains in its vertical position the outer drum made of 25 mm thick aluminum plates rotates continuously around it. This creates a temperature difference, warm at the top and cold at the bottom, of only about 0,002 ^{o}K per the 50 cm of height of the inner drum. Now the inner drum with the test assemblies can be turned by rotating it on its head by 180° without interrupting the temperature measurements or the quality of the insulation.

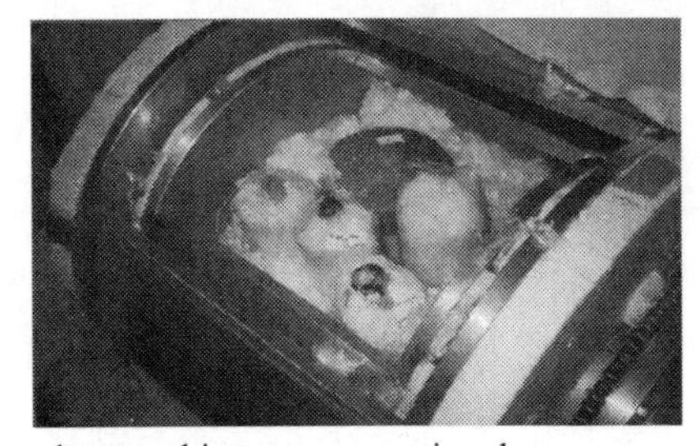

FIGURE 3: Apparatus with outer rotating drum and inner non-rotating drum

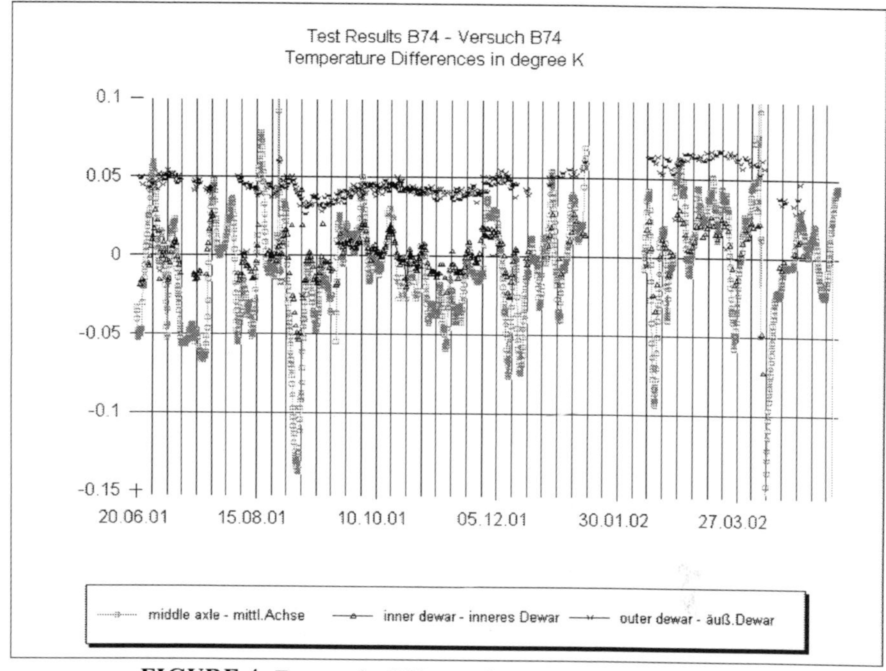

FIGURE 4: Test results B74 over time span of half a year in °K

TEST RESULTS

It is very difficult to give meaningful values for the accuracy of the temperature measurements. I believe that values given are accurate to about ± 0,002 °K. Such a low value is possible because we do not measure absolute temperatures but temperature differences. For measuring small values we use five leg thermopiles instead of single thermocouples which increase the accuracy. For those measurements where the absolute values are not as important as the direction of the temperature gradient, each temperature was measured with switched polarity.

The temperature differences shown by the three thermocouples were recorded every 4 minutes over a time of half a year. Figure 4 shows a printout where each point represents the average of a 4-hour time span equalling the average of 60 temperature measurements.

Any point above the zero line means that the upper junction of the thermocouple is warmer than the lower junction.

It can easily be seen that the the outside of Dewar insert 2 is always warmer at the upper junction than at the lower one by about 0,040 °K corresponding to 0,174 °K/m. This is the result from the temperature gradient in the surrounding room air, which is warmer at the top and colder at the bottom, typically by 1 °K per meter of room height. The insulation around the Dewar inserts and especially the thick copper plates are the reasons for reducing this gradient for the Dewar inserts.

Contrary to this, at the inner axle most of the time the temperature of the upper junction is lower than at the lower one.

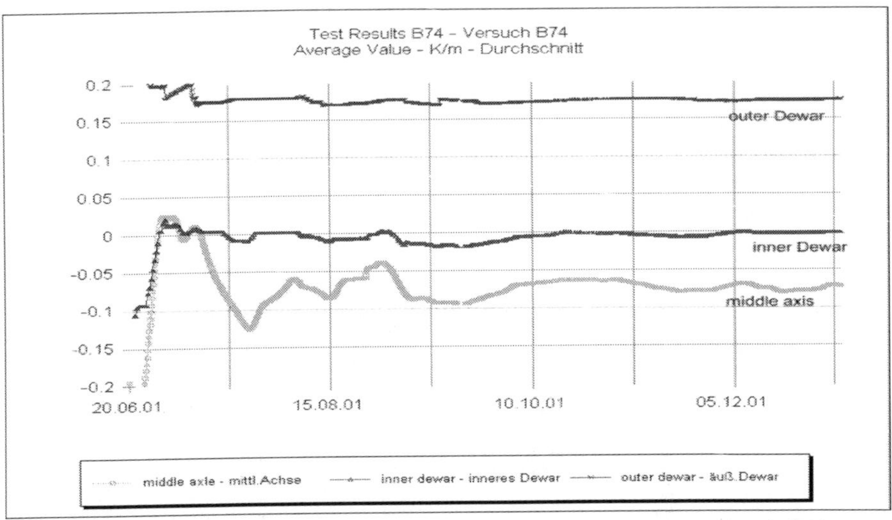

FIGURE 5: Test results B74 over time span of half a year: Average values in °K/m

This result is more clearly demonstrated in figure 5 showing average values from the beginning of the test until the date in question. Table 1 shows the average values for the half year test span for the actual distance between junctions of the thermocouples and normalized for a height of 1 m.

TABLE 1. Average Temperature Differences over 6 Months:

	from figure 5:	distance between junctions:	for 1 m:
Dewar 2	+0,040 °K	0,23 m	+0,174 °K
Dewar 1	-0,001 °K	0,18 m	-0,0055 °K
Middle axle:	-0,012 °K	0,17 m	-0,0705 °K

The average value over the test span of half a year for Dewar 2 shows clearly a positive value, warm at the top and cold at the bottom. The temperature difference for Dewar 1 was practically zero. The average for the middle axle shows a negative value meaning that in the average the top was colder than the bottom.

INTERPRETATION OF TEST RESULTS

That the middle axle shows, on the average, a temperature gradient, cold at the top and warm at the bottom, cannot be explained by applying the normal formulas for the conduction of heat. Under these laws, heat can travel only from an area with a higher temperature to an area with a lower temperature. But my test results can be explained by assuming that the molecules traveling vertically are accelerated by gravity in their downward path and decelerated in their way upwards. On the average the molecules will hit the upper wall of the enclosure with a lower speed than the lower wall. Elastically bouncing off the walls with the temperature of the wall, under equilibrium conditions, the upper wall would show a lower temperature than the lower one. This temperature difference due to the influence of gravity I call T(Gr).

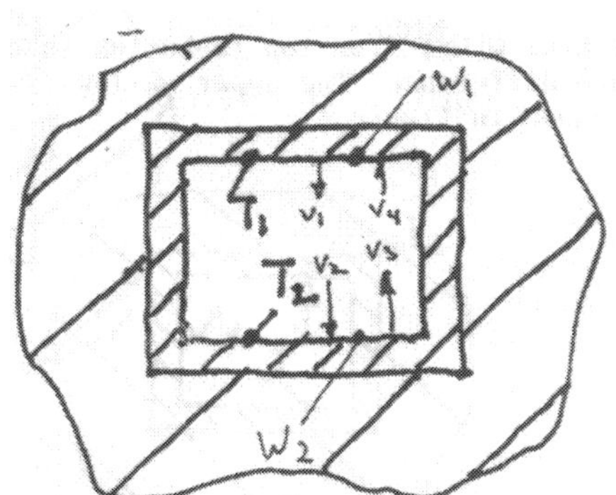

The value of T(Gr) can be calculated by equating the potential energy of the molecules with the increase of their speed which represents their temperature. At least for a rarified gas no compression work has to be performed on the downward path. Therefore calculating with the specific heat c_p, for air we get

$$T(Gr) = -0{,}01\ ^{\circ}K/m$$

This value is typically found in the atmosphere [6]. With higher gradients a column of gas would be instable. Therefore the innermost measuring area was filled with fine powder in order to avoid any convection currents. Furthermore the higher values found in the experiments can be explained only by assuming that acceleration by gravity affects only the vertical speed component of a molecule negating the equipartition of energy

It is believed that this argument for the conditions of a rarified gas is equally valid for dense gases. Details are discussed in [7].

SUMMARY

Measuríng the temperature distribution in isolated spaces filled with a gas and a powder a vertical temperature gradient was found, cold at the top and warm at the bottom as argued by J. Loschmidt. [3].

This result seems to be a contradiction to the 2nd law of thermodynamics. If correct, it would make possible the creation of work out of a heat bath.

ACKNOWLEDGEMENTS

Extremely helpful were numerous discussions with Dr. Alexander Haltmeier which covered many aspects of the experiments and especially their theoretical interpretations. I am thankful to Mrs. Gisela Hoffmann who supervised the data collection systems and generated the graphs presented here.

REFERENCES

[1] Boltzmann, L., *Wissenschaftliche Abhandlungen, Vol.2, edited by F.Hasenoehrl,* Leipzig, 1909

[2] Maxwell, J.C., *The London, Edinburgh, and Dublin Philosophical Magazine and Journal of Science*, **35**, p. 215 (1868)

[3] Loschmidt, L., *Über den Zustand des Wärmegleichgewichts eines Systems von Körpern mit Rücksicht auf die Schwerkraft,* Sitzungsberichte der Mathematisch-Naturwissenschaftl. Klasse der Kaiserlichen Akademie der Wissenschaften **73.2** (1876), p. 135

[4] Trupp, A., Physics Essays, **12**, **No. 4**, 1999

[5] Graeff, R.W., *Gravity Machine,* Web Site <*firstgravitymachine.com*>

[6] Bohren, C.F. *Atmospheric Thermodynamics*, pp. 167 ff., Oxford University Press, New York, 1998

[7] Graeff, R.W., *Gravity Machine, my Search for Peaceful Energy*, to be published September 2002

My Path to Peaceful Energy
Appendix 5

VIEWING THE CONTROVERSY LOSCHMIDT – BOLTZMANN/MAXWELL THROUGH MACROSCOPIC MEASUREMENTS OF THE TEMPERATURE GRADIENTS IN VERTICAL COLUMNS OF WATER

Roderich W. Graeff

International Licensing
Prof.-Domagk-Weg 7, D-78126 Königsfeld, Germany
102 Savage Farm Drive, Ithaca, NY 14850-6500, USA
E-mail rwgraeff@yahoo.com

Abstract

In order to clarify the dispute between Loschmidt and Boltzmann/Maxwell concerning the existence of a temperature gradient in insulated vertical columns of gas, liquid or solid, macroscopic measurements of the temperature distribution in water were performed. A negative temperature gradient, cold at the top and warm at the bottom, is found in insulated vertical tubes, while the outside environment has a reverse gradient. This is explainable by the influence of gravity. These test results strengthen the suggestions of Loschmidt, and contradict the statements of Boltzmann and Maxwell

In the original the graphs, especially Figure 2, 3, and 4, are printed in color with the result that they are much better understandable. The original paper in color can be downloaded from the webpage www.firstgravitymachine.com.

VIEWING THE CONTROVERSY LOSCHMIDT – BOLTZMANN/MAXWELL THROUGH MACROSCOPIC MEASUREMENTS OF THE TEMPERATURE GRADIENTS IN VERTICAL COLUMNS OF WATER

Roderich W. Graeff

Private Scholar

Prof.-Domagk-Weg 7, D-78126 Königsfeld, Germany

102 Savage Farm Drive, Ithaca, NY 14850-6500, USA

E-mail rwgraeff@yahoo.com

Abstract

In order to clarify the dispute between Loschmidt and Boltzmann/Maxwell concerning the existence of a temperature gradient in insulated vertical columns of gas, liquid or solid, macroscopic measurements of the temperature distribution in water were performed. A negative temperature gradient, cold at the top and warm at the bottom, is found in insulated vertical tubes, while the outside environment has a reverse gradient. This is explainable by the influence of gravity. These test results strengthen the suggestions of Loschmidt, and contradict the statements of Boltzmann and Maxwell.

Key words: temperature gradient, gravity, water, Second Law, isolated system, energy production, heat bath, Maxwell, Boltzmann, Loschmidt

INTRODUCTION

Late in the 19th century J. Loschmidt believed that a vertical column of gas or a solid in an isolated system would show a temperature gradient under the influence of gravity, being cold at the top and warm at the bottom. L. Boltzmann and J. C. Maxwell disagreed. Their theories and understanding of the Second Law supported an equal temperature over height. This historical discussion between J. Loschmidt, L. Boltzmann and J. C. Maxwell is covered in [1], [2], and [3]. A. Trupp gives a good summary in [4].

The author reported for the first time in [5] and [8] actual measurements of the temperature gradient in gas columns in isolated systems. The value found for air with -.07 K per meter of height seems to strengthen the position of Loschmidt. These experiments are critically discussed by Sheehan [10].

In trying to reach more stable results, the measurements were extended to vertical columns filled with a liquid. A first report was published in [9]. Water was selected, because of its high density. This way temperature fluctuations of the environment affect the temperature gradient less than when measuring gases.

It is known that temperature gradients in gases and liquids are stable only up to the adiabatic lapse rate [6]. Higher negative values are not possible, because the column of a gas or a liquid becomes instable. Lower temperature at the top than at the bottom create higher densities at the top resulting in convection currents which would diminish the temperature gradient to values below the adiabatic lapse rate. In order to make greater values possible, the author tried various convection-suppressing designs. It was found that the use of fine powders, like glass powder, largely eliminated convection currents. It had the added advantage that it prevented any heat exchange by radiation within the test setup.

The historical dispute between J. Loschmidt, L. Boltzmann and J. C. Maxwell

In trying to formulate and understand the Second Law, Boltzmann calculated in 1868 that a column of gas should have the same temperature at the top and at the bottom, but his calculations were limited to ideal gases.

Loschmidt disagreed with some of his conclusions and assumptions. He thought that gravity would create a temperature gradient, cold at the top and warm at the bottom, especially in solids. He felt that this would not contradict the Second Law and had the following vision for the future:

"Thereby the terroristic nimbus of the Second Law is destroyed, a nimbus which makes that Second Law appear as the annihilating principle of all life in the universe, and at the same time we are confronted with the comforting perspective that, as far as the conversion of heat into work is concerned, mankind will not solely be dependent on the intervention of coal or of the sun, but will have available an inexhaustible resource of convertible heat at all times" [3] .
Loschmidt never explained, why a temperature gradient would not contradict the Second Law. He believed that only measurements could decide this dispute but knew that improved sensors and instruments would be needed to measure the small gradients he expected.

Maxwell expected equal temperatures at the top and bottom and in his book "Theory of heat", published in London in 1877, he writes (p. 320):

"... if two vertical columns of different substances stand on the same perfectly conducting horizontal plate, the temperature of the bottom of each column will be the same; and if each column is in thermal equilibrium of itself, the temperatures at all equal heights must be the same. In fact, if the temperatures of the tops of the two columns were different, we might drive an engine with this difference of temperature, and the refuse heat would pass down the colder column, through the conducting plate, and up the warmer column; and this would go on till all the heat was converted into work, contrary to the Second Law of thermodynamics. But we know that if one of the columns is gaseous, its temperature is uniform. Hence that of the other must be uniform, whatever its material."

THEORETICAL VALUE FOR TEMPERATURE GRADIENT T(Gr)

No published treatise is known to the author for calculating the vertical temperature gradient T(Gr) in solids or liquids under the influence of gravity. But the value of T(Gr) can be calculated by equating the potential energy of the molecules to the increase of their speed on their downward path. Their speed is related to their temperature. When bouncing off the bottom wall their kinetic energy is zero at the moment of impact. Though the loss of potential energy on their downward movement their energy is totally converted to an increase of their average "temperature". A heat transfer takes place between water molecules and the upper and the lower walls of the tube, until the wall temperatures are equal to the "temperature" of the impinging water molecules and equilibrium has been reached.

The potential energy is

$$E_p = -M \cdot g \cdot H$$

with M = mass; g = constant of gravity; H = height difference

(negative, because g and H are measured in opposite directions)

We equate this potential energy E_p with the amount of energy available for a temperature increase of this mass

$$E_{avail} = M \cdot c_{Gr} \cdot T$$

with c_{Gr} = effective specific heat; T = Temperature difference

We now can equate E_p wit E_{avail} or

$$E_p = E_{avail} = M \cdot g \cdot H = M \cdot c_{Gr} \cdot T$$

or

$$T = \frac{g \cdot H}{c_{Gr}} = T_{Gr}$$

c_{Gr} is not the normal specific heat of the liquid in question, because the acceleration through g affects only the vertical speed component of the molecule. The potential energy is converted only into an increase of their speed in their lateral downward direction. No energy is added to the other degrees of freedom, like the remaining two lateral directions (left to right and front to back) or to the rotational energy in molecules consisting of more than one atom. Therefore,

$$c_{Gr} = \frac{c}{n}$$

with c = specific heat; n = number of degrees of freedom

We therefore get

$$T_{Gr} = \frac{-g \cdot H}{c_{Gr}} = \frac{-g \cdot H \cdot n}{c}$$

With this formula for a height of 1 meter and taking the number of degrees of freedom for water as 18, we obtain

$$T(Gr) = -0.04 \text{ K/m}$$

EXPERIMENTAL SETUP

The column used in the experiments reported here has a height of 850 mm. It is chosen as a compromise between a greater height, allowing a greater temperature gradient, which is easier to measure, but having the difficulty to create a good insulation against the temperature fluctuations in the environment, and a smaller height with the opposite advantages and disadvantages.

Great care has to be taken to improve the accuracy of the temperature measurements. Temperature gradients are measured primarily with thermocouples. Often they are used as thermopiles connecting 5 thermocouples in serie. Critical values are measured twice with switched polarity correcting for any zero offset.

The experimental results are generated by three methods. In addition to hourly values an average value over time is calculated using the so called “future average” eliminating the initial time periods when equilibrium had not been reached yet. The third method seeks the values of the measured gradients at times of unchanging temperatures in the test column indicating periods of no heat flow in or out.

FIGURE 1

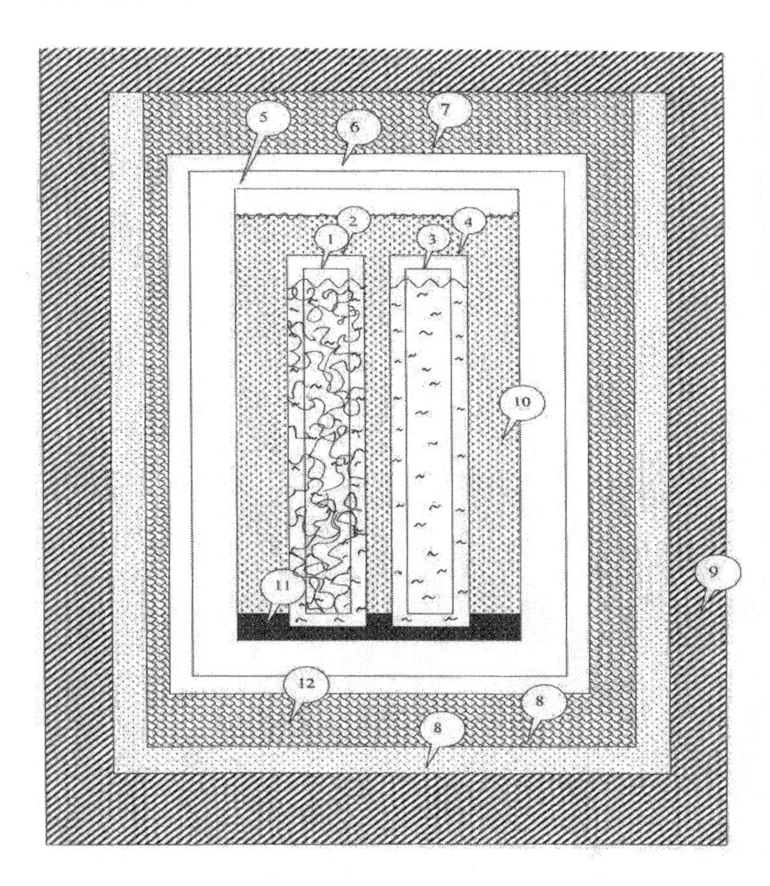

1: Glass tube 1, filled with water and glass powder, L= 850 mm, D= 40 mm
2: PVC tube 1, L= 910 mm, D= 50 mm .
3: Glass tube 2, filled only with water, L= 850 mm, D= 40 mm
4: PVC tube 2, L= 910 mm, D= 50 mm.
5: PVC tube 125 mm, L= 1000 mm, D= 125 mm
6: Aluminum tube 150 mm, L= 1100 mm, D= 150 mm
7: Aluminum tube 220 mm, L= 1200 mm, D= 220 mm
8: Double wall housing L= 1500 mm, D= 500 mm
9: Glass fiber insulation 100 mm
10: Glass foam, balls 1 mm
11: Brass shavings
12: PET fibers

Test B 372, as shown in Fig 1, measures the vertical temperature gradient in two identical glass tubes of 40 mm diameter and 850 mm length. Each glass tube is individually surrounded by a PVC tube of diameter 50 mm and length 910 mm. Tube 1 **(1)** and its surrounding PVC tube **(2)** are filled with water and fine glass powder, while tube 2 **(3)** and its PVC tube **(4)** only with clear water.

These are arranged in a PVC tube of diameter 125 mm and 1000 mm length **(5).** The remaining space is filled with small balls of glass foam of 1mm diameter **(10)**. The bottom part is filled with small brass shavings **(11)** in order to try to equalize the bottom temperatures of the two 50 mm PVC tubes.

The assembly is inside a 150 mm diameter and 1100 mm long aluminum tube of wall thickness 5 mm **(6)**. This in turn is placed into another aluminum tube of 220 mm diameter, 1200 mm length and a wall thickness of 5 mm **(7)**. Each of these is closed at the top with round aluminum plates of the same thickness.

The aluminum tubes containing the test assembly are standing in the center of a double walled aluminum housing of height 1500 mm and an inner diameter of 500 mm with 50 mm between the two walls **(8).** This space is filled with water. The whole assembly is insulated on the outside with 100 mm of glass wool **(9).** The space between the larger aluminum tube and the inner aluminum housing, i.e. between **(7)** and **(8),** is filled with fine PET fibers **(12).**

The temperatures inside the test setup are measured by thermocouples and by thermistors. These are mounted at the tops and at the bottoms of the inner axes of the two glass tubes. Additional sensors are mounted on the outside of these glass tubes and on the outside of the two PVC tubes. The temperatures of the double wall aluminum housing are measured 3 cm below the top and above the bottom.

EXPERIMENTS AND MEASUREMENT RESULTS

The test setup B372 was installed in May 2006. All sensors were connected to DMM Multimeter Keithley model 2700 and the data fed into a computer. Measurement results are reported from December 2006 through March 2007, a time period long after the setup, so that it can be expected that equilibrium conditions had been reached.

B372: H_2O in two glass tubes, one with and one without glass powder

FIGURE 2: Temperature gradients curve 1 – 4 and Thermistor temperatures 5 and 6

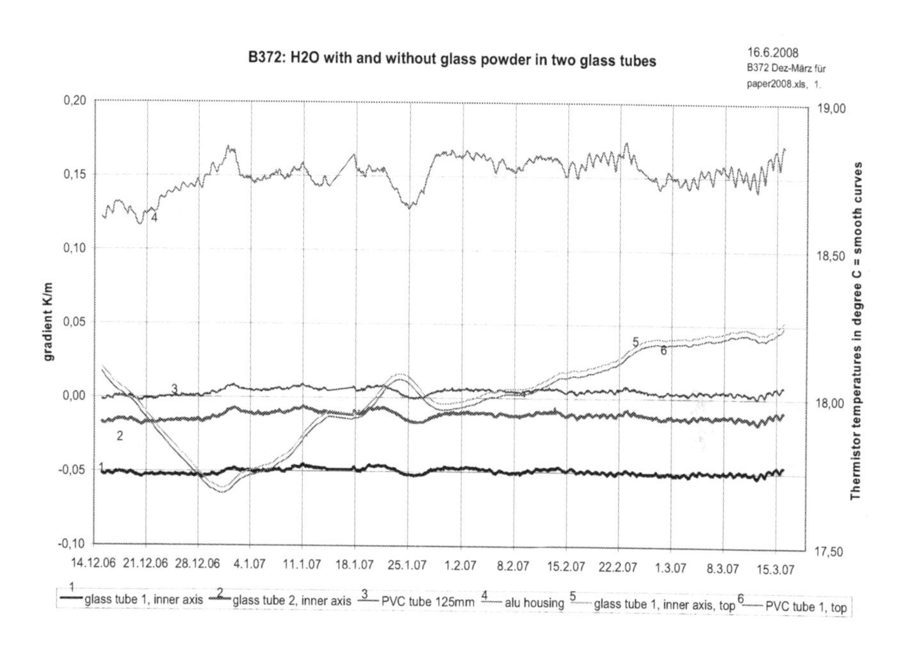

Fig. 2 shows 4 temperature gradients of the 8 values measured by thermocouples as temperature differences from December 14 through March 15. Each point of the curve represents a 10 value average (of a ten times repeated reading of the same sensor) measured every hour, using the scale on the left side of the graph. The smooth lines represent the 2 temperatures measured by thermistors of the 6 values measured hourly in centigrade, using the scale on the right side.

Environmental influences

Ideally, the measurements would take place in an isolated system not allowing the exchange of matter or energy across the boundaries. While the exchange of any matter can be eliminated, the exchange of energy cannot be avoided even with an optimal insulation. Because the temperature on the outside will always fluctuate to some degree, some energy will always pass in and out through the boundaries and influence the measurements.

The temperatures, measured by the thermistors at various locations within the test setup, give an indication of the amount of energy entering or leaving the system. From initial values around 18.10 C the temperatures all declined to about 17.50 C during the first 17 days (winter) and rose to a peak around 18.25 C during the following 50 days (spring). This gives a maximum change of only 0.75 C in a 13 week period that is caused by the temperature fluctuations of the environment.

Even an air-conditioned room can have such fluctuations. These experiments were carried out in a basement without air-conditioning, but with a thermostat-controlled heating system during the winter. The smooth parallel temperature curves (see curves 5 and 6 in Figure 2) indicate that the heat transfer took place uniformly in all parts of the test setup, not significantly disturbing the temperature differences that we tried to measure.

First method to establish a value for the vertical temperature gradient

The most important result is shown in curve 1 of Figure 2 (lowest blue curve), the temperature gradient of the inner axis of glass tube 1, filled with water and glass powder. It is quite stable around a value of about -.05 K/meter; the minus sign indicating a lower temperature at the top than at the bottom.

The glass tube filled only with water , tube 2, curve 2, (lowest red curve) with its value of about -.01 K/m, has a less negative gradient than tube 1. This is plausible, because tube 2 contains only water containing no convection-hindering glass powder like tube 1.

Next further out is the PVC tube of diameter 125 mm, curve 3, enclosing both the PVC tubes 1 and 2 and the glass tubes 1 and 2. It shows a slightly positive gradient close to zero , which means that the top is warmer than the bottom.

Also very important is the gradient on the inner wall of the aluminum housing, curve 4 with a value of +.15 K/m. It is always positive, warm at the top and cold at the bottom. Only under these conditions with positive gradients at the outside locations the negative gradients at the inner axis of tube 1 or 2 -- cold at the top and warm at the bottom -- become meaningful.

Temperatures within the experimental setup

As already discussed in the section "Environmental influences" the smooth curves (5 and 6 in Figure 2) represent temperatures measured by the thermistors. In comparing the measurements at different locations, one has to consider that the precision of a thermistor amounts to only +/- .1 C. But the measurements are very constant over time, as indicated by the smoothness of the curves, whereby the temperature change over time is measured to a much greater precision than the absolute values.

This fact becomes very important, when one looks at long time periods, during which the temperatures in a tube do not change. During these times one can find out, whether a temperature gradient T(Gr) exists under equilibrium conditions.

Second method to determine the temperature Gradient T(Gr) as a long term average

While in Figure 2 the curves show some variations over time, Figure 3 provides a better resolved picture in the form of long term averages.

FIGURE 3

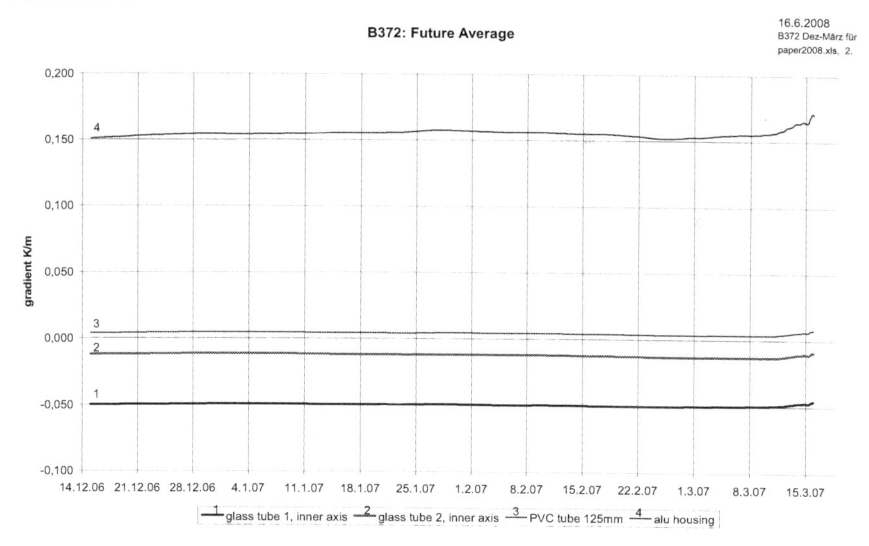

Fig. 3 "**B372: Future Average**" shows average values of all gradients over time. Each point on a curve represents the average value calculated for this gradient from that time through the last point of the curve to the right. For example, the values shown on December 14 are, therefore, the averages for the time from December 14 through March 15, while the last points on the right represent the values on March 15. Thus we can ignore the right end of the curves, where too few measurements are included in every point and the values are not meaningful. For the inner axis of tube 1 we get a steady average gradient of -.05 K/m.

Third method to determinate T(Gr) through a regression analysis at equilibrium

The measured values of T(Gr) fluctuate over time, because even the best insulation can not prevent small temperature changes in the test setup. In a regression analysis T(Gr) can be found very efficiently, when all measured temperature gradient values are plotted as a function of the rate of the temperature change (Figure 4). We measured these rates both for the top and at the bottom of the tubes, and found very similar results. The parallel nature of the actual temperature changes over time at different locations in the system were already observed in Figure 2.

In Figure 4 the x-axis stands for the rate of temperature changes measured only at the top of the tubes in question. The correct value of T(Gr) can be obtained, whenever the rate of temperature change is zero. At these times no heat is flowing into or out of the system and we have equilibrium conditions with no temperature change over time.

Trend lines are calculated as least squares regression lines for the scattered values. The trend line for the blue triangles for inner tube 1 (water with glass powder) show an intercept with the vertical zero line , where the rate of temperature change is 0, at -0.05 K/m. The red markers give a T(Gr) value of -0.12 K/m for inner tube 2 (only water). Both of these values agree well with the long term average values, seen on curves 1 and 2 in Figure 3. The error bars correspond to +/- 1 SDs for all measured values.

FIGURE 4

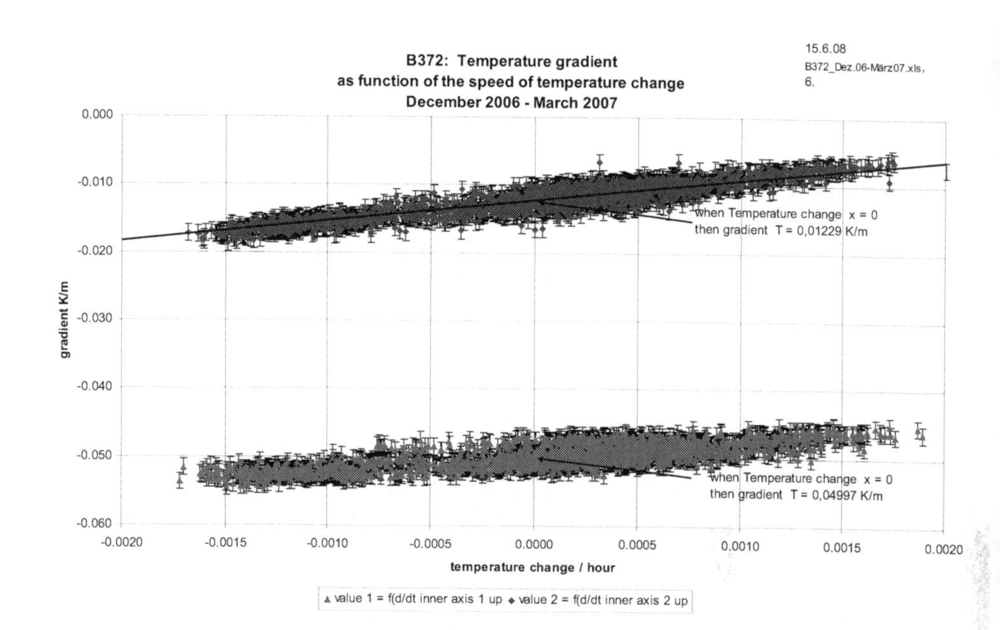

DISCUSSION

Precision of measurements

Thermistors have a precision of only about +/- 0.1 K, not sufficient to measure gradients of 0.01 K/m without making difficult additional calibrations. In the above measurements thermistors are used not for reliably measuring gradients, but in order to establish the changes of temperatures over time at different locations. The precision of measuring these temperature changes is better than 0.001 K/hour.

Type E thermocouples are used to measure the temperature gradients as the difference of the voltages between two thermocouple points. Connecting these in series, one obtains thermopiles. The values reported here are more than 10 times the precision of individual thermocouple measurements.

Actually, the precision of the absolute value of T(Gr) (20 or 30% higher or lower) is not as important as finding out, whether the direction of the temperature gradient is positive or negative. But this direction can be decided upon to a very great precision, because the zero offset of the instrument can be determined measuring each value twice, the second time with switched polarity.

Contradicting possible explanations for the negative temperature gradient, cold at the top and warm at the bottom, within today's accepted laws of thermodynamics

a) Exothermic process

By adding glass powder to the water of tube 1 an exothermic process could possibly start. The generated heat would be transported to the outside through an increase of the internal temperatures. If the lower part of tube 1 is better insulated than the upper part, the top of tube 1 would be more intensively cooled than the bottom. As more heat would leave the upper parts the consequence would be of a lower temperature in that area, creating a negative temperature gradient in the tube.

Response:
In this case the upper temperature of tube 1 would have to be higher than the upper temperature of tube 2. But as a measurement of this difference shows (not shown in graph 2.) the opposite is being measured.

b) Water evaporation:
Water could possibly evaporate from the surfaces at the top of tube 1 and tube 2, cooling the upper part of the water.

Response:
On the top of the water in tube 1 and 2 are layers of a synthetic motor oil with a thickness of about 8 mm eliminating the possibility of water evaporation.

c) Convection currents creating an adiabatic gradient:
The fluctuations of the water temperatures in tube 1 and 2 could create slow vertical movements of the water in the tubes creating an adiabatic temperature gradient, cold at the top and warm at the bottom.

Response:
Any convection current should hardly be possible in tube 1 where the water is filled with glass powder. In tube 2, filled only with water, one can imagine small convection currents which could explain an gradient of up to -.002 K/m, which is the adiabatic gradient for water, but the measured values of -.01 to -.05 K/m are an order of magnitude larger.

d) Precision of measurements:
The precision of the instruments measuring the resistance values of the thermistors and the voltages of the thermocouples are not sufficient to measure these small gradients.

Response:
Thermistors have a precision of only about +/- .1 K, not sufficient to measure gradients of .01 K/m without making difficult additional calibrations. In the above results thermistors are used not for measuring gradients but to establish the changes of the temperatures over time in different locations. The precision for these temperature changes is better than .001K /hour.

Thermocouples are used to measure the temperature gradients as the difference of the voltages of two thermocouple points, the thermocouples often used in the form of thermopiles. The values reported are more than 10 times the precision of these measurements including the effects of the instruments.

e) Equilibrium
As long as equilibrium conditions have not been reached the measured gradients might change and disappear.

Response:
Fig. 4. shows the temperature gradient for the inner axis being negative for times when there is no heat inflow or outflow to the test column which constitutes equilibrium conditions, meaning no temperature changes in the test column over time.

CONSEQUENCES OF THE MEASURED TEMPERATURE GRADIENTS FOR THE SECOND LAW

Provided the experimental results get reconfirmed by other investigators this temperature difference of tube 1 between the top and the bottom with an average of -.05 K/m could be used to create work by supplying electric power created by a thermocouple. This is actually continuously taking place during the experiments described here. The amount of energy so produced is, of course, extremely small. It does not affect the equilibrium condition of the experiment, because this small amount of energy taken out of the system is easily replenished from the heat bath of the environment. But the surprising observation is that heat flows under the influence of gravity from a cold reservoir to one with a higher temperature.

The Second Law of Thermodynamics, as stated by Clausius in 1854 (11) says: *"No process is possible for which the sole effect is that heat flows from a reservoir at a given temperature to a reservoir at a higher temperature."*
It is assumed that the process takes place within an isolated system with no exchange of matter and energy across its bounders. It also implies, like any other presently used statement of the Second Law, that the isolated system might be exposed to a force field, like gravity, and in spite of this, the assertion remains valid.

Contrary to the statement by Clausius, the reported results show that *in an isolated system under the influence of a force field like gravity heat can flow from a reservoir at a given temperature to a reservoir at a higher temperature.* This leads to the need of a new general statement of the Second Law:

In isolated systems – with no exchange of matter and energy across their boundaries AND WITH NO EXPOSURE TO FORCE FIELDS - initial differences of temperature, densities, and concentrations in assemblies of molecules will disappear over time, resulting in an increase of entropy.

Conversely:
In isolated systems - with no exchange of matter and energy across its borders - FORCE FIELDS LIKE GRAVITY can generate in macroscopic assemblies of molecules temperature, density, and concentration gradients. The temperature differences can be used to generate work.

SUMMARY

Measurements of the temperature gradient in insulated vertical tubes, filled just with water or with water and small glass beads, show a negative temperature gradient, cold at the top and warm at the bottom.

These gradients appear in spite of positive temperature gradients in the environment. They are not explainable by today's accepted laws of heat transport in liquids, gases and solids, because positive temperature gradients in the environment would allow only positive gradients within the test setup.

If the experimental results are confirmed, the temperature differences created in vertical, isolated columns of water under the influence of gravity allow the production of work out of a heat bath using only the effect of gravity. Therefore basic statements of the Second Law of Thermodynamics would have to be restated to reflect the effects of force fields like gravity.

REFERENCES

1. Boltzmann, L., *Wissenschaftliche Abhandlungen, Vol.2, edited by F.Hasenoehrl,* Leipzig, 1909
2. Maxwell, J.C., *The London, Edinburgh, and Dublin Philosophical Magazine and Journal of Science*, 35, p. 215 (1868)
3. Loschmidt, L., *Über den Zustand des Wärmegleichgewichts eines Systems von Körpern mit Rücksicht auf die Schwerkraft,* Sitzungsberichte der Mathematisch-Naturwissenschaftl. Klasse der Kaiserlichen Akademie der Wissenschaften 73.2 (1876), p. 135
4. Trupp, A., Physics Essays, 12, No. 4, 1999
5. Graeff, R.W., *Gravity Machine,* Web Site <*firstgravitymachine.com*>

6. Bohren, C.F. *Atmospheric Thermodynamics*, pp. 167 ff., Oxford University Press, New York, 1998
7. Graeff, R.W., *Gravity Machine, my Search for Peaceful Energy*, to be published autumn 2009
8. Graeff, R.W., *Measuring the temperature distribution in gas columns*, CP 643, Quantum Limits to the Second Law, New York 2002, American Institute of Physics, AIP Conference Proceedings, Vol. 643
9. Graeff, R.W.,*Produktion von nutzbarer Energie aus einem Wärmebad*, NET-Journal, ISSN 1420-9292, Jupiter-Verlag, Zürich, February 2003
10. Čápek,V., and Sheehan, D.P., *Challenges to the Second Law of Thermodynamics*, pp 203 ff., Springer, 2005
11. Clausius, R., Abhandlungen ueber die mechanische Waermetheorie, Vol. 1, (F.Vieweg, Braunschweig, 1864); Vol. 2, (1867).

My Path to Peaceful Energy
Appendix 6

Kreisfähigkeit – Circability

Roderich W. Graeff

International Licensing
Prof.-Domagk-Weg 7, D-78126 Königsfeld, Germany
102 Savage Farm Drive, Ithaca, NY 14850-6500, USA
E-mail rwgraeff@yahoo.com

December 2007

CIRCABLE ACTION means in view of every single individual and every animal and plant species, to perform only those actions which within one generation return to their original condition on its own or can be returned to it.
CIRCABLE ACTION allows all activities as long their consequences make it not impossible for future generations to recreate their original conditions.

CIRCABILITY (KREISFÄHIGKEIT)

CIRCABLE ACTION means in view of every single individual and every animal and plant species, to perform only those actions which within one generation return to their original condition on its own or can be returned to it.
CIRCABLE ACTION allows all activities as long their consequences make it not impossible for future generations to recreate their original conditions.

DIRECT CONSEQUENCES:
This statement results in 5 direct consequences, which lead, to use the words of Prof Kueng, to 5 directives. Each of these consequences eliminates actions whose results could not be rectified and returned to their original conditions by future generations.

Physical hurting a human being or killing one is not retractable, therefore:

The killing or physical hurting of a human being can not take place anymore: Killing is out, killing is passé.

In order to fulfill this directive we get automatically the following consequences:

- Peace preservation instead of preparing for war
- Dialog instead of condemnation
- Development of violence-free methods for the negating of violence by un-peaceful people
- Changing the military to peace preserving police and trained assistance giving specialists
- Defending against threatening strangers not by building walls or preparing for the use of force but by building bridges and offering local help, education and assistance.
- Making possible local CIRCABLE LIVING instead of population movements
- Building of schools and hospitals instead of prisons
- No more sending of ones troops to secure ones own energy supply.

Not retractable are the changes in the atmosphere due to the burning of fossil fuels or the creation of radiating atomic waste materials through the generation of atomic energy, therefore

The burning of fossils fuels or the use of atomic energy is off limits.

Consequences:

- Speedy switch from the use of atomic energy or the burning of fossil fuels to an energy production based on only renewable energy sources.
- Learning, acting, creating and helping instead of consuming.
- Striving for education, creativity and sharing instead for an increased consumption.

The increase of the number of people on this earth is not retractable, therefore:
III. Families have at the most two children.

Consequences:

- Striving for families with maximal 2 children instead advertising for more children.
- Instead of asking for more children avoidance of child mortality
- Equal rights and equal education for women world wide.

The elimination of an animal or plant species is not retractable, therefore:
IV. Actions which endangers the survival of any animal or plant species are off limits.

Consequences:

- Strengthening all actions which allow the survival of all animal and plant species.

Without responsible actions of all people within the limits of CIRCABLE ACTIONS non retractable actions would take place, therefore:
V. Responsible personal action.

Consequences:

- Every one tries out of a feeling of responsibility for other human beings, animal and plants and especially for the next generations to plan and execute all his actions within the idea of CIRCABILITY.

My Path to Peaceful Energy
Appendix 7

How to Build a Gravity Machine

Roderich W. Graeff

International Licensing
Prof.-Domagk-Weg 7, D-78126 Königsfeld, Germany
102 Savage Farm Drive, Ithaca, NY 14850-6500, USA
E-mail rwgraeff@yahoo.com

1. Introduction

Why do you look at this addendum? Do you want to build a Gravity Machine? What do you expect?

There are a number of definitions for "machine". One common one is: a machine is a device that uses energy to perform some activity, for instance, some type of work. My Gravity Machines can produce work out of a heat bath.

A heat bath is a system whose heat capacity is so large that when in formal contact with some other system its temperature remains constant. The heat bath is effectively an infinitive large reservoir of energy. In real life a heat bath could be the water in a lake, or an ocean, or the soil in the ground, or the atmosphere around us. The Gravity Machine, which we will discuss and I will tell you about how to build one, would be a machine which can create work out of the energy of a heat bath without the addition of any other energy, just by using the effect of the gravity field.

Our Gravity Machine creates a temperature difference in connection with a heat bath. This temperature difference can then be used to drive a machine for instance; a Stirling motor, which in turn can be connected to a saw mill to cut wood, or to an electric generator producing electricity. More directly we can use our temperature difference in connection with a thermopile and produce electricity.

So, if your interest is to produce electricity out of a heat bath and to build machines for this purpose, then keep on reading. If you follow my description you really can build such a machine. But if you think that the electricity you will be producing is enough to light a lamp or to produce hot water on the one side and cold air for your fridge on the other, then you are mistaken. The machine which you can build in your home will be big enough to demonstrate the principle, which is to show that without introduction of outside energy your machine will produce a temperature difference and a very small amount of electricity. If your aim is to make a machine big enough to light an electric bulb you migh have to wait for another five or ten years. Or you will have to do the development work still needed yourself, investing quite a bit of your time and financial resources.

Already today you can demonstrate by building your own machine the sensational feature of our present machines: You will be able to produce electricity out of a heat bath. You should be able to create a graph like this one:

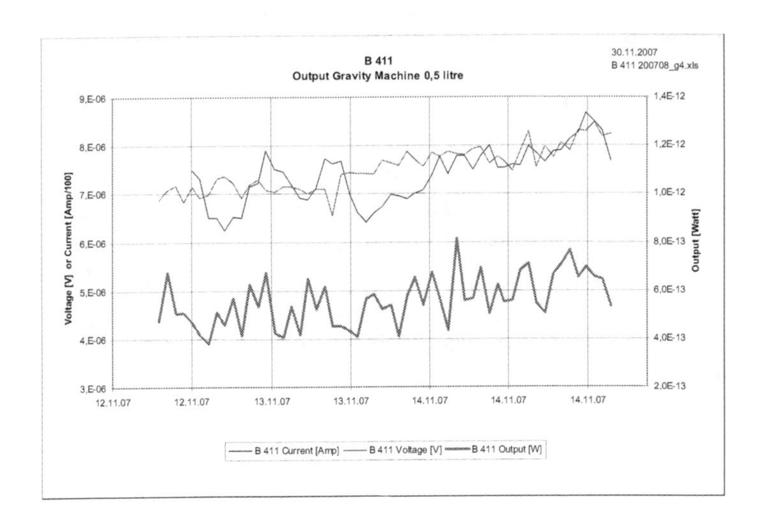

This graph shows measured values of the amount of electricity produced by a Gravity Machine. You can build or purchase such a machine. We will describe what steps you have to take.

STEP 1. Choose what type of machine do you want to build.

I will discuss the building of a machine using the design of B 76 described in chapter 16. This is a machine using air filled with glass powder

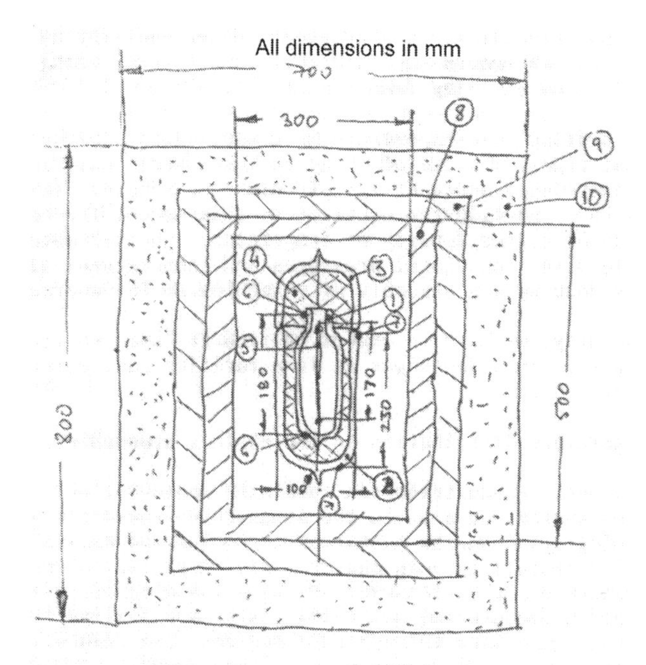

The innermost Dewar (1) of ½ liter is filled with a fine powder in order to eliminate convection currents and radiation between the inner wall surfaces. A thermocouple (5) is arranged in the middle axle with a height distance between junctions of 170 mm. A second thermocouple (6) is mounted on the outside of the Dewar insert (1) with a vertical distance of 180 mm. A third thermocouple (7) is mounted on the outside of Dewar insert 2 with a vertical distance of 230 mm.

The Dewar inserts are held in place by fine PET fibers within a box (8) fabricated out of 40 mm thick polystyrene foam panels. This box is surrounded by 50 mm thick panels (9) consisting of pressed copper wires. The whole setup is insulated against the room air by 100 mm thick polystyrene foam panels (10).

The diameter of the thermocouple wire is kept as small as possible in order to reduce any heat conduction through these wires. The measuring instrument measures the voltage of the thermocouples with a resolution of 1×10^{-7} V corresponding to 0,0003 °K.

STEP 2. Make a descision:

a. Purchase a complete machine from us, or
b. Purchase a set of all needed materials from us and assamble it yourself, or
c. Start from scratch.

STEP 3: Select a building and demonstration space.

STEP 4: Decide if you want to be able to turn your machine on its head

.STEP 5: Select the type and size of the heart, the inner axle, of the machine.

STEP6: Select the type, size and arrangement of the inner insulation

STEP 7: Select type and size of the temperature equalizer

STEP 8: Select type, size and arrangement of the outer insulation and temperature equalizers.

STEP 9: Select and assemble the temperature sensors

STEP 10: Produce scetsch of macine und photos

STEP 11: Start to assemble from the inside out

STEP 12: Select your data collection system and instrumentation

At present we are putting the finishing touches on our latest edition of our booklet

How to build a Gravity Machine.

If you want to continue to plan the building of your own Gravity Machine please send us an E-mail to

rwgraeff@yaho.com

and we will send to you free of charge an up to date edition describing in detail all the steps just listed, including a list of all materials necessary to build a Gravity Machine Type B 76.

And then you can go ahead with

a. Purchase a complete machine from us, or
b. Purchase a set of all needed materials from us and assamble it yourself, or
c. Build your machine start from scratch.

Good Luck! I am sure you will find it very exiting and enjoyable!

My Path to Peaceful Energy
Appendix 8

Gravity Machines for Sale

Roderich W. Graeff

International Licensing
Prof.-Domagk-Weg 7, D-78126 Königsfeld, Germany
102 Savage Farm Drive, Ithaca, NY 14850-6500, USA
E-mail rwgraeff@yahoo.com

Dated October 14, 2010

Gravity machines for sale

My company

Dr. Roderich W. Graeff,
CONSULTING.

with offices in

102 Savage Farm Drive
Ithaca NY 14850
USA

and

Professor Domagkweg 7
D 78126 Koenigsfeld
Germany

Tel 49 7725 91191
E-mail rwgraeff@yahoo.com

offers for sale

a. a set of all needed materials for assembling a Gravity Machine
b. complete Gravity Machines.

These gravity machines are are primarily used as demonstration and research machines. They can contain as its power producing element a gas, a liquid or a solid. Details of availability are described in a leaflet available under www.rwgraeff@yahoo.com

The following picture shows the smallest standard machine type 1-2010. Within the insulation box it contains an inner vertical arranged active element in which a temperature gradient developes under the influence of gravity. This difference can be used to generate very small amounts of electricity. This ability of the generation of electricity develops without any imput of power from any source, i.e. from a battery.

Gravity Machine, smallest standad size Type 1-2010

Gravity Machine Type 1-2010

This Gravity Machine Type e 1-2010 contains air with glass powder within a Dewar flask. A cardboard box protects the vacuum panels which surround the test setup. A 5cm thick polystyrene box protects the panels from the outside and provide an additional 5cm of insulation all around.

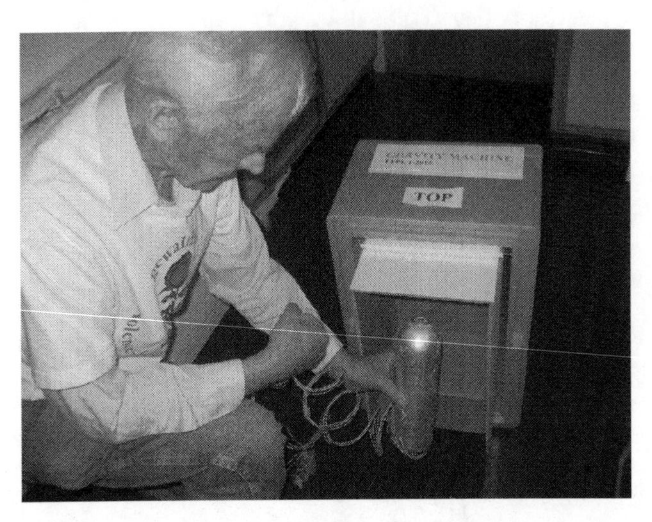

This Gravity Machine contains a half liter aluminium can filled with Xenon gas.

Gravity machines are covered by

EUROPEAN PATENT APPLICATION Number 03 002 000.2-2315.

REFERENCES

1. Boltzmann, L., Wissenschaftliche Abhandlungen, Vol.2, edited by F.Hasenoehrl, Leipzig, 1909
2. Maxwell, J.C., The London, Edinburgh, and Dublin Philosophical Magazine and Journal of Science, 35, p. 215 (1868)
3. Loschmidt, L., Über den Zustand des Wärmegleichgewichts eines Systems von Körpern mit Rücksicht auf die Schwerkraft, Sitzungsberichte der Mathematisch-Naturwissenschaftl. Klasse der Kaiserlichen Akademie der Wissenschaften 73.2 (1876), p. 135
4. Trupp, A., Physics Essays, 12, No. 4, 1999
5. Graeff, R.W., Gravity Machine, Web Site <firstgravitymachine.com>
6. Bohren, C.F. Atmospheric Thermodynamics, pp. 167 ff., Oxford University Press, New York, 1998
7. Graeff, R.W., Gravity Machine, my Search for Peaceful Energy, to be published autumn 2010
8. Graeff, R.W., Measuring the temperature distribution in gas columns, CP 643, Quantum Limits to the Second Law, New York 2002, American Institute of Physics, AIP Conference Proceedings, Vol. 643
9. Graeff, R.W., Produktion von nutzbarer Energie aus einem Wärmebad, NET-Journal, ISSN 1420-9292, Jupiter-Verlag, Zürich, February 2003
10. Čápek,V., and Sheehan, D.P., Challenges to the Second Law of Thermodynamics, pp 203 ff., Springer, 2005
11. Clausius, R., Abhandlungen ueber die mechanische Waermetheorie, Vol. 1, (F.Vieweg, Braunschweig, 1864), Vol. 2, (1867)
12. Chuanping Liao, Temperature Gradient Caused by Gravitation, International Journal of Modern Physics B, Vol. 23, No. 22 (2009) 4685-4696 or at http://www.worldscinet.com/ijmpb/23/2322/S0217979209052893.html
13. R. C. Tolman, Phys. Rev. 35, 904 (1930)
14. R. C. Tolman, and P. Ehrenfest, Phys. Rev. 36, 1791 (1930)
15. W. M. Suen and K. Young, Phys. Rev. 35, 406 (1987)
16. C. A. Coombes and H. Laue, Am. J. Phys. 53, 272 (1985)
17. F. L. Roman, J. A. White and S. Velasco, Eur. J. Phy. 16, 83 (1995)
18. S. Velasco, F. L. Roman and J. A. White, Eur. J. Phy. 17, 43 (1995)
19. B. F. Whiting, Physica A 158, 437 (1989)
20. Ro. O. Esposito, M. Castier and F. W. Tavares, Chem. Eng. Sci. 55, 495 (2000)
21. F. Montel, Fluid Phase Equilibria 84, 343 (1993)
22. J. Perez, C. R. Phys. 7, 406 (2006)
23. J. Coulomb and G. Jobert, the Physical Constitution of the Earth, Translated by A.E.M. Nairn (Oliver & Boyd Ltd., Edinburgh and London 1963)

24. Quantum Limits to the Second Law, New York 2002, American Institute of Physics, AIP Conference Proceedings, Vol. 643
25. Schmidt, Klaus, Die Brandnacht, Reba-Verlag GmbH Darmstadt (1964), bzw. Schmidt, Klaus: Die Brandnacht, Dokumente von der Zerstörung Darmstadts am 11. September 1944, Darmstadt, Verlag H. L.Schlapp, 2003. ISBN 978-3-87704-053-9, 22,00 €
26. Caidin, Martin, The Night Hamburg Died, Ballantine Books, New York (1960)
27. Gräff, Siegfried, Tod im Luftangriff – Ergebnisse pathologisch-anatomischer Untersuchungen, H.H.Nölke Verlag, Hamburg (1955), pages 40, 42
28. Anna Cataldi, Letters from Sarajevo - voices of a besieged city, translation by Avril Bardoni, USA (1994)
29. Scholl, Inge, Die weiße Rose, Fischer Taschenbuch Verlag, Frankfurt/M. (1981)
30. Eddington, Sir Arthur Stanley, The Nature of the Physical World (1927), The Gifford Lectures 1927, Cambridge University Press: Cambridge UK, 1933, reprint, pp.74-75.
31. Memo, Heft Nr. 1, Museum für Hamburgische Geschichte, (1993)
32. Kelvin, Lord (Thomson, W.), Mathematical and Physical Papers, Vol.I, Cambridge University Press, Cambridge, 1882
33. Planck, M., Vorlesungen über die Theorie der Wärmestrahlung (Barth, Leipzig,, 1906); Treatise of Thermodynamics 7th edition, translated by Ogg, A. (Dover, New York 1945)
34. Xu Yelin, A Trial and Study on Obtaining Energy from a Single Heat Reservoir at Ambient Temperature, Science Press, Beijing, (1988)
35. Xu Yelin, Missiles and Space Vehicles 3, 53 (2000)
36. Chiatt, Leonardo, Has the 2nd Law of Thermodynamics relly been Violated? (Postal Address: L. Chiatti, ASL VT Physics Laboratory, Via San Lorenzo 101, 01100 Viterbo, Itly, fisica1.san@asl.vt.it
37. Xin Yong Fu , Realization of Maxwell's Hypothesis, An Experiment against the Second Law of Thermodynamics, Xin Yong Fu, Zi Tao Fu, Shanghai Jiao Tong University, xyfu@sjtu.edu.cn
38. Clausius, R., Abhandlungen über die mechanische Wärmetheorie, Vol. 1, (F.Vieweg, Braunschweig, 1864); Vol. 2 (1867); The Mechanical Theory of Heat (Macmillan, London, 1879); Phil. Mag. 2 1, 102 (1851); 24, 201, 801, (1862)
39. Maxwell, J.C., Theory of Heat (Longmans, Green, and Co., London, 1871)
40. Ernst Schmidt: Einführung in die technische Thermodynamik, 3rd edition, 1945, Springer-Verlag, Berlin

END OF BOOK!

Made in the USA
Charleston, SC
09 January 2013